T0314057

BIOPHYSICAL METHODS FOR BIOTHERAPEUTICS

BIOPHYSICAL METHODS FOR BIOTHERAPEUTICS

Discovery and Development Applications

Edited by

Tapan K. Das

Center of Excellence for Mass Spectrometry and Biophysics
Bristol-Myers Squibb
Hopewell, NJ

WILEY

Published by John Wiley & Sons, Inc., Hoboken, New Jersey.
Published simultaneously in Canada.

For general information on our other products and services or for technical support, please contact our Customer Care Department within the United States at (800) 762-2974, outside the United States at (317) 572-3993 or fax (317) 572-4002.

Wiley also publishes its books in a variety of electronic formats. Some content that appears in print may not be available in electronic formats. For more information about Wiley products, visit our web site at www.wiley.com.

Library of Congress Cataloging-in-Publication Data:

Biophysical methods for biotherapeutics: discovery and development applications / edited by Dr. Tapan K. Das.
      p. ; cm.
   Includes bibliographical references and index.
   ISBN 978-0-470-93843-0 (cloth)
   I. Das, Tapan K., editor of compilation.
   [DNLM: 1. Biopharmaceutics–methods.   2. Drug Discovery–methods.   3. Technology, Pharmaceutical–methods. QV 35]
   RM301.5
   615.7–dc23

                                                            2013038071
Printed in the United States of America.

10 9 8 7 6 5 4 3 2 1

CONTENTS

PREFACE

The biotechnology industry, emerging since 1970s, is enjoying robust growth with estimated equity investments in the tune of $400 billion today. The growth is fueled by opportunities of biologic drugs in unmet medical-need areas including cancer and immunology. Currently there are over 270 approved biotech drugs for a wide range of therapies, and hundreds of candidates are in clinical development. The biotherapeutic class of drugs encompasses a range of biologic-based compounds including monoclonal antibodies, enzymes, antibody fragments, glycosylated proteins, other recombinant proteins, peptides, conjugated or fused peptides, antibody drug conjugates (ADC), protein-based vaccines, oligonucleotides, protein–lipid complexes, and carbohydrates. Protein- and peptide-based biologics dominate the list of approved as well as clinical development candidates.

In contrast to small molecule-based therapeutic candidates, biotherapeutic molecules present a much higher level of complexity arising from several degrees of structural elements that are required for appropriate biological function. For protein-based biologics, in addition to the amino acid sequence (referred to as primary structure), there are various forms of secondary structures (helical, β-sheet, β-turn, unordered, etc.) that often coexist in a protein. When the higher order structural builds (tertiary, quaternary) are added to the secondary structure, a composite and complete structure is formed that is often referred to as protein conformation. Maintenance of the integrity of protein conformation requires a great deal of attention in all stages of development—from early discovery through clinical phases to commercial development. A good understanding of protein structure and function and its sensitivity to a variety of solution, interface, and environmental conditions is critical to ensure an active and nondegraded form is preserved through the development stages. Additionally, protein drug substance is most often a heterogeneous mixture of closely related species that add complexity to efforts toward maintaining original conformation(s) and ensuring drug stability.

It is well recognized that in addition to using appropriate analytical techniques to monitor stability and integrity of a biologic candidate, employing a wide range of biophysical methods is paramount in the development process. To deal with the numerous degradation issues known to occur with biologic molecules, biotech researchers have innovatively adapted physical and chemical technologies from across diverse fields in addition to using the classical biophysical/biochemical methods. However, there is no comprehensive textbook available that discusses application diversity of biophysical methods and the type of information sought as a function of the phases in biotherapeutic development.

This book focuses on systematic applications of biophysical technologies and methods in stages of biotherapeutics development. Four areas are emphasized in this book: (1) novel applications of traditional biophysical techniques, (2) emerging technologies and their applications, (3) biophysical applications relevant to stage-wise development of a clinical biotherapeutic candidate—from discovery through clinical phases to commercial, and (4) focused discussion of some of the thermodynamic, conformational, and stabilization concepts aided by biophysical research.

The chapters in this book are laid out in a theme (Sections 1–3) based on clinical phases of development (Section 1: Early discovery stages and biotherapeutic candidate selection; Section 2: First-in-human and up to proof-of-concept; Section 3: Phase III and commercial development). This gives a comprehensive view to a biotherapeutics development scientist of what biophysical studies are needed for which phase and what purpose, how to apply some of the biophysical techniques for what type of information sought, what type of orthogonal biophysical characterization may be expected by regulatory agencies, and very importantly the limitations of each technique and its applications—"myth versus truth." The editor believes that this book will be a good guide to biophysics experts as well as beginners to help them with the big picture of biotherapeutics development and for developing an organization's short- and long-term strategies for resource investment in biophysical research.

The contributing authors in this book are prominent researchers with proven track records. The authors added excellent CASE STUDIES and discussed results with literature data and concepts. *My deepest gratitude to all authors for making outstanding contributions to make this book possible.* My sincere thanks to many individuals for guidance, help with reviewing, and the publishing process. In particular, I thank Drs. Kevin King and Sandeep Nema of Pfizer; Dr. Andy Vick of Wilresearch; Dr. Michael Hageman of Bristol-Myers Squibb; and Jonathan Rose and Amanda Amanullah of Wiley.

This book is dedicated to my parents who unconditionally nurtured and supported my passion for science and technology and my wife Paramita for inspiring and supporting me to complete the book.

Tapan K. Das
Hopewell, NJ, USA

ABOUT THE EDITOR

Dr. Tapan K. Das is Director of Center of Excellence, Mass Spectrometry and Biophysics in the Biologics Development group, Bristol-Myers Squibb. Prior to this position, he was an Associate Research Fellow and Head of Biophysical Center of Excellence in Pharmaceutical R&D, Biotherapeutics Pharmaceutical Sciences, Pfizer.

His key interest areas include structural characterization of biotherapeutic candidates progressing through clinical development toward commercialization, pharmaceutics, and development of novel biophysical and biochemical applications. He is actively engaged in fostering collaborations and sharing learning through internal and external networks.

Dr. Das began his career at Albert Einstein College of Medicine, New York. He currently serves as the Chair of the Biotechnology Section of American Association of Pharmaceutical Scientists (AAPS), USA. He has published 61 articles in professional scientific journals, books, and patents and presented numerous talks.

LIST OF CONTRIBUTORS

Ahmad M. Abdul-Fattah, Pharmaceutical Development and Supplies, Pharma Technical Development Biologics EU, F. Hoffmann-La Roche Ltd., Basel, Switzerland

Alptekin Aksan, Biostabilization Laboratory, Mechanical Engineering Department and BioTechnology Institute, University of Minnesota, Minneapolis, MN

Angela W. Blake-Haskins, BioPharmaceutical Development, Human Genome Sciences, Inc., Rockville, MD

James W. Bryson, Protein Science and Structure Department, Bristol-Myers Squibb Research and Development, Princeton, NJ

Christine P. Chan, Biologics R&D, Genzyme—A Sanofi Company, Boston, MA

Michael L. Doyle, Protein Science and Structure Department, Bristol-Myers Squibb Research and Development, Princeton, NJ

Yiqing Feng, Lilly Research Laboratory, Indianapolis, IN, USA

Gary L. Gilliland, Janssen R&D, LLC, Spring House, PA

Boris Gorovits, Department of Pharmacokinetics, Pharmacodynamics, and Metabolism, Pfizer, Andover, MA

Carol Hirschmugl, Department of Physics, University of Wisconsin, Milwaukee, WI, and Synchrotron Resource Center, University of Wisconsin, Stoughton, WI

Lisa M. Jones, Department of Chemistry, Washington University in St. Louis, St. Louis, MO, and Department of Chemistry and Chemical Biology, Indiana University-Purdue University, Indianapolis, IN

Judy E. Kim, Department of Chemistry and Biochemistry, University of California at San Diego, La Jolla, CA

Corinna Krinos-Fiorotti, Department of Pharmacokinetics, Pharmacodynamics, and Metabolism, Pfizer, Andover, MA

Virginie Lafont, Protein Science and Structure Department, Bristol-Myers Squibb Research and Development, Princeton, NJ

Brian S. Leigh, Department of Chemistry and Biochemistry, University of California at San Diego, La Jolla, CA

Yen-Huei Lin, BioPharmaceutical Development, Human Genome Sciences, Inc., Rockville, MD

Zheng Lin, Protein Science and Structure Department, Bristol-Myers Squibb Research and Development, Princeton, NJ

Hanns-Christian Mahler, Pharmaceutical Development and Supplies, Pharma Technical Development Biologics EU, F. Hoffmann-La Roche Ltd., Basel, Switzerland

Paul E. Morin, Protein Science and Structure Department, Bristol-Myers Squibb Research and Development, Princeton, NJ

Hendrik Neubert, Department of Pharmacokinetics, Pharmacodynamics, and Metabolism, Pfizer, Andover, MA

Melissa D. Perkins, BioPharmaceutical Development, Human Genome Sciences, Inc., Rockville, MD

Vishard Ragoonanan, Department of Pharmaceutics, University of Minnesota, Minneapolis, MN

Christopher J. Roberts, Department of Chemical Engineering, University of Delaware, Newark, DE

Bonita Rup, Department of Pharmacokinetics, Pharmacodynamics, and Metabolism, Pfizer, Andover, MA

Nazila Salamat-Miller, Rare Disease Business Unit, Shire, Department of Pharmaceutical and Analytical Development, Lexington, MA

Diana E. Schlamadinger, Department of Chemistry and Biochemistry, University of California at San Diego, La Jolla, CA

Lumelle A. Schneeweis, Protein Science and Structure Department, Bristol-Myers Squibb Research and Development, Princeton, NJ

Zahra Shahrokh, Rare Disease Business Unit, Shire, Department of Pharmaceutical and Analytical Development, Lexington, MA

Li Shi, Shanghai Zerun Biotechnology Co., Ltd., Member of Wison Group, Shanghai, China

Justin B. Sperry, Analytical Research and Development, Biotherapeutics Pharmaceutical Sciences, Pfizer, Inc., Chesterfield, MO

Thomas M. Spitznagel, BioPharmaceutical Development, Human Genome Sciences, Inc., Rockville, MD

Glen Spraggon, Genomics Institute of the Novartis Research Foundation, San Diego, CA

John J. Thomas, Rare Disease Business Unit, Shire, Department of Pharmaceutical and Analytical Development, Lexington, MA

Sheng-Jiun Wu, Janssen R&D, LLC, Spring House, PA

Zhuchun Wu, BioPharmaceutical Development, Human Genome Sciences, Inc., Rockville, MD

Aaron P. Yamniuk, Protein Science and Structure Department, Bristol-Myers Squibb Research and Development, Princeton, NJ

Joseph Yanchunas, Jr., Protein Science and Structure Department, Bristol-Myers Squibb Research and Development, Princeton, NJ

SECTION 1

EARLY DISCOVERY STAGES AND BIOTHERAPEUTIC CANDIDATE SELECTION

1

BIOPHYSICAL METHODS APPLIED IN EARLY DISCOVERY OF A BIOTHERAPEUTIC: CASE STUDY OF AN EGFR-IGF1R BISPECIFIC ADNECTIN

Michael L. Doyle, James W. Bryson, Virginie Lafont, Zheng Lin, Paul E. Morin, Lumelle A. Schneeweis, Aaron P. Yamniuk, and Joseph Yanchunas, Jr.

Protein Science and Structure Department, Bristol-Myers Squibb Research and Development, Princeton, NJ, USA

Biophysical Methods for Biotherapeutics: Discovery and Development Applications, First Edition. Edited by Tapan K. Das.
© 2014 John Wiley & Sons, Inc. Published 2014 by John Wiley & Sons, Inc.

1.1 INTRODUCTION

Biophysical characterization of protein therapeutics and associated reagents in drug discovery is critical to selection and optimization of molecules that have the desired biological activity and to selection of drug candidates that can be efficiently developed and manufactured. Protein therapeutic molecules are larger and more complex than small-molecule drugs. Consequently, analytical strategies for determining whether a protein therapeutic is pure, stable, and homogeneous require that a larger number of physical properties be investigated, including characterization of tertiary and quaternary structures. Furthermore, several physical properties of protein therapeutics, for example, aggregation state, require multiple, orthogonal methods to confidently define them (Table 1.1).

In addition to production and characterization of hundreds or thousands of drug candidates during drug discovery, a large number and diversity of protein reagents must also be produced and characterized. To begin with, the biological target must be produced in a form that is well behaved and representative of the functional form to be targeted *in vivo*. There are a multitude of other protein reagents needed to run the program as well (e.g., multiple affinity-tagged forms of the target for use in a variety of assays, truncated forms of the target for structural studies, counter-targets, co-targets, and nonhuman species ortholog variants of the target; Figure 1.1; see also Kim and Doyle [1] for a detailed listing). Target reagents that are aggregated or misfolded confound the drug discovery process during hit identification and downstream assays. The famous admonition "garbage in, garbage out" is often cited as a reminder that biophysically well-behaved reagents generally lead to higher success rates during lead identification and optimization of protein therapeutics. Biophysical methods thus play a wide variety of roles in the characterization of biotherapeutic candidates and protein reagents during the early discovery stages of biotherapeutics.

Biophysical characterization is a central part of the selection and optimization process. But how much biophysical characterization is optimal for each type of reagent or biotherapeutic candidate molecule, and how does the extent of biophysical characterization change during each stage of the discovery process? The goals of this chapter are to describe the types of biophysical methods that are used in a stage-dependent manner throughout discovery for reagent and drug candidate production of protein therapeutics and to discuss how the application of these methods in discovery help to de-risk the potential costly challenges later in the development and manufacturing phases.

TABLE 1.1. Biophysical and biochemical methods used to characterize targets, reagents, and drug candidates for protein therapeutic discovery programs in terms of identity, purity, stability, oligomeric status, binding activity, and molecular binding mechanism

Method[a]	Molecular information	Targets	Reagents	Hits	Leads	Final candidates
Analytical SEC	Self-association	1	1	1	1	1
Thermal melt	Thermal stability			1		
Biosensor	Confirm binding	1	1	1		
SDS PAGE	Purity, approximate mass	1	1		1	1
LC-MS	Identity, primary structure, purity	1	1	2	1	1
SEC-MALS	Self-association, absolute mass	1	2		1	1
DSC	Thermal stability	2	2		1	1
Biosensor	Binding affinity, kinetics, epitope discrimination	1	1		1	1
AUC—sedimentation equilibrium	Self-association, absolute mass, dimerization constant	2			2	2
X-ray crystallography	Define epitope, define atomic binding interactions	2			1	1
ITC	Solution binding affinity, molar ratio	2			2	1
KinExA	Very tight solution binding affinity					1
Thermal stability profiling	Thermal stability over diverse set of conditions	2	2		2	1
Accelerated degradation	Indicator of manufacturability				2	1

Cases where a method is frequently used are designated by 1, and cases where the method is less frequently used but recommended are designated by 2.
[a]SEC, size-exclusion chromatography; SDS PAGE, sodium dodecyl sulfate polyacrylamide gel electrophoresis; DSC, differential scanning calorimetry; MALS, multiple angle light scattering; ITC, isothermal titration calorimetry; KinExA, kinetic exclusion assay. Accelerated degradation refers to a set of biophysical methods (see text).

The discovery process is described in this chapter by several stages: target generation, hit evaluation, lead selection, lead optimization, lead formatting, and final lead candidate selection of a molecule to progress into development. We note that the types and extent of biophysical characterization will depend to some degree on the molecular class of the protein therapeutic (monoclonal antibody, Adnectin, antibody fragments,

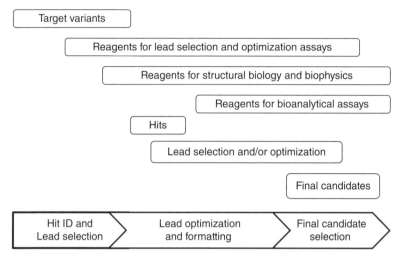

Figure 1.1. Scheme showing the different classes of protein reagents and drug candidates produced and characterized by biophysical methods from the initiation of a drug discovery program through selection of a final molecule for subsequent progression into development. Initially the protein production and biophysical characterization efforts are focused on the target(s) and reagents. As the program progresses, the amount of protein chemistry increases and shifts toward the production and characterization of protein therapeutic candidate molecules. The type and extent of biophysical characterization done for each class of protein and for each stage of discovery is different as described in Table 1.1.

non-antibody fragments, etc.) and the technology used for selecting lead candidates (immunizations, phage display, RNA display, etc.). The purpose here is to present a case study of biophysical applications during the discovery of a bispecific Adnectin against epidermal growth factor receptor (EGFR) and insulin-like growth factor-1 receptor (IGF1R). Many of the details for this system have been reported elsewhere [2].

1.2 TARGET IDENTIFICATION

Identification of a drug's biological target is a critically important part of a biotherapeutic discovery program. One of the expanding areas in biotherapeutics research is the design of bispecific biotherapeutics that bind to two different, already validated biological targets. The proposed benefits for the bispecific-targeting approach include improved efficacy and lower cost of goods than developing two drugs independently.

Drug targets may also be identified from genetic validation studies (correlation between mutation of target and disease state) or pharmacological validation studies (utilizing a surrogate molecule such as a natural ligand to demonstrate efficacy in a non-clinical setting). The Holy Grail for identification of completely novel targets is to utilize the growing information from genomic, proteomic, and interactomic studies to draw correlations between specific drug targets, or sets of drug targets, and treatment of disease.

This chapter describes a case study for discovery of an Adnectin [3] bispecific biotherapeutic that targets inhibition of both EGFR and IGF1R (Emanuel et al. [2]). EGFR is a clinically validated target for cancer therapy, and there are both small-molecule kinase inhibitors and biotherapeutic inhibitors of the extracellular domains presently available as marketed drugs. IGF1R is also an attractive target for cancer therapy and there are several small-molecule and biotherapeutic inhibitors in preclinical and clinical studies [4].

1.3   TARGET GENERATION

Once a target has been identified, it is usually produced recombinantly to provide suffi-cient material to enable selection of biotherapeutic candidate "hits" through a screening or selection process. There are several technologies commonly used for generating bio-therapeutics hits, including in vivo immunization, phage display, mRNA display, and yeast display [5, 6]. All of these technologies rely on the production of biophysically well-behaved target molecule. Biophysical methods thus play a critical role as "gate-keeper" at this phase of discovery, to ensure the quality of the target being used for screening or selections is suitable for generating the best candidates.

The first step in producing the target reagent is to engineer a form of the target molecule that will be expressed well and has acceptable biophysical behavior when puri-fied. Sometimes the design is fairly straightforward. For instance, the construct design, expression, and purification for some targets may be well described in the literature. Construct design may also be straightforward if the protein target itself is structurally small and simple. An example would be a soluble target such as a cytokine. The construct design of a simple small protein could be as straightforward as expressing the entire native protein. On the other hand, construct design of large membrane-spanning protein targets can be much more challenging since the membrane-spanning and intracellular regions usually need to be deleted in order to make well-behaved soluble extracellu-lar fragment(s) of the target. Whether or not some or all of the extracellular domains extracted from the full-length protein can be expressed, purified, and well behaved biophysically is often not known in advance.

1.3.1   Multiple Constructs Strategy

Given significant uncertainties and risk surrounding the production of critical target molecules, it is prudent to approach the problem with the design of multiple constructs in parallel, at least through DNA expression vector or small-scale expression screening stages. There are several reasons for designing multiple constructs up front for a target molecule. First, most target molecules need to be produced as fusions with a variety of affinity tags (e.g., His tags, Flag tag) to facilitate purification and development of different types of downstream assays. These non-native sequences may in turn alter the native functional or biophysical behavior of the target. Thus, different types of tags, each having different linker sequences joining them to the target molecule, may need to be made and tested for suitable functional and biophysical behaviors by trial and error. Second, different domain regions, or fragments, of a target protein will have different intrinsic expression and biophysical properties, some of which will have acceptable biophysical and functional behaviors and others will not. As a general rule, the more novel is the target, the less is known about its expression and biophysical and functional properties and the greater the risk is of making it in useable form. Novel targets thus deserve more upfront engineering of multiple constructs. Finally, different forms of a target protein may generate different types of epitope families of lead drug candidates from the high-throughput screening or selection process, for reasons that may not be obvious. In order to obtain a sufficient diversity of initial drug candidates to evaluate during discovery, it is therefore useful to screen against multiple forms of the target molecule. For all these reasons, it is prudent to carefully plan out the target design strategy and backup strategies at the beginning of the target generation process, since the cycle time from construct engineering through biophysical and functional assessments is measured in weeks to months.

In the cases of EGFR and IGF1R, there are extensive precedences in the literature for making a variety of extracellular fragments. Moreover, there are three-dimensional crystal structures for some of these fragments, showing where the self-contained domain regions are at atomic resolution. We designed multiple variants of the extracellular regions of EGFR and IGF1R target proteins. The variants included different purification tags, different expression hosts, and different length variants of the extracellular regions. A subset of the constructs designed were expressed, purified, and characterized with biochemical and biophysical methods as described in Table 1.1.

Production of the target molecule, and multiple variants thereof, is only a subset of the total number of reagents needed to support a protein therapeutics drug discovery program. The scheme in Figure 1.1 describes the various classes of additional reagents needed, as well as protein therapeutic drug candidates, that must be produced and characterized during the discovery phase. Ideally one would like to have all the variants of the target, co-targets, counter-targets, and species ortholog targets upfront in the early phase of a discovery program in order to facilitate selection of leads with the optimal diversity and cross-reactivity profiles. However, producing all these reagents upfront is very time consuming and it is not uncommon for a program to move forward as soon as an adequate amount of the human target protein is available, and then to produce the other reagents for optimizing cross-reactivity and potency later in the program.

1.4   HIT EVALUATION

In the earliest stage of drug candidate biophysical assessment, many potential lead candidate molecules need to be evaluated in high-throughput mode (typically on the order of hundreds or thousands, or more, depending on the hit identification technology being used). The purification methods used at this stage are high throughput and must be robust and simple enough to generate large numbers of candidates within a reasonable period of time, but do not need to yield proteins that are as high in purity or quantity as will be needed in the later stages of discovery. The biophysical assessment at this stage must also be rapid and simple and be able to distinguish the higher-quality lead candidates from the lower-quality leads. Some of the key biophysical methods used for hit identification include analytical size-exclusion chromatography (SEC), biosensor analysis, and thermal stability fluorescence (TSF; Table 1.1). These methods provide information about the self-association, binding affinity, and conformational stability properties of the hit molecules, respectively, and can be conducted in high-throughput mode using small quantities (sub-milligram) of protein sample.

1.4.1 Qualitative and Rapid Self-Association Check

Figure 1.2 shows example analytical SEC data [7, 8] for a well-behaved homogeneous candidate protein therapeutic in comparison to one that is heterogeneous and contains high molecular weight (HMW) species. Here we assume the homogeneous profile reflects a monomeric drug candidate. This assumption will be more rigorously tested at later stages of discovery using the more rigorous methods in Table 1.1. The presence of aggregates or HMW species suggests that production and storability of the molecule will likely involve more challenges during discovery than the molecule that exhibits homogeneous, monomeric behavior. Furthermore, the heterogeneity observed at the hit stage signals a risk that the poorer behavior might be retained during the later stages of

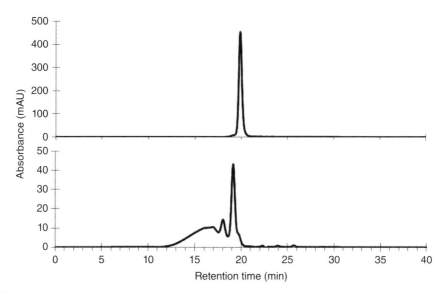

Figure 1.2. Analytical size-exclusion chromatography data showing examples of the elution profiles of early-stage Adnectin drug candidates. The top panel shows a homogeneous, monomeric drug candidate, and the bottom panel shows a candidate that has high molecular weight (HMW). Data of this type is used to select the most promising drug candidates for advancement.

discovery and during development. Barring any other exceptionally redeeming properties of the candidate having the HMW species present (such as being one of the very few hits having unique cellular activity or potency), one would normally select the homogeneous molecule to progress into the subsequent stages of discovery.

1.4.2 Qualitative and Rapid Thermal Stability Check

The conformational (or folding) stability of a protein is broadly used as a general measure of stability. This is because the partially or fully unfolded species of proteins are usually more prone to physical and chemical mechanisms of degradation (e.g., aggregation, proteolytic clipping, deamidation) than are the natively folded species. Thermal denaturation of proteins can be measured by many different technologies. One commonly used method that is rapid and requires only microgram amounts of protein is TSF [9]. This method goes by several different names such as thermofluor, thermal stability perturbation, and thermal shift assay. Here we refer to it as thermal stability fluorescence or TSF. Figure 1.3 shows an example of thermal stability for an Adnectin as measured by TSF. In this experiment, the temperature of the protein sample in the presence of an extrinsic fluorophore is increased, while the fluorescence of the sample is monitored. When the protein unfolds, there is an increase in exposed hydrophobic surface area which then binds to the extrinsic fluorophore and causes an increase in fluorescence. In principle, one can monitor the extent of unfolding from the extent of the change in fluorescence shown in the figure. A convenient measure of the thermal stability that can be used to rank-order the relative thermal stabilities of a series of closely related

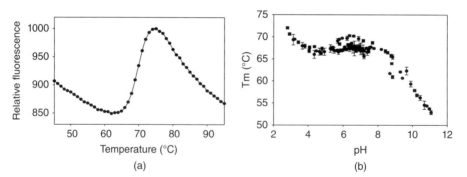

(a)   (b)

Figure 1.3. Example of thermal stability of a biotherapeutic candidate molecule as measured by thermal stability fluorescence. (a) Fluorescence of an Adnectin candidate in the presence of the extrinsic fluorophore anilinonaphthalene sulfonic acid (ANS) as a function of temperature. As the protein unfolds, hydrophobic regions are exposed to solvent, bind ANS, and cause an increase in fluorescence. The midpoint of the transition (Tm) is obtained by curve fitting and is used as a qualitative measure of thermal stability. The Tm for the curve shown is 70.2°C. (b) Tm values measured in high-throughput mode for the same Adnectin in many different buffer pH conditions. The experiment was done in 384-well format and demonstrates the ability to rapidly screen buffer conditions that may influence the thermal stability of the drug candidate.

protein drug candidates is the temperature at which half the protein is unfolded, also called the midpoint temperature and denoted by Tm [10]. Generally speaking, a higher Tm is preferred, as it implies the conformational stability is higher. All other parameters being equal, one would prefer to progress drug candidates that have higher thermal stability, with the anticipation that they may be easier to produce, handle, and store. However, it is also important to recognize that the Tm by itself does not always predict shelf life or manufacturability of a protein therapeutic. In some cases, aggregation can be initiated by the solubility limit of the natively folded protein or a chemically modified folded form of the protein [11].

1.4.3   Confirmation of Binding

One of the most important factors used to evaluate hit candidates is to determine whether or not they bind the target molecule, and if so, how tight the interaction is. Biosensor is a biophysical method often used at the hit evaluation stage because they can be run in higher throughput mode, while consuming very little of the hit molecules [12–14]. Biosensor is a workhorse technology for all phases of protein therapeutics drug discovery and more will be described about this technology later in this chapter and throughout the book. Because the purity values of the hit molecules may not be accurately understood, analysis of the association kinetics is difficult to interpret quantitatively (the association kinetics are dependent on an accurate knowledge of the active concentration of reactant in solution phase which is usually the drug candidate). Instead, the main goal for biosensor work at the hit evaluation stage is to confirm the hit molecules bind to the target. This would normally be done at concentrations of reactants high enough to allow detection of binders that have an acceptable affinity, but low enough to reduce potential nonspecific interactions with the surface. For example, the hits could be tested at a single concentration of 1 µM to discern if they bind with equilibrium dissociation constants of at least approximately 1 µM. If binding is not detected at 1 µM, then the hit molecule either does not bind the target or its affinity is much weaker than 1 µM and perhaps of little interest as a lead molecule. The rate of a hit dissociating from the target may also provide useful information for comparing between hits. Hits having unusually long dissociation rates likely indicate they are binding either with higher affinity or by distinct binding modes compared to hits with much faster dissociation rates.

1.5   LEAD SELECTION

The next stage of discovery is the selection of lead families of candidates for optimization and progression into the later stages of discovery. The decisions about which candidate molecules to advance have long-term consequences for the success and challenges that will be encountered by the program, including whether the binding epitopes are able to elicit suitable biological efficacy from the target and whether there are any chemical or physical liabilities associated with the lead candidate or family. Ideally one would like to select multiple lead families that bind to a diversity of epitopes, to maximize likelihood of favorable biological activity, and have favorable biophysical properties, to increase the chances of ultimately producing candidates that have superior stability and manufacturability attributes.

The biophysical properties that are used as part of the selection criteria include self-association, conformational stability, binding affinity, and binding epitope. In order to measure these biophysical properties rigorously, it is necessary to produce the potential lead molecules at the milligram scale and to purify them to a higher purity standard (e.g., sample is at least 95% molecule of interest). The biophysical methods themselves are also more rigorous at this stage. Prior to biophysical analysis the candidates usually undergo an evaluation of purity and identity by SDS PAGE and LC/MS. SDS PAGE and the LC part of LC/MS provide information about purity, and the mass spectrometry data provide mass information of sufficient accuracy to confirm the identity of the protein candidate to its expected amino acid sequence.

1.5.1   Self-Association

Prior to selecting a lead candidate it is important to obtain an accurate understanding of the self-association properties in a standard biological buffer system such as phosphate buffered saline (PBS) or histidine buffer. Ideally, one would like to evaluate the self-association properties in more than one buffer in order to minimize the risk of buffer-specific anomalous behavior. The analytical SEC assessment done at the hit identification stage provides a qualitative measure of self-assessment but can sometimes be obscured by interactions with the column matrix or non-candidate impurities [8]. These obstacles can be overcome to a large extent by coupling the SEC method with multiple angle light scattering (SEC/MALS) [15]. The MALS detector system allows one to measure the absolute mass of the protein sample across the elution peak(s), irrespective of elution time. Figure 1.4 shows SEC/MALS data for an Adnectin lead molecule. In this case the protein elutes with a homogeneous profile as measured by absorbance at 280 nm. The dotted curve drawn across the elution peak represents the weight-average MW of the sample measured at many individual time points during peak elution. The average value of the measurements across the main peak is 11 kDa and is within error equal to the mass of a homogeneous monomer of the protein (theoretical mass = 10.9 kDa). At time points earlier than the main peak elution, the light-scattering signal detects MW species for very small amounts (1% or less) of HMW material that are of a size approximately that of a dimer.

Another method for rigorous analysis of protein self-association behavior is sedimentation equilibrium analytical ultracentrifugation (SE-AUC). Like SEC/MALS, SE-AUC is a method that measures the absolute mass of the protein sample [16]. It is significantly more time consuming than SEC/MALS but has the advantage that it can measure self-association equilibrium constants for simple equilibrium systems such as monomer–dimer equilibria. This is an important advantage for lead molecules against targets that are influenced by dimerization. In such cases, the dimerization constants for a series of lead molecules can be used as a criterion for selection of progressible candidates and provide insight into the final format of the drug molecule (e.g., monomeric or dimeric). Figure 1.4c shows AUC data for the same lead Adnectin in Figures 1.4a and 1.4b. The data are from a sedimentation equilibrium experiment and provide strong confirmation that the protein is a homogeneous monomer over the

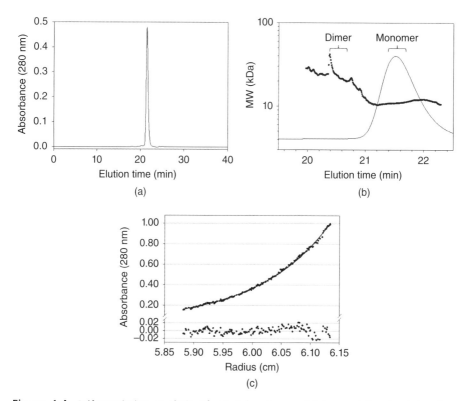

Figure 1.4. Self-association analysis of an Adnectin candidate as determined by size-exclusion chromatography combined with multiple angle light scattering shown in (a) and (b) and analytical ultracentrifugation shown in (c). The Adnectin eluted from a size-exclusion column (a) with a major peak (99% of 280 nm signal) at 21.6 min and a minor peak (1%) at 20.5 min. From light-scattering data collected during the run, the MW versus elution time plot (b) shows that the main peak eluted with a MW consistent with monomeric protein, and the shoulder likely contained dimer. (c) shows the sedimentation equilibrium analysis of the absolute mass of the Adnectin as measured by analytical ultracentrifugation. The best-fit curve shown is the one representative from a set of multiple centrifugation speeds fit globally to a single mass species. The best-fit from the global analysis yielded a mass of 10.3 kDa. This agrees well with the theoretical mass of 10.9 kDa.

concentration range shown (A280 from about 0.1 up to 1) based on the goodness of fit to the single-exponential curve-fitting analysis.

1.5.2 Thermal Stability

At the lead selection stage, the preferred method for measuring thermal stability of the lead candidates is by differential scanning calorimetry (DSC). DSC is the gold standard method for measuring thermal unfolding. It measures the excess heat capacity of the protein as temperature is scanned and directly monitors unfolding from the change in heat for the reaction. One reason why DSC is the preferred method is that it is not susceptible to fluorescence or other optical artifacts that sometimes occur with TSF or other optical methods. Another reason is that DSC instrumentation offers high-precision and high-accuracy temperature control.

DSC is in principle a rigorous way to also measure the thermodynamics of the unfolding–folding equilibrium, including the free energy, enthalpy, and heat capacity changes. These parameters describe the conformational energy of the protein in detail. That said, DSC is oftentimes subject to artifacts such as scan-rate dependence of the unfolding curve, lack of unfolding–folding reversibility, or artifactual heats originating from side reactions such as aggregation. Thus, in practice, DSC data are mainly used in drug discovery as a semi-quantitative measure of stability. Even so, it is more direct than optical methods and can be controlled more precisely.

Figure 1.5 shows DSC data for two Adnectin candidates. The one shown in the lower part of the figure is superior in two ways. First, the unfolding begins at a higher temperature and the Tm is several degrees higher. This indicates the candidate has higher conformational (or folding) stability. Second, the unfolding reaction of the lower one is reversible. That is, after thermally unfolding, it can be cooled, refold in the calorimeter,

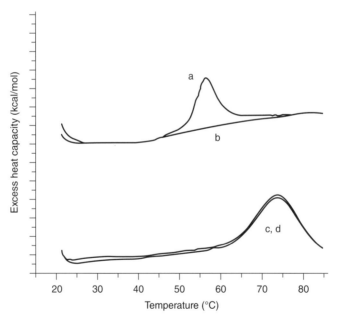

Figure 1.5. Differential scanning calorimetry (DSC) of two different Adnectin drug candidate molecules. The top two traces are indicative of a thermally irreversible protein system. The Adnectin in the top portion of the panel denatured in the first thermal scan (a) shows no evidence of regaining structural integrity in the time frame allotted for the second thermal scan (b). The lower half of the panel displays a different Adnectin molecule that displays essentially complete thermal reversibility under the conditions tested. The first thermal scan (c) and second scan (d) are observed to overlay, indicating the protein melted in the first scan has recovered structural integrity and behaves identically when thermally scanned a second time.

and be thermally unfolded again. The ability to unfold–fold reversibly is in some cases an indicator for improved expression levels [17] and reduced aggregation tendency [18]. This is presumably due to the ability of the refolding reaction to compete against aggregation side reactions of the unfolded form(s).

1.5.3   Binding Affinity, Kinetics, and Epitope

The primary objective of most protein therapeutic programs is to identify a molecule that binds to a preferred epitope on the target and with high affinity. A preferred epitope is one that yields one or more of the following outcomes: strong or partial antagonism of target function, strong or partial agonism, selective modulation of some target functions but not others, presence or absence of target degradation, cross-reactivity to the same epitope on the target from nonhuman species used in critical program assays, cross-reactivity to closely related human co-targets, lack of cross-reactivity to human liability targets, and so on.

Unfortunately, it is usually not possible to know if a hit molecule binds to a preferred epitope, and it is therefore necessary to select a diversity of lead candidates that bind to different epitopes for further studies and affinity optimization. A common method for selecting lead molecules that bind to distinct epitopes on a target is biosensor technology. To accomplish this, lead molecules are examined in pairs to determine if they can bind to the target molecule simultaneously or not. Ideally one should have a sound understanding of the binding affinities, kinetics, and concentration ranges used in such studies, as described by Yamniuk et al. [19]. Figure 1.6 depicts an example

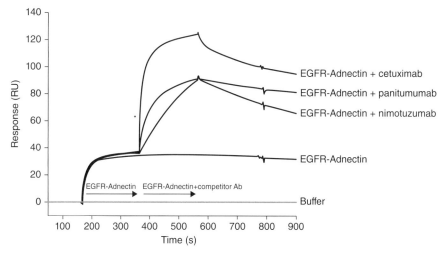

Figure 1.6. Biosensor data, measured on a Biacore T100 surface plasmon resonance instrument, showing an EGFR Adnectin does not compete with clinically approved mAbs for binding to EGFR. EGFR was immobilized by amine coupling and experiments were conducted as described elsewhere [2]. Briefly, the EGFR Adnectin was flowed over the EGFR surface alone at 450 nM, and then either alone at 450 nM or together with 450 nM mAb as shown. Reproduced with permission from Reference 2.

epitope discrimination study with a lead EGFR Adnectin in comparison to three anti-EGFR monoclonal antibodies. The studies were conducted using biosensor technology and demonstrate that the EGFR Adnectin binds to the target molecule EGFR at an epitope that is sterically distinct from EGFR antibodies cetuximab, panitumumab, and nimotuzumab. Details about these experiments are provided by Emanuel et al. [2].

At the lead selection stage, the protein candidates should be purified to high standards and in hundreds of microgram to milligram amounts. They are therefore suitable for enabling a more rigorous analysis of binding affinity and kinetics using biosensor technology. Generally speaking, the higher the affinity the more attractive the lead molecule is for advancing further into the later stages of discovery. The kinetics may also provide clues for discriminating between different modes of binding to target. Slower association kinetics may be due to rate-limiting conformational change in the target (or the drug candidate) and thus could reflect a novel mode of binding the target that could correspond to a novel biological outcome. Example biosensor data showing the association and dissociation curves for EGFR and IGF1R Adnectins binding their targets are discussed in Sections 1.6 and 1.8.

1.6   LEAD OPTIMIZATION

Once lead candidates have been selected they are optimized in terms of their binding affinity, cross-reactivity, potency, and biophysical stability attributes. Again, biophysical technologies play a central role in guiding the optimization to generate advanced lead candidates that have the desired biological activity and are likely to be manufacturable. The biophysical methods used at this stage are very similar to those used in the lead selection stage, but the extent of characterization is increased.

During lead optimization, assays are often performed that require the drug candidates to bind to the target from nonhuman species such as mouse or rat. In order to help validate these types of assays biosensor technology is often used to demonstrate that the leads bind to the nonhuman targets and that upon affinity optimization to the human target, the cross-reactivity toward the nonhuman targets is maintained (or not). For the EGFR part of the EGFR–IGF1R bispecific molecule, we produced extracellular fragments of EGFR from multiple species to verify with biosensor technology that the leads bind to EGFR from species relevant to preclinical efficacy and toxicology studies. Similarly, for the IGF1R part of the molecule, we also conducted biosensor studies to assess the nonhuman species cross-reactivity. Figure 1.7 shows biosensor data for binding the leads to human, monkey, mouse, and rat IGF1R. The results demonstrated very tight and nearly indistinguishable affinities for human and monkey IGF1R. Binding to rat and mouse was also observed, and in an affinity range acceptable for downstream studies involving those targets. This is a case study with a favorable outcome. However, not all programs are as fortunate. Some programs could require a parallel discovery effort to create a species-specific surrogate biologic in order to conduct critical studies needed to progress the program.

Structural biology methods such as X-ray crystallography and NMR are also important biophysical tools for lead optimization. Solving the three-dimensional structure of a

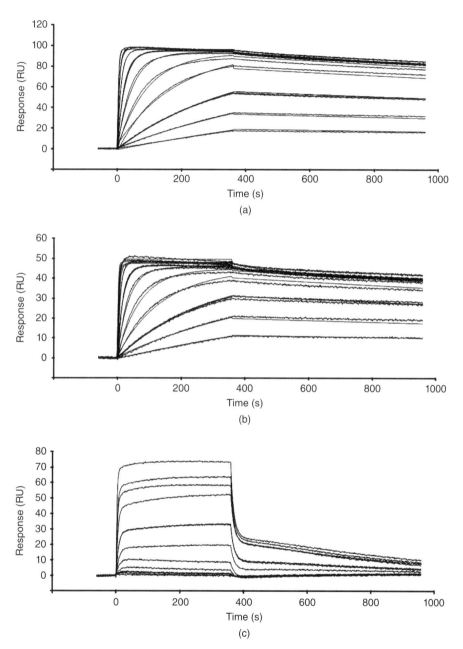

Figure 1.7. Biosensor data for anti-IGF1R Adnectin binding to IGF1R-Fc from different species: human (a), monkey (b), mouse (c and d) and rat (e and f) IGF1R. In each panel, the IGF1R-Fc target was captured on the surface by protein A and the Adnectin was flowed across the surface at multiple concentrations from 1 to 500 nM. The data for human and monkey IGF1R were well described by a simple Langmuir fitting model. In contrast, the kinetic data for mouse and rat (c, e) could not be well described with a simple model.

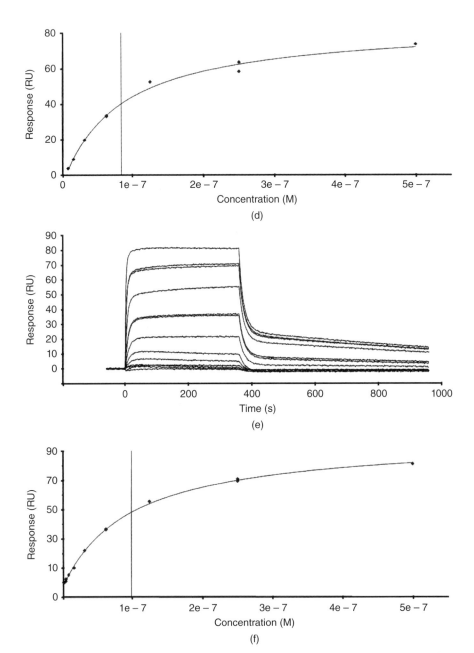

Figure 1.7. (*Continued*) Instead, the affinities for mouse and rat IGF1R were obtained from a steady-state equilibrium analysis (d, f) by fitting to a simple equilibrium model [20] based upon the amount of Adnectin bound versus total concentration of Adnectin in solution. Binding equilibrium dissociation constants for the four interactions were thus determined as 0.3 nM (a, human), 0.2 nM (b, monkey), 84 nM (d, mouse), and 99 nM (e, rat).

lead molecule in complex with its biological target provides an atomic-level understanding for designing improved binding affinity or cross-reactivity to species ortholog forms of the target or co-targets. An example where crystallography was used to determine the three-dimensional structure of an Adnectin lead bound to its biological target is reported by Ramamurthy et al. [21]. In their report the authors describe in molecular detail how the Adnectin lead molecule binds the target. The unique ways the Adnectin binds offer unexpected opportunities to optimize the lead molecule–target interaction.

1.7 LEAD FORMATTING

Most of the so-called next-generation antibody fragment and non-antibody fragment discovery programs offer novel ways to format the drug candidates to provide multivalent and/or multispecific target binding [22,23]. Many of these programs also include molecular formatting, such as fusion to the Fc fragment of an antibody or by covalent conjugation to large inert moieties such as PEG, to create extended pharmacokinetics. Any of these formatting modifications may alter the biophysical stability or target-binding ability of the lead molecules, and biophysical methods are required to determine which formats yield final candidates with acceptable stability and binding profiles. In the present case, mono-Adnectin leads against EGFR were formatted into single-chain bispecific leads by in-line fusion with a high-affinity, stable mono-Adnectin against IGF1R. Furthermore, in order to extend the pharmacokinetics of the candidates, they were conjugated to 40 kDa branched PEG. A full analysis of the biochemical, cellular, and biophysical properties of the component mono-Adnectins and the formatted versions is given by Emanuel et al. [2]. In some cases formatting by these or other mechanisms can decrease (or increase) binding affinity and/or change self-association, aggregation, solubility, or thermal stability of the lead. Biophysical analysis of formatted lead candidates is critical to guide optimization of the formatting itself (size or location of PEG, orientation and linkers of in-line fusions, etc.), and selection of the preferred combination of lead candidate and format. In developing a bispecific candidate, for example, the physiological mechanism might require that a single bispecific candidate molecule be able to bind simultaneously and with high affinity to two different targets. This would be tested at the lead optimization and formatting stages by biosensor or perhaps ITC or AUC studies.

1.7.1 Solubility

As the discovery program moves into the later stages of discovery, more emphasis is placed on selecting candidates that have a better chance of having cost-effective manufacturability. A critical part of manufacturability assessment is the determination of solubilities of the drug candidates. In some programs an estimate for the dosing amount and concentration is known at this stage of discovery, and in other programs the dosing information may not be well understood yet. In either case, selecting candidates with better solubility profiles will facilitate the downstream process development and formulation work.

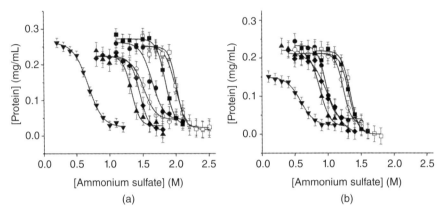

Figure 1.8. Determination of relative solubility values for seven EGFR mono-Adnectins and an anti-IGF1R Adnectin (a), and the same seven when formatted as a bispecific Adnectin with the single anti-IGF1R Adnectin (b). The IGF1R Adnectin in (a) is shown as stars and is the furthest curve to the right. Relative solubility values were measured by an adaptation of the ammonium sulfate precipitation method reported by Trevino et al. [24]. The data in this figure show the concentration of drug candidate remaining soluble as a function of molarity of ammonium sulfate. The higher the solubility of a protein, the more ammonium sulfate required to cause precipitation. Thus, the more left-shifted the curve, the less soluble the protein therapeutic candidate. The results provide a way to rank-order the solubilities of the drug candidate molecules.

Normally the dosing concentration is quite high for a typical protein, and reaching the high concentrations can be a substantial challenge for many programs, as described in the other chapters of this book. There are several commonly used methods for measuring protein solubility. Typically they require larger amounts of protein and can be time consuming to conduct. A method that has become popular recently for rapid assessment of the solubilities of several candidates at a time, and thus suitable for use during discovery, is PEG precipitation or ammonium sulfate salting-out studies [24, 25]. The method is conducted at conditions where the protein is in the native form, and the objective is to gain information about the solubility of the native form. We conducted ammonium sulfate salting-out studies on the EGFR mono-Adnectins to help select the candidates with the most favorable solubility behavior. The results are shown in Figure 1.8. Most of the candidates had similar solubility behavior, but one lead candidate in particular had much lower solubility. The rank-order of estimated solubilities obtained by this higher throughput method was in good agreement with solubility limits measured by centrifugal spin concentration (data not shown), although the most soluble molecules could not be fully characterized by the spin method due to a lack of sufficient amounts of the candidate molecules to reach their upper limit solubility values. We also tested the same EGFR candidates after formatting to the bispecific format using a common IGF1R mono-Adnectin with all of them. The rank-order of solubilities was in good agreement with the individual EGFR mono-Adnectins, indicating that for this system the mono-Adnectin behavior was predictive of the behavior in the bispecific format. The same trends were also observed after PEGylation (not shown).

1.7.2 Thermal Unfolding Behavior

Although the solubility of the natively folded protein is an important property to understand, oftentimes aggregation of a drug candidate is related to the insolubility of partially or fully unfolded forms of the protein [11, 26]. Thus it is important to conduct careful thermal unfolding studies on late-stage protein therapeutic candidates. As discussed above in the Section 1.5, the DSC analysis should include an assessment of the onset temperature for unfolding (the temperature when unfolding can first be detected), the Tm or midpoint temperature for unfolding, and the reversibility of unfolding. In general higher unfolding temperatures and greater reversibility are desirable properties to select for, since the less exposure that a protein has for its unfolded form(s), the less likelihood for reacting along unfolding-based aggregation pathways.

1.8 FINAL DEVELOPMENT CANDIDATE SELECTION

In the final stage of discovery a candidate molecule is chosen to progress into development. At this stage, there is an increased emphasis on the use of biophysical methods to select lead candidates that will have favorable manufacturability. That is, how stable are the candidates? Can a process be developed to make the final candidate at larger manufacturing scale? Will the final candidate have a long shelf life? How concentrated can the candidate be made and remain stable? Answering these questions directly in the discovery stages is impractical due to the large number of candidates that must be made and characterized and the small scale of protein production typically done in discovery. However, there are biophysical approaches that allow for the selection of a final candidate based on its broad biophysical behavior in many conditions and by multiple methodologies. The assumption made is that a candidate that exhibits superior biophysical properties under a broad range of conditions will stand the greatest chance for performing well in development.

To address these issues most of the biophysical methods in Table 1.1 are needed. The number of biophysical methods used during final candidate selection is expanded to increase confidence in understanding the biophysical properties of each of the candidates. For example, to determine the binding affinity with high confidence, orthogonal methods such as ITC [27] or KinExA [28] are used. These methods measure affinity in solution and avoid potential immobilization-dependent or complex kinetics artifacts that are sometimes present in biosensor studies. Similarly, a thorough understanding of the aggregation properties of a protein requires multiple methods [29].

Additionally, the number of solution conditions studied must also be expanded to understand the generality of the drug candidate's stability profile. Studies at this stage include broad screening of each candidate's thermal stability against pH, salt, and excipients. Figure 1.3b shows the Tm value obtained by TSF over a wide range of pH. The results show that the Adnectin candidate is quite stable from low pH up to around pH 8, but exhibits a drop in thermal stability at pH values above pH 8. Knowledge about specific conditions that stabilize or destabilize a candidate is useful for production and handling during discovery phases and also serves to inform the design of production and formulation work done later in development. These data provide a rapid and insightful view of the general stability behavior [30] of the protein candidate and can identify conditions or excipients that may facilitate or obstruct the manufacture or storage of the candidate. This type of data is useful for selecting the final candidate for development and it also provides the development teams with fundamental thermal stability information that can be used to jump-start formulation and process development.

In order to rapidly assess the chemical and physical liabilities of the candidates, accelerated degradation studies are conducted. The candidates are thus exposed to conditions of pH, temperature, protein candidate concentration, time, and so on that challenge a protein's stability to reveal any degradation reactions that may exist [31, 32]. Many of the biophysical methods listed in Table 1.1 are used as part of the accelerated studies, and several others are used, such as ion exchange chromatography or hydrophobic interaction chromatography, to investigate for the production of charge or conformational species.

1.9 CONCLUDING REMARKS

The discovery phase of a protein therapeutic is a protein chemistry-intensive environment that includes the production of a large number of drug candidates and a large number of diverse protein reagents. Biophysical methods are necessary to ensure the fidelity of the drug discovery process. Protein reagents that are made in discovery and relied upon for critical assays must be characterized to ensure that they are properly folded and active. Biophysical methods also play a prominent role in characterizing the lead candidates and for selecting well-behaved leads for progression through the later stages of discovery. Finally, biophysical methods play a central role in the selection of biotherapeutics that will be cost-effective to manufacture and safe as drugs.

ACKNOWLEDGMENT

We thank Thomas Palm for use of the differential scanning calorimeter for initial characterization of some of the EGFR Adnectin leads.

REFERENCES

1. Kim, Y.J. and Doyle, M.L. (2010) Structural mass spectrometry in protein therapeutics discovery. *Anal. Chem.*, **82**, 7083–7089.
2. Emanuel, S.L., et al. (2011) A fibronectin scaffold approach to bi-specific inhibitors of epidermal growth factor receptor and insulin-like growth factor receptor. *mAbs*, **3**, 38–48.

3. Lipovsek, D. (2011) Adnectins: engineered target-binding protein therapeutics. *Protein Eng. Des. Sel.*, **24**, 3–9.

4. Scartozzi, M., et al. (2011) State of the art and future perspectives for the use of insulin-like growth factor receptor 1 (IGF-1R) targeted treatment strategies in solid tumors. *Discov. Med.*, **11**, 144–153.

5. Rothe, A., Hosse, R.J., and Power, B.E. (2006) Ribosome display for improved biotherapeutic molecules. *Expert Opin. Biol. Ther.*, **6**, 177–187.

6. Mondon, P., Dubreuil, O., Bouayadi, K., and Kharrat, H. (2008) Human antibody libraries: a race to engineer and explore a larger diversity. *Front. Biosci.*, **13**, 1117–1129.

7. Goetz, H., et al. (2004) Comparison of selected analytical techniques for protein sizing, quantitation and molecular weight determination. *J. Biochem. Biophys. Methods*, **60**, 281–293.

8. Arakawa, T., Ejima, D., Li, T., and Philo, J.S. (2010) The critical role of mobile phase composition in size exclusion chromatography of protein pharmaceuticals. *J. Pharm. Sci.*, **99**, 1674-1692.

9. Cummings, M.D., Farnum, M.A., and Nelen, M.I. (2006). Universal screening methods and applications of thermoFluor. *J. Biomol. Screen*, **11**, 854–863.

10. Eftink, M.R. (1995) Use of multiple spectroscopic methods to monitor equilibrium unfolding of proteins. *Methods Enzymol.*, **259**, 487–512.

11. Philo, J.S. and Arakawa, T. (2009) Mechanisms of protein aggregation. *Curr. Pharm. Biotechnol.*, **10**, 348–351.

12. Bronner, V., et al. (2010) Therapeutic antibodies: discovery and development using the ProteOn XPR36 biosensor interaction array system. *Anal. Biochem.*, **406**, 147–156.

13. Safsten, P., et al. (2006) Screening antibody-antigen interactions in parallel using Biacore A100. *Anal. Biochem.*, **353**, 181–190.

14. Leonard, P., et al. (2007) High throughput ranking of recombinant avian scFv antibody fragments from crude lysates using the Biacore A100. *J. Immunol. Methods*, **323**, 172–179.

15. Wen, J., Arakawa, T., and Philo, J.S. (1996) Size-exclusion chromatography with on-line light-scattering, absorbance, and refractive index detectors for studying proteins and their interactions. *Anal. Biochem.*, **240**, 155–166.

16. Doyle, M.L. and Hensley, P. (1997) Experimental dissection of protein-protein interactions in solution. *Adv. Mol. Cell. Biol.*, **22A**, 279–337.

17. Gerber, E. and Demarest, S.J. (2007) A broad range of Fab stabilities within a host of therapeutic IgGs. *Biochem. Biophys. Res. Commun.*, **355**, 751–757.

18. Remmele, R.L., Bhat, S.D., Phan, D.H., and Gombotz, W.R. (1999) Minimization of recombinant human Flt3 ligand aggregation at the Tm plateau: a matter of thermal reversibility. *Biochemistry*, **38**, 5241–5247.

19. Yamniuk, A.P., et al. (2012) ABRF-MIRG benchmark study: molecular interactions in a three component system. *J. Biomol. Tech.*, **23**, 101–114.

20. Morelock, M.M., Ingraham, R.H., Betageri, R., and Jakes, S. (1995) Determination of receptor-ligand kinetic and equilibrium binding constants using surface plasmon resonance: application to the lck SH2 domain and phosphotyrosyl peptides. *J. Med. Chem.*, **38**, 1309–1318.

21. Ramamurthy, V., et al. (2012) Structures of Adnectin/protein complexes reveal an expanded binding footprint. *Structure*, **20**, 259–269.

22. Beck, A., Wurch, T., Bailly, C., and Corvaia, N. (2010) Strategies and challenges for the next generation of therapeutic antibodies. *Nat. Rev. Immunol.*, **10**, 345–352.

23. Carter, P.J. (2011) Introduction to current and future protein therapeutics: a protein engineering perspective. *Exp. Cell Res.*, **317**, 261–1269.

24. Trevino, S.R., Scholtz, J.M., and Pace, C.N. (2008) Measuring and increasing protein solubility. *J. Pharm. Sci.*, **97**, 4155–4166.

25. Gibson, T.J., et al. (2010) Application of a high-throughput screening procedure with PEG-induced precipitation to compare relative protein solubility during formulation development with IgG1 monoclonal antibodies. *J. Pharm. Sci.*, **100**, 1009–1021.

26. Brummitt, R.K., et al. (2011) Nonnative aggregation of an IgG1 antibody in acidic conditions: part 1. Unfolding, colloidal interactions, and formation of high-molecular-weight aggregates. *J. Pharm. Sci.*, **100**, 2087–2103.

27. Doyle, M.L. (2001) Titration microcalorimetry. *Curr. Protoc. Protein Sci.*, **20.4**, 1–24.

28. Darling, R.J. and Brault, P.A. (2004) Kinetic exclusion assay technology: characterization of molecular interactions. *Assay Drug Dev. Technol.*, **2**, 647–657.

29. Philo, J.S. (2006) Is any measurement method optimal for all aggregate sizes and types? *AAPS J.*, **8**, E564–E571.

30. Mezzasalma, T.M., et al. (2007) Enhancing recombinant protein quality and yield by protein stability profiling. *J. Biomol. Screen*, **12**, 418–428.

31. Li, Y., Mach, H., and Blue, J.T. (2011) High throughput formulation screening for global aggregation behaviors of three monoclonal antibodies. *J. Pharm. Sci.*, **100**, 2120–2135.

32. Brummitt, R.K., Nesta, D.P., and Roberts, C.J. (2011) Predicting accelerated aggregation rates for monoclonal antibody formulations, and challenges for low-temperature predictions. *J. Pharm. Sci.*, **100**, 4234–4242.

2

X-RAY CRYSTALLOGRAPHY FOR BIOTHERAPEUTICS

Glen Spraggon

Genomics Institute of the Novartis Research Foundation, San Diego, CA, USA

Biophysical Methods for Biotherapeutics: Discovery and Development Applications, First Edition. Edited by Tapan K. Das.
© 2014 John Wiley & Sons, Inc. Published 2014 by John Wiley & Sons, Inc.

2.1 INTRODUCTION TO X-RAY CRYSTALLOGRAPHY

2.1.1 Early X-Ray Crystallography for Biologics

The advances made in macromolecular crystallography over the last four decades, like other disciplines in protein science, have been correlated with the growth of the production of macromolecules for biotechnology and scientific applications. In its infancy, X-ray crystallographers were reliant on plentiful supplies of pure protein that could be obtained and purified from natural sources. By coincidence, but also because of biological importance, two protein families involved in early protein crystallography studies also turned out to be immensely important as biotherapeutics. The first of these was the small peptide insulin. Insulin has a number of firsts in the protein world: the first protein sequenced [1, 2]; the first commercially available protein biotherapeutic, for the treatment of type I diabetes (Eli Lilly in 1923); and also the first industrially produced recombinant biologic [3]. This availability coupled with biomedical interest also initiated early crystallographic studies and as such insulin was also one of the first proteins whose X-ray structure was solved [4]. This seminal work paved the way for the use of structure in biotechnology; indeed crystallization is still used in the industrial manufacture of insulin [5] and methods used to crystallize insulin have contributed to the formulation of long-acting insulins. Structure-based design is still being used to improve the therapeutic properties of the peptide.

The other biotherapeutic family that quickly assailed itself to crystallographers were the immunoglobulins, more specifically immunoglobulin G (IgG), a large (150 kDa) heterodimeric protein. High-resolution structure studies of this molecule have been ongoing since the influential work of Rodney Porter who separated fragments of the antibody "Fab" (fragment antigen binding) and Fc (the fragment crystallizable) and obtained crystals of the Fc region of a naturally occurring rabbit γ-globulin antibody in 1958 [6]. Subsequent studies on the Fc fragment in the 1960s by Humphrey, Poljak, and Goldstein aided in further characterizing the crystals [7] but it was not until the mid-1970s that Deisenhofer and coworkers [8, 9] determined the structure of the Fc domain, with further studies extending the structure to higher resolution in the early 1980s. The Fab portion of the IgG, because it is the binding portion of the antibody, has attracted the majority of interest from the X-ray crystallographic and biotherapeutics community. First solved in 1972, and with subsequent structures of paratope/epitope complex structures, initially for lysozyme, the IgG Fab fragment has provided a model system for how shape and biophysical diversity can be generated by the complementarity determining regions (CDRs) of the immunoglobulin variable domains [10–12], enabling the immune system to bind a very diverse array of pathogen antigens. Again, advances in biotechnology were the major determining factors for studying these molecules at high resolution. The ability to produce monoclonal antibodies against a protein of choice [13] coupled to expression technologies has made structure determination relatively routine. There are currently dozens of Fab structures in the Protein Databank (PDB) [14], as well as a handful of full IgG antibody structures illustrating the considerable conformational diversity between the fragments around the flexible Y-shaped scaffold [15]. Similar advances have been seen in the biotechnology field. While it was not until 1986 that the first monoclonal antibody therapy based on IgG was approved, muromonab-CD3, IgG-based antibodies are now the mainstay of biologics with greater than 20 IgG-based biologics approved, and hundreds now taking part in clinical trials.

Many aspects of early stages of biotherapeutic discovery and development mirror the processes of macromolecular crystallography, the need for large quantities of well-behaved, homogenous protein being primary to both. As such, all of the major families of biotherapeutics have representative structures solved by crystallography. This work, which started primarily to define the three-dimensional folds of these various proteins, is rapidly developing into a discipline, using the structure from crystallography as part of the design cycle in biotherapeutic applications.

The science of macromolecule X-ray crystallography is now mature and major advances in all aspects of the technology have led to the possibility of a rapid turnover of results. The chapter details the basic techniques that can be utilized to maximize the effort spent and the use of structure for the design and analysis of biologics.

2.2 MODERN X-RAY CRYSTALLOGRAPHY

2.2.1 Construct Design and Protein Production

The most important aspect of any macromolecular crystallography project is the production of relatively large quantities of the pure macromolecule. While the levels of homogeneity and purity necessary to produce crystals are open to debate, the dogma is that the more monodisperse a protein is, the better the chance of success [16]. On many levels, the amount and ability to produce a homogenous material is influenced by the starting DNA sequence, referred to as the construct, and often only small variations in the sequence can lead to startling changes in the molecular properties.

2.2.1.1 Bioinformatics and Construct Design. The basic starting point for structural biology projects is a one-dimensional sequence, either protein or DNA. Bioinformatics tools for doing analysis on biotherapeutics are plentiful and useful for the analysis of sequences (Figure 2.1). The first important methodology is to use homology

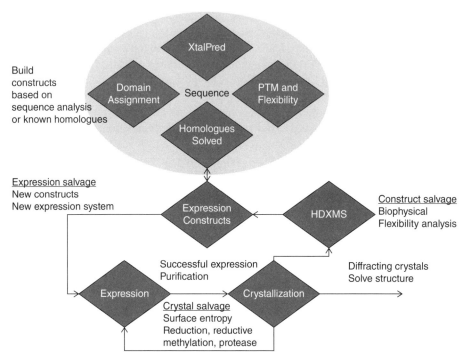

Figure 2.1. Flowchart of the construct and crystallization process. PTM, posttranslational modifications; HDXMS, high-resolution deuterium exchange mass spectroscopy

with proteins of known structure to design the construct. A number of useful tools for biotherapeutics can be used to derive structural information on the construct. Generally the first procedure for biologics is to locate structural domains within the construct, which can be done with one of the various domain analysis databases such as Pfam, Interpro, SCOP, CATH, and Conserved Domain Database (CDD) [17–23]. These tools often cross-reference one another, so can be used in tandem as a starting point to develop a model of the domain boundaries and possible linkers between them. This may be incredibly useful in later stages of the project where it may be necessary to solve the high-resolution structure of the protein in pieces and then model the overall conformation with hybrid methods such as electron microscopy (EM) or small angle X-ray scattering (SAXS).

Using the Basic Local Alignment Search Tool (Blast) algorithm [24] to search the primary sequence against the PDB is a rapid method to locate similar structural homologues already solved. After running the search, a Blast hit with a sequence identity match of greater than 30% and covering a significant portion of the search sequence should provide a good place to start with construct boundaries. Failure to get a significant match over more than 20% of the protein will likely reduce the chance of solving the structure by molecular replacement (see Section 2.4.1). These structural homologues, and the literature associated, will also provide a template for the various methods to express or solve the structure of the target protein, although the means used to express the protein may not be the only or the most convenient way.

Further to the design of the boundaries of the construct a wide variety of other primary sequence analysis tools can be explored on the Expert Protein Analysis System (ExPASy) site [25]. These tools provide the means to identify features such as sites of posttranslational modification, transmembrane (TM) spanning regions, the presence of a signal sequence, or the presence of disordered regions of the proteins.

Of the posttranslational modifications, the most important class for crystal generation are the presence of glycosylation sites which can sometimes hinder crystallization, by inhibiting the ability of the protein to form sufficient contacts to organize in three dimensions. Prediction of N-linked glycosylation sites is relatively trivial and follows a set motif, either Asn-X-Ser or Asn-X-Thr; O-linked sites are less predictable [26] but also less common in secreted proteins. Many N-linked glycosylated proteins have been crystallized, suggesting that leaving the asparagine residue in the construct may be acceptable. In cases where this is not the case, the standard procedures are to mutate the asparagine residue of the motif to a glutamine or to express in a system that suppresses glycosylation [27]. Other posttranslational modifications may also be important in both biotherapeutic development and crystallization and can be panned for with the various one-dimensional tools available within ExPASy site (http://www.expasy.org).

Many of the proteins of interest in biotherapeutics are cell-surface receptors, attached to the cell surface by one or more TM spanning helices. Crystallizing the exo-region in the presence of these TMs has so far proven intractable, so the general method is to remove the TM from the construct. The distinguishing feature of TM segments is the hydrophobic character of the residues which transverse the membrane. Kyte and Doolittle exploited this information to predict TM helices in protein sequences. To do this they used a hydrophobic index for each of the amino acids and a sliding window against the primary sequence [28] to define a "GRAVY" index at each point of the sequence. Derivatives of this index are the cornerstone of all subsequent TM prediction algorithms [29].

Other secondary structure prediction methods can provide insight into the construct design stage [30]. It is often disadvantageous to cut a sequence in the middle of a secondary structure element as these are generally ordered regions that contribute to the three-dimensional topology of the protein. Conversely, many tools can also predict flexibility or intrinsic disorder within a protein [31–34]. When this disorder is located at the N- or C-terminus of the construct these regions can be easily deleted by simple truncation; if they occur in the middle of the sequence, the path forward is more challenging to determine. Recently biophysical means to experimentally measure the flexible components of constructs using mass spectroscopy have been determined [35, 36]. These can be incredibly useful for construct design when even a small amount or protein is available for analysis and such experimental procedures are becoming the standard in many crystallization endeavors.

Producing constructs for crystallization is a multivariate process and the number of variables needed to produce a viable construct that expresses and then crystallizes is large. Predicting these criteria *a priori* can often be problematic. As our empirical knowledge of crystallization experiments grows, our ability to correlate success from the statistics generated should also increase. Interfaces such as XtalPred [37], which utilize this information, are now beginning to be applied to predict the likelihood of crystallization of a primary sequence. While such servers are, by nature, probabilistic, their use can often weigh the success in favor of the crystallographer.

2.2.1.2 Construct Design for Expression and Purification. Molecular biology techniques allow the insertion or addition of sequences into proteins, which can act as purification tags. This ability to express proteins to order has revolutionized the crystallographic field and such approaches are now fundamental to producing the necessary constructs for crystallization. Methods continue to improve from the earliest molecular biology experiments [38,39] and techniques such as the PIPE cloning method [16, 40], which remove the need for restriction enzymes, have not only increased this efficiency but also lowered the cost of clone production.

Tags are inserted at the N- and C-termini of the designed construct and while some can increase expression most are designed to enable affinity purification of the protein. The most common is the histidine tag (between 5 and 10 histidines attached to either the N- or C-terminus) but other larger protein tags such as maltose binding protein (MBP), glutathione-S-transferase (GST), and fragment crystallizable (Fc) are also in common use [41–43]. In general, linkers and short tags do not have an ordered structure, so the longer the tag, the more degrees of freedom it generates, hindering the formation of crystal contacts and thus crystallization. There has long been debate about whether the presence of a histidine tag hinders crystallization [44]; in many cases it does not encumber crystallization, but in most cases no structure can be assigned to the histidine tags. There are, however, very few examples of large fusion proteins that have crystallized where a flexible linker has been inserted between the tag and the protein of interest, although rigidifying the linker can sometimes improve the crystallizability of such constructs [45]. To circumvent these processes it is often necessary to insert a cleavage sequence between the tag and the protein so that removal of the fusion can be achieved. This scheme can also be used to improve the purity of the resultant protein, by going through an affinity/cleavage/reverse cycle where the protein is purified by an affinity column followed by cleavage and subsequently passing the protein over the column again to collect the flow-through of cleaved protein [16, 40].

2.2.2 Macromolecular Crystallization

2.2.2.1 Crystals. The need for crystals to solve the high-resolution structures of macromolecules stems from the fact that X-rays interact extremely weakly with electrons. Put simply, a crystal aligns a vast number of molecules into a lattice, acting as an amplifier of this weak interaction to produce a measurable signal. Recent advances in the generation of higher intensity X-rays is beginning to reduce the size of the crystal needed [46], but currently the only reliable way to obtain a diffraction pattern from a molecule is from a macroscopic crystal.

While protein crystals need to be in an aqueous state to maintain structural integrity [47], they still follow all of the rules of three-dimensional solid-state crystals established from mathematic principles in the late nineteenth century [48–50]. The space group is defined by a set of symmetry elements, known as point groups, of which there are 30, further classified into 7 distinct crystal systems combined with a set of 14 Bravais lattices. These combinations ultimately yield 230 possible space groups in three dimensions. Proteins and nucleic acid crystals can only form in 65 of these due to the chiral nature of the amino acids and nucleotides, excluding space groups with mirror symmetry. To the macromolecular crystallographer, the important aspects of the crystal are the unit cell, corresponding to the size of the box that when repeated along the three translational axes will reproduce the crystal, and the asymmetric unit which is the portion of the unit cell which is unique and not dependent on the application of the space group's symmetry operators. The space group provides a description of the symmetry of the crystal and is important in that it represents how the molecules in the crystal are packed and how much data need to be collected from the crystal to reconstruct the asymmetric unit. However, in modern crystallography, these details are invariably encoded algorithmically and as such, appreciation of the contents of the asymmetric unit is generally all that is required. The interested reader is referred to the International Table for Crystallography [51] for further information.

2.2.2.2 Modern Approaches to Protein Crystallization.
Protein crystallization is an old technique and originated as a method for the purification of proteins, exemplified by the presentation of the 1946 Nobel prize for chemistry to James Sumner, John Northrop, and Wendell Stanley for the purification and crystallization of proteins [52]. X-ray studies of protein crystals in 1934 by Bernal and Hodgkins [47] on pepsin crystals revealed that diffraction could only be achieved by maintaining the fluid environment around the crystals, thus initiating macromolecular crystallography as a science. The basic methods for crystallization have not changed since and involve bringing a protein into an ordered solid state by inducing a molecular supersaturation [53]. Achieving this is brought about by screening high concentrations of macromolecules against arrays of polymers, salts and organic solvents at various pHs and temperatures to induce this supersaturated state and prompt crystal nucleation. Various methods are available to carry this out; vapor or liquid diffusion methods are the most common. Here, the concentration of crystallization reagents is slowly increased over time by equilibration against the solution at a different solute potential (Figure 2.2) by either a vapor or liquid interface. Currently, our inability to predict suitable conditions for crystallization has resulted in many high-throughput coarse screen approaches for crystallization whereby suitable conditions are sampled from those already shown to work. Initiated in 1991 by Jancarik and Kim [54], these sparse matrix techniques have proven extremely successful and are frequently updated via further data mining of successful conditions as more data become available from structural genomics and other initiatives [55]. Typically crystallization screens now consist of hundreds of different conditions performed at varying temperatures with solutions of proteins between 1 and 100 mg/mL.

Having to screen over increasing arrays of conditions means that relatively large amounts of pure, homogenous protein has to be generated. Over the last decade this has led to a continued shrinking of the volume of individual crystallization experiments, to increase efficiency and cope with the growing demands for structure. Nano-volume miniaturization of vapor diffusion experiments has reduced the need for protein by

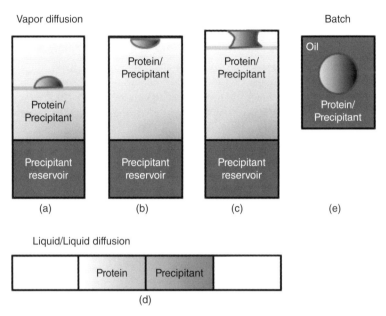

Figure 2.2. Schematic representations of the most common macromolecular crystallization techniques: (a) sitting drop vapor diffusion, (b) hanging drop vapor diffusion, (c) sandwich drop vapor diffusion, (d) free interface diffusion, and (e) batch crystallization under oil

an order of magnitude and also potentially speeds equilibration and thus crystal generation [56]. This technology has largely replaced traditional hand setup plates and, when combined with high-throughput robotic systems and computational laboratory information management, provides an excellent means to catalogue and analyze the screening results for macromolecular crystallization experiments. In the last few years a number of commercial modular systems for nano-volume crystallization, imaging, and classification of crystallization experiments have become readily available (Figures 2.3 and 2.4).

Other crystallization techniques involving different methods for achieving nucleation have also evolved; batch crystallization [57] can now effectively screen many thousands of conditions (Figure 2.2e) with relatively little protein. When protein is extremely scarce then liquid/liquid or free interface diffusion microfluidic techniques pioneered by Quake et al. [58] can provide distinct advantages. Here micro-hydraulic systems can often perform hundreds of experiments and produce viable crystals with as little as 10 μL of protein [59] (Figures 2.2 and 2.3). As experiments get smaller so do the crystals they generate, imposing limitations on crystallization technology. However, with technologies for the generation of X-rays [46], the collection of data and handling of crystals is also advancing, and while these bounds have not at present been fully explored, the current limit for successful data collection of a crystal at a micro-focus synchrotron beamline is a crystal with dimensions in the region of ~5 μm on a side (Figure 2.5).

(a)

(b)

Figure 2.3. Examples of current high-throughput crystallization platforms: (a) nano-volume dispenser apparatus used to set up 96 well sitting drop vapor diffusion plates at submicroliter volumes (Art Robbins Instruments Sunnyvale CA, USA, Phoenix); (b) a microfluidic system used to prepare liquid/liquid diffusion plates (left), a plugmaker liquid/liquid interface plate capable of setting up screen with as little as 10 μL of protein (right) (Emerald Bio Seattle WA, USA)

2.3 X-RAY DATA COLLECTION

To obtain atomic resolution structures of macromolecules, diffraction data must be collected with light at a wavelength close to the size of covalent bonds. This falls within the spectrum of electromagnetic hard X-rays (typically on the order of 1.0 Å wavelength). In-house X-ray generators utilize the characteristic radiation of various metals, the most common being copper; striking a rotating copper anode with accelerated electrons produces radiation at a wavelength of 1.54 Å. However, modern third-generation synchrotron X-ray sources [60] can be orders of magnitude more intense and generate

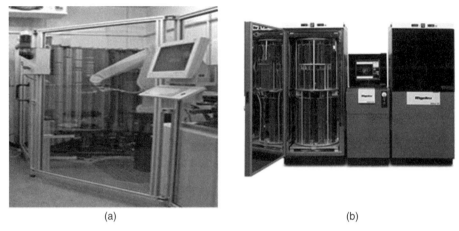

(a)   (b)

Figure 2.4. Automated imaging systems for recording crystallization images: (a) custom-built automated imaging systems for the storage and recording of crystallization images (GNF Systems, San Diego, CA); (b) Rigaku's Automation Carlsbad CA, USA minstrel automated imaging system incorporating UV imaging for the detection of crystals

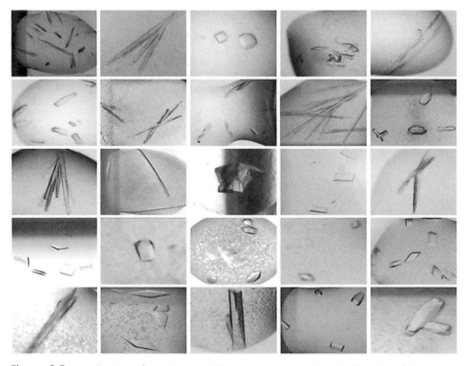

Figure 2.5. A collection of protein crystal images demonstrating the diversity of shape and sizes of crystals that have produced viable X-ray data sets

highly parallel beams, ideal for X-ray data collection. In the future, many advances in the field will be made as X-ray beams become more intense and instrument precision increases. This will enable data collection on more complex targets with lower collection time and on increasingly smaller crystals.

2.3.1 Crystal Mounting

As was shown in 1934 by Bernal and colleagues, collecting diffraction data from a crystal must be done in the presence of a hydrated environment to maintain the structural integrity of the crystal. Typical protein crystals are extremely small, between 5 and 100 μM on an edge, and can rarely be seen with the naked eye. This necessitates the use of a microscope to enable crystal harvesting. This process has evolved over the years from one where the crystals were gently sucked into a capillary, excess crystallization removed, and then sealed with wax for room temperature data collection to the use of cryo-crystallography techniques. Cryo-crystallography procedures involve lassoing of the crystal into thin nylon loops prior to vitrification at liquid nitrogen temperatures (100 K) for data collection. This cryo-protection of samples has been necessitated by the increase of intensity of modern X-ray sources which greatly increases the radiation damage experienced by the sample [61], and cryo-crystallographic techniques are now the standard, often allowing collection of a complete data set from a single crystal without significant radiation damage. Cryo-preservation of a crystal requires the prevention of ice formation within the crystal, which can damage the fragile structure of the crystal. While many crystallization conditions in standard screens contain conditions that prevent the nucleation of ice, most do not. Basic cryo-cooling involves adding enough of a cryoprotectant, typically 10–25% of glycerol, ethylene glycol, or an organic solvent such as MPD [62], or coating the crystal in an oil, such as Paratone-N or perfluoro-polyether, thus removing ice-forming liquids from the sample prior to quick vitrification in liquid nitrogen. The rapid cooling can be achieved by either dipping the specimen in liquid nitrogen or placing it directly in the stream of liquid nitrogen vapor on the goniostat prior to data collection.

2.3.2 Collecting a Data Set

X-ray data collection is analogous to photographing an object without a lens. The image macromolecular crystallographers collect is the intensity of scattering of the X-rays diffracted from the object. If a lens could be inserted between these scattering waves and the detector, the waves can be combined and a real-space object formed. For hard X-rays, no such lens currently exists, so experiments are confined to measuring the intensities of these scattered waves and computationally providing the lens by means of a Fourier transform. This measurement is carried out in a "reciprocal space" where distances that are far apart correspond to close distances in real space. This is compounded by the fact that the molecules are contained within a crystal, which, as previously described, acts as an X-ray amplifier but only at certain points in the reciprocal space. If we continue the analogy with a photograph our data collection is the observation of the scattering intensities from an object, after filtering through an opaque grid of points; known in crystallography as reflections.

Standard data collection is performed by the oscillation method [63] and takes place by mounting the crystal on a goniometer and aligning the crystal perpendicular to the X-ray beam. For cryo-crystallography applications the crystal is maintained at liquid nitrogen temperatures by streaming liquid nitrogen vapor over the crystal. Collection can then be achieved by rotating the crystal by a small oscillation width around one or more axis for a set number of degrees until enough data are collected to reconstruct the crystal. Modern data collection strategies digitally capture data on an imaging plate or CCD detector. Synchrotron X-ray sources are so intense that radiation damage to the crystal [64, 65] can be a major issue in data collection, which limits the exposure that can be given to crystals. Recently, pixel single-array detectors have become available at a number of synchrotrons allowing the collection of data in a matter of seconds [66].

The strategies employed to collect a data set depend on a number of quantities intrinsic to the crystal, the detector and beamline geometries [63]. The mosaicity of a crystal is a measure of how many degrees the crystal needs to be rotated to collect a full reflection. It is based on a model of a crystal whereby a single crystal is decomposed into a number of blocks orientationally displaced relative to one another. As such, to collect a complete reflection one has to rotate through the maximum displacement. The resolution of the data is a measure of the amount of detail we expect in the electron density. The reciprocal nature of the data collection dictates that the farther the reflections are from the center of the detector the better the resolution that can be obtained. More importantly the resolution is defined by the signal of the data at positions away from the position of the central beam. Convention generally measures this resolution as the highest resolution data shell collected with a signal to noise above 1.0 (although many authorities pick 2.0 as a cutoff). Resolution is typically measured in Angstroms ($\text{Å}\ 10^{-10}$ m). Table 2.1 details the features that can modeled into a typical crystal structure at different resolution ranges.

2.3.3 Data Reduction

The raw data from a set of images needs to be reduced to a set of reflections. These reflections, or indexes, are the three-dimensional integer coordinates of the diffraction pattern and each reflection has an associated value which is the intensity of the reflection and a standard deviation representing the error in the measurement of the data. A number of X-ray data reduction programs are commonly used and all go about data processing in a similar way. Data reduction first proceeds by indexing the raw data to determine the orientation of the crystal relative to the beam and measure the unit cell. Knowing these, the location of every reflection on the images can be computed and raw images can be processed by integrating a spot to reduce to a three-dimensional integer coordinate with an associated value and percentage of the recording reflection that was on the image. Finally, reflections are scaled together to yield a data set of intensities by merging like reflections, bringing them onto a common scale. This process yields statistics given in resolution shells from low to high resolution. Because there is a drastic reduction in intensity as resolution increases, statistics are much better at low resolution and become rapidly less significant as the reflections approach atomic resolution. Of the measures of the quality of data, the most common are R_{merge} and $R_{symm,}$ which measure the

TABLE 2.1. Information that can be derived from a crystal structure at a given atomic resolution

Resolution (Å)	O/P^a	Available modes of refinement	Information that the X-ray structure can provide
>6.0	<0.1	• Rigid body refinement of domains • Overall displacement refinement • TLS refinement of domains	Fitting of high-resolution structures into low-resolution model to define quaternary structure of complex
>3.5	0.5	• Rigid body refinement • Constrained atomic refinement if NCS • Group displacement parameter refinement of residues	Low-resolution structure, secondary structure, some side chain information Dynamic information on movement of domains and individual residues
3.0	0.75	• Individual atomic positional refinement • Group displacement parameter refinement	Details of side chain interactions
2.5	~1.0	• Individual atomic positional refinement • Group or ADP refinement	Most atomic positions well defined, addition of some core waters
2.0	3.0	• Individual atomic positional refinement • Individual ADP refinement	Atomic positions well defined, waters, occupancy refinement where applicable
<1.5	>9.0	• Individual atomic positional refinement • Individual anisotropic ADP refinement • Occupancy refinement	Full anisotropic ADP refinement Occupancy refinement where applicable

NCS, non-crystallographic symmetry; TLS, translational, libration, and screen refinement; ADP, atomic displacement parameter.

$^aO/P$: observations to parameters ratio when taking into account four parameters for individual atoms, three positional and one atomic displacement parameter.

agreement between like reflections. R_{merge} and R_{symm} are often used interchangeably but in fact refer to cases when data are combined from multiple and single crystals, respectively. Along with this statistic, which is generally below 10% overall, a number of other statistics are also useful in determining the quality of data. The most important of these are I/sigma, the signal to noise of the data in resolution shells, the redundancy or how many times on average a reflection is measured, and the completeness in resolution shell. All of these factors can be utilized in assessing the quality and the resolution limit of the diffraction.

Finally the data are used to unambiguously determine the space group of the structure; this can be done by analysis of the scaling statistics in related point groups and examination of the systematic absences of the data. With the exception of a number of polar space groups where two possibilities are possible, analysis of these factors will ultimately provide the means to unambiguously determine the space group of the crystal.

2.4 SOLVING THE STRUCTURE OF THE CRYSTAL

The final product of a crystallographic study is the structure, a set of three-dimensional coordinates representing the various atoms contained within the asymmetric unit of the crystal. To derive these coordinates a number of stages have to be traversed, the first being the derivation of an electron density map corresponding to the density of the electrons within the crystal, by application of a Fourier transform. The impact of recombinant DNA technologies, along with miniaturization of crystallization technologies, has greatly increased the ability to generate crystals for data collection technologies. Unfortunately, as was previously stated, the inability to focus hard X-rays limits current technology to capturing the intensities in reciprocal space of a sampled grid of the data. These intensity data are actually only half of the data necessary to reconstruct the electron density and only provides information on the vectors between atoms [67] but none on the positions of atoms relative to an origin. To derive the positional electron density information in real space a Fourier transform of the structure factors of the reflections must be performed. The structure factor consists of a vector of amplitude given by the square root of the intensity and angle called the phase of a reflection. This phase of a reflection is lost in normal data collection and gives rise to the "phase problem" in X-ray crystallography. Information, must therefore be provided either by physical techniques where the information is provided via additional data or by purely computational techniques whereby relations between the amplitude data are used to derive the phases; such techniques are known as direct methods and are reviewed in papers by Sheldrick and Uson [68, 69]. For the most part, the use of direct methods has limited use for macromolecule structure, because of its requirement for data better than 1.2 Å resolution. This generally leaves the biotherapeutic crystallographer with two options to solve structures: molecular replacement where known structures are used to derive phases and heavy metal techniques; both will be briefly outlined.

2.4.1 Molecular Replacement

Currently the study of structures of biotherapeutics is limited to a relatively small number of protein folds and domains. As such, the most prevalent technique for solving structures of biotherapeutics is molecular replacement [70, 71]. The process is conceptually very simple and relies on the fact that similar structures can be used to derive the position of the molecule in the crystal and thus derive initial phases by back calculation of data from correctly positioned molecules. In the general case, where one molecule is contained within the asymmetric unit, this involves a search over six-dimensional space, three rotations and three translations. To cut down the search time this positioning can be separated into rotational and translational searches. At each position in the search, calculated data are generated from the positioned probe model and compared with the experimentally derived data to derive a score. Over the half century that molecular replacement has been performed a number of target functions have been developed to aid in determining the correct molecular orientation of the probe molecule; these functions can be extremely sensitive to resolution ranges or to slight perturbations of the orientation of the molecule. Recently the increase in computational power and development of more robust target functions such as the maximum likelihood function in the program PHASER [72,73] are increasing the robustness of these methods. However, programs will generally produce some sort of solution, so it is always necessary to confirm that the solution generated is the correct one. This assessment can be done in a number of ways and is discussed in Section 2.4.4.

2.4.2 Heavy Atom Techniques

The first method of deriving phases from a macromolecular crystal was the "multiple isomorphous replacement" method. Here, datasets are collected from a native crystal and crystals generated by the addition of multiple heavy atoms, such as gold or mercury. Because the addition of heavy atoms to certain points in the crystal is a vector sum of the native and added atom, comparison of the data sets can be used to derive phase information [74]. Experimentally, this method can be cumbersome and has largely been superseded by the multi- and single-wave anomalous dispersion techniques (MAD/SAD). For MAD/SAD methods, phases are derived by the comparison of data, collected at one or more characteristic wavelengths, from a single crystal incorporating heavy atoms [75]. These specific wavelengths are characteristic to the absorption edges of the heavy atom incorporated and, as such, generally require the tunability of modern synchrotron radiation sources. The major issue with MAD and SAD data sets is collection of data with a high enough signal to noise to locate the positions of the inserted heavy atoms by direct or Patterson methods [68,69]; this signal typically only varies by a few percentage points of the total intensity of the reflection. Generally, selenium is the atom used due to the ability to incorporate it at the expression level via selenomethionine [76] but other atoms can be soaked into crystals in identical ways to that used for isomorphous derivatives.

The errors in phases generated from the above techniques can still be quite large, resulting in electron density maps that are far from optimal. Many different techniques have been developed to refine these phases which take place by iterative application of real-space filtering of data, solvent flattening histogram matching, and skeletonization interspersed with phase combination [77].

2.4.3 Confirming the Validity of a Solution

Once initial phases are derived, combination with experimental amplitudes provides the crystal structure factors, from which electron density maps can be generated. As the name suggests, electron density maps are distributions of the electrons in the unit cell. Generation of initial phases from either molecular replacement or isomorphous replacement techniques often requires some interpretation as to whether the solution is correct; algorithms for both techniques will always produce solutions. In the case of molecular replacement, the phases provided by the model will ensure that the electron density will always look like the protein model; because of this "phase bias," investigation of the solution is always a good idea; a good solution tends to be many standard deviations above the next highest and this measure of Z-value can provide confidence that the solution is correct. Looking at the molecular packing can also be insightful; a correct solution should not have too many overlapping regions when the symmetry of the crystal is applied. However, the best confirmation is to look at an electron density map generated from the model and the experimental data. Such maps generated from maximum likelihood functions [78] can reduce model bias and identify features within the electron density map which are not present in the model. Electron density identifying features such as different side chains or known ligands provides good confidence that the solution is correct. As this additional density is generated from experimental amplitudes it is highly unlikely to appear from random.

For heavy atom solutions a number of other parameters are available such as the figure of merit and phase error. Both are good indicators of the correctness of the solution, but inspection of density is also critical to see that a sensible protein structure is visible in the density. Over the last decade, the advent of auto-building algorithms such as ARP/wARP [79, 80] and SOLVE [77, 81] has greatly reduced the burden of solution validity. These procedures allow the automatic combination of building with refinement, starting from the primary sequence of the molecule and initial phases, and are applicable for resolutions better than around 2.8 Å. Success in building and refining a significant portion of the protein model, for either heavy atom techniques or molecular replacement, generally ensures that the initial solution is correct.

2.4.4 Building and Refining the Structure

After initial electron density maps and starting models are available, building and refining of the structure can be completed. The refinement process involves fitting a number of parameters to best match the observations provided by the experimental data. What can be fitted depends on the number of observations to parameters used to model the crystal and this in turn relies on the resolution of the data and characteristics of the crystal such as the solvent content and the number of copies of the molecule present in the asymmetric unit. It is important to note that in the absence of ultra-high-resolution data <1.2 Å all molecular refinements include geometrical restraints or constraints defined by parameters derived from small-molecule crystal structures [82]. Table 2.1 provides a coarse guideline of what features can be fitted to a model at a particular resolution. Central to this process of completing the structure is the use of molecular graphics software to view the model and manually into the electron density. A number of molecular graphics programs are currently available for this, the most extensively used being O [83], XFit [84], and Coot [85]. All programs possess the ability to display electron density and add, remove, or alter the model by manual or real-space fitting of this model to the electron density. Building and refinement cycles are done iteratively, whereby in-between model building, reciprocal space refinement of the three-dimensional coordinates and atomic displacement parameters against the experimental data is performed. In a refinement cycle, the data calculated from the atomic coordinates are compared against the experimental data, usually the amplitudes, to minimize a target function.

Modern refinement programs, Refmac [86], Phenix [87], and Buster [88], universally incorporate a maximum likelihood derived target function. Here a small number of reflections, known as the reference-set, typically 5%, are left out of the refinement procedure to derive an estimate of the correctness of the refinement [89]. By performing this evaluation in resolution shells, this has the general effect of down-weighing the higher resolution until the model becomes sufficiently accurate to adequately represent the data. Assessment of the quality of the structure is generally performed by a number of metrics relating to the agreement of the experimental and observed data and the geometry of the model.

The main statistic used to measure the correctness of the solution is the R-factor, a measure of the agreement between experimental and calculated amplitudes. As would be expected the nonlinear and multivariate nature of the refinement procedure has many global minimums; this is compounded by the need to add supplementary data in the refinement process with macromolecule chemical information derived from small-molecule data, the Engh–Huber parameters [82]. How to weigh the relative contributions of the geometric terms with that of the X-ray data is something that can generally be achieved by the use of Free-Rfactor as defined by Brunger [89]. When refinement is converged, for a good structure the gap between the R-factor and the Free-Rfactor is typically a few percentage points and the R-factor is at or below 20%. Concomitantly, the geometry should also be in good agreement with known values for bond distances and angles and in accord with the Ramachandran plot of the structure. The Ramachandran plot is a graph of the phi–psi torsion angles of the protein main chain relating the torsion angles of the backbone carboxy and amide nitrogen [90]; it is an excellent cross-validation tool at the end of refinement. A good structure should have very few residues that lie outside of the favored region of the Ramachandran distribution along with geometric features that confer expected values; programs such as PROCHECK [91] and its derivatives are very useful to compile this information from the final structure.

2.5 UNDERSTANDING THE TARGET THROUGH STRUCTURE

2.5.1 The Model

The purpose of X-ray crystallography for biotherapeutics is to derive an atomic representation of the macromolecular structure of the crystal for viewing and investigation via display software explicitly for the use of discovery and design purposes. The final model is a set of three-dimensional (x,y,z) coordinates of the nuclei of the various atoms. These data should also contain other information, such as the crystal dimensions, space group, disorder present in the crystal, and the methods used to produce and solve the crystal structure. The prevalent format that is used for macromolecular structure is the PDB format, a text-based fixed-field representation of the crystal structure. Individual lines that contain coordinate information are detailed by a beginning "ATOM" line for macromolecule entries or HETATM for non-macromolecule entities such as ligands, ions, or waters. Scheme 2.1 displays the format of this line with a description of the 11 fields below.

```
ATOM 32 N AARG A 3 11.281 86.699 94.383 0.50 35.88 N
ATOM 33 N BARG A 3 11.296 86.721 94.521 0.50 35.60 N
      1 2  3  4 5    6       7       8     9    10  11
```

1. The beginning of a starting protein or nucleic acid is represented by ATOM. A HETAM in this field specifies the beginning of a non-macromolecular entity such as a water, ion, ligand or in some cases carbohydrate.
2. The number of the atom is generally a sequential number starting at 1; this is unique.
3. The Atom Name: atom names within a residue are unique and are a maximum of three characters long. Ions and metals are displaced one field to the left.
4. The designation of an alternative coordinate: generally this field is blank indicating there is only one conformation of this atom—then beginning at A,B,C other conformations are designated.
5. The residue name: standard three-letter designation of a residue.
6. The monomer chain: a one-letter description of the chain of the macromolecule, typically beginning at "A" but in many cases is given a nonsequential value. For example, Fab structures often have chain designations "L" and "H" for the light and heavy chains, respectively.
7. The residue number: a sequential number corresponding to the numbering of the macromolecule residue in the chain. A combination of the chain and residue number should be sufficient to uniquely identify a residue in the structure.
8. The next three fields assign the three-dimensional coordinates of an atom, represented as Cartesian coordinates with units of angstroms. This is the most common coordinate system for atoms, but it is possible to embody coordinates in unitless "fractional" coordinates which refer to the coordinates relative to their fractional position along axes of the unit cell.
9. The Occupancy: a measure of the probability of finding an atom in a particular crystal unit cell at this position. For macromolecules, where resolution is limiting, typically the occupancy takes a value of 1.0. In higher resolution the presence of alternative conformations cause an atom to be in two or more distinct positions in the crystal in different unit cells. The relative fractional occupancy of the atomic positions in the unit cells is this occupancy value which must add up to one over the various conformations of the atoms.
10. Atomic Displacement Factor: this column is the final numeric value and goes by various names, atomic displacement parameter (ADP), temperature factor, or B-factor, the most descriptive being atomic displacement factor. This provides a uniform harmonic model of the displacement of the atomic nucleus around its coordinates; the higher the number the more the mean displacement from the coordinate position.
11. The final column is the element that the atom is. This is the symbol of the atom in the periodic table.

Scheme 2.1. Protein Databank format.

An additional line can appear below an atom, starting with the key word ANIS. This is a further model of the dynamics of the atom, supplementary to field 10—the atomic displacement factor—and appears in two situations. In the first situation, when data resolution becomes very high (<1.5 Å), it becomes possible to model the motion of individual atoms via an anharmonic model, whereby the motion of the atom is characterized by nine numbers, representing a thermal ellipsoid, following the ANIS key word. The second use of the ANIS line is used for lower resolution structures to model rigid body domain movements by translational, libration, and screw (TLS) refinement. This model or motion can be particularly useful for biotherapeutics, which are often multi-domain structures. In these cases it is possible to quantify the inter-domain movement and flexibility by TLS refinement [92].

2.5.2 The Protein Databank and Related Resources

One of the great resources in the field of protein structure is the PDB [93]. It was originally established in the early 1970s as a repository of all structures published and has since been growing at an exponential rate. In the modern era it is now impossible to publish a structure without first depositing the structure and its accompanying data in the PDB (www.rcsb.org) which now contains well over 80,000 macromolecular structure entries. This resource has allowed a large number of databases and search tools to grow, using the PDB as the central source of structure data, to characterize protein domains SCO [94], CATH [95], Pfam, and InterPro [17, 19]. These tools along with the Protein Database itself are essential to mining information on a solved structure.

2.5.3 Information Provided by X-Ray Crystallography

The final product of an X-ray crystallography study is the three-dimensional coordinates detailed above. Other methods such as NMR and EM can be used to provide similar or complementary information but currently, X-ray crystallography provides the quickest and highest resolution way to structurally characterize a macromolecule at the atomic level and provide a template for the design and improvement of biotherapeutic properties. A number of levels of information can be obtained by analysis of the structure far beyond the primary sequence. Traditionally macromolecular structural information is classified into secondary, tertiary, and quaternary kinds; the most basic of these is the correct secondary structure. Secondary structure can only be estimated to an accuracy of up to 80% from the primary sequence alone [96]. For proteins, secondary motifs can be assigned from the X-ray structure via various pattern-matching techniques [97–99] and fall into three categories, periodic, non-periodic and random. The periodic motifs, known as helices and sheets, result from repeated values of backbone dihedral angles, the phi–psi angles, from the Ramachandran plot [90]. The helices are self-contained units, and hydrogen-bond main chain carboxyls with amides a small number of residues apart. They can be further subdivided into motifs according to the number of residues per turn: α-helices (3.6), 3-10 helices (3.0), and π-helices (4.4). By far the most common of these motifs is the α-helix. The sheets, as the name suggests, form fairly flat blankets of structure with hydrogen bond patterns occurring between carboxyl and amide atoms of adjacent strands arranged in either a parallel or antiparallel topology. The second type of secondary structure is the turn [100]; these motifs are non-periodic in their phi–psi angle arrangement but have a defined number of residues with a distinct hydrogen bond pattern between the residues. The direction of the chain is always reversed, hence the name turn. There are five distinct forms of turns (α, β, γ, δ, π) which form turns with four, three, two, one, and five residue bonds, respectively.

The final category of secondary structure is the loop. These are regions of the structure which do not fall into the periodic or turn motifs, and cannot be categorized easily into distinct patterns of hydrogen bond structure. The detection of loops within structures can be extremely informative; structures with a high percentage of defined secondary structure tend to be stable, while those with a high percentage of loops which can adopt multiple conformations due to the lack of hydrogen bond restraints are often less stable. Both of these situations may be useful in extending the properties of a given biotherapeutic.

The tertiary structure of a protein is the overall topology that the secondary structure of the chain folds into. Despite the near-infinite variety of the sequence space of proteins, the number of distinct topologies, or folds, that these proteins collapse into appears to be limited in number, with many seemingly unrelated sequences at the primary structure level sharing the same topology. For current biologics, the vast majority of therapeutics are represented by a relatively small number of folds, all of which have some sort of structural representation in the PDB (Figure 2.6). Biotherapeutics are almost always extracellular proteins of sorts which have evolved naturally as duplications and combinations of similar, often disulfide-rich repeating domains; for a review see Bork [101]. However, the actual molecule in therapy frequently contains some sort of chemical modification such as pegylation or are chimeras of molecules fused together to enhance pharmacokinetics. Amongst these folds, the immunoglobulins, four-helix bundles, β-trefoils, and insulin-like growth factor are extremely common but with an extremely diverse repertoire of extracellular function (Figure 2.6). The tertiary structure therefore defines the nature of these folds and allows the X-ray crystallographer and protein engineer to compare and contrast the structure with others at the three-dimensional level, with structure comparison tools such as DALI [102, 103]. The way that individual secondary structure motifs and domains interact with each other on the three-dimensional level provides valuable insight into the design of chemical modifications or modifications of the proteins themselves to yield new function or pharmacokinetic properties.

This final level of protein structure interpretation, known as the quaternary structure, is of particular importance in the biotherapeutics field and details how individual chains of proteins interact with one another. This can define how an individual molecule homo-oligomerizes or how two proteins interact with one another on the biological level. The crystal structure is an extremely accurate way of determining these interactions and because computational prediction of quaternary structure from three-dimensional coordinates is still largely inaccurate, crystallography still remains one of the few ways to define these protein–protein interactions. Interpretation must, however, be qualified by some warnings.

The information provided from X-ray diffraction work is in the context of a crystal, which is a solid-state combination of unit cells held together often by many small interactions between the proteins in the asymmetric unit, termed crystal contacts [104]. The crystal can be made up of a mixture of biologically significant protein–protein interactions and crystal contacts which only occur in the context of the crystal. As

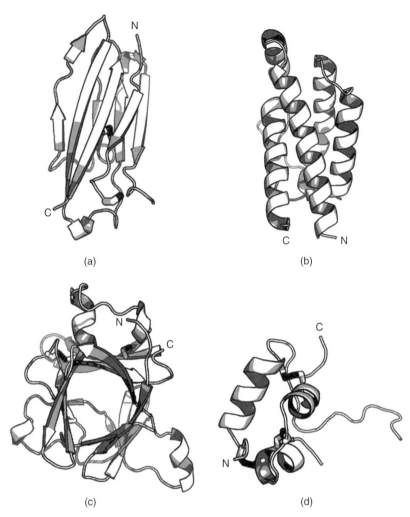

Figure 2.6. Common protein folds encountered in biotherapeutics. Ribbon diagrams of structures of common folds found in biotherapeutics: (a) structure of a typical immunoglobulin domain taken from the heavy chain of a Fab fragment (PDB ID 1MJU [168]), this two-layer β-sandwich consisting of between 6 and 9 β-strands is the most common domain in extracellular proteins. (b) A four-helix bundle (interleukin-2, PDB ID 1M47 [169]); the four helices in this domain are arranged in an antiparallel fashion with this motive occurring in many cytokines. (c) A β-trefoil fold is a 12-stranded β-strand structure with pseudo threefold symmetry represented in the diagram by interleukin 1 receptor antagonist (PDB ID 1IRA [170]). (d) The small cysteine bridge-mediated insulin-like fold is characterized by insulin-like growth factor (PDB ID 1IMX [171])

such, crystal contacts can sometimes be misinterpreted to infer oligomeric state or a biologically relevant protein–protein interaction. There are, however, a number of studies that aid in the evaluation of the meaningfulness of contacts in a biological context [105–108] and these studies should be taken into account when biological information is interpreted from a structure without the support of supplementary experimental evidence.

Crystals can also provide some information of the dynamics of a macromolecule, at the atomic level in the form of atomic displacement parameters and at the domain level with TLS parameters. Again all of these details must be viewed in the context of a crystal, with its supporting lattice which may constrain the movement of atoms relative to the biological situation. In addition to this, the resultant structure is an average of the electron density of all of molecules in the unit cells of a crystal, which for a typical crystal is extremely large. Therefore, any part of the structure that does not have one or more discrete conformations will not be apparent in the electron density maps, as this density will be smeared out over all of the space that the region occupies. These regions of the macromolecule are often termed disordered and while this is a mischaracterization of the state, it refers to areas in the primary sequences that have a high degree of conformational variability. In this context, the lack of structural information can provide qualitative information on the flexibility of loops, linker regions, and termini. When combined with computation and biophysical experimental data from NMR and HDXMS [35, 36] it can be a powerful tool for increasing stability, decreasing aggregation, or designing better crystallizing proteins.

2.6 APPLICATIONS OF X-RAY CRYSTALLOGRAPHY TO BIOTHERAPEUTICS

2.6.1 Antibody-Based Biotherapeutics

As previously mentioned, crystallographic and biotherapeutics have largely focused on the archetypal IgG molecules, which are extremely abundant in plasma. Normal IgG antibodies are glycoproteins, composed of four chains, two light (L) chains (~23 kDa) and two heavy (H) chains (~60 kDa), giving a total molecular weight of approximately 160 kDa (Figure 2.7a). All of these chains are composed of immunoglobulin domains, one of the most common domains in the human genome (Figure 2.6a) [109]. The L chains in human are further subdivided into two types, λ and γ, while immunoglobulin heavy chains are partitioned into five distinct chains, γ, μ, α, δ, and ε, corresponding to different antibody isotypes, IgG, IgM, IgA, IgD, and IgA, respectively. The heavy chain variations are associated not only with sequence but also to either additional constant immunoglobulin domains (IgE) or different oligomeric states (IgA, dimer; IgM, pentamer). Further, there are four different human IgGs naturally expressed relating to slight differences in the constant domain sequences.

For antibodies, immunoglobulin domains are assigned a standard nomenclature partitioning the domains into variable (V) and constant domains (C) with a number defining the position of the C domain relative to the N-termini. Thus a light chain is denoted V_L, C_L1 and a heavy chain V_H, C_H1, C_H2, C_H3 (Figure 2.7a). The variable

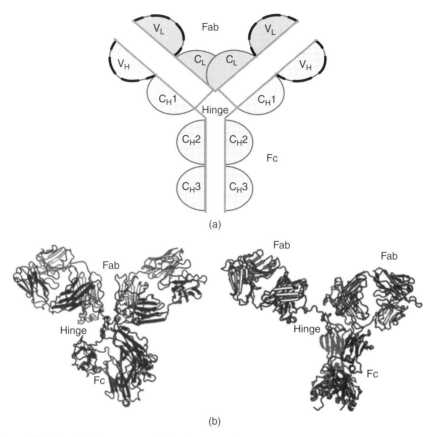

Figure 2.7. Antibody structure. (a) Schematic of IgG showing domain nomenclature and regions of interest in the IgG; complementarity determining regions highlighted as bold lines. (b) Ribbon representations of full-length IgG structure, heavy chains shown as black, light chains as light gray. Left: PDB ID 1IGY [172] showing the expected symmetric Y-shaped nature. Right: a more asymmetric full-length antibody (PDB ID 1HZH [15]) demonstrating the inherent flexibility in the quaternary structure

domains are further subdivided into invariant regions and CDRs which are the six loops of the variable domains whose hypervariability allows specific targeting of an antigen (Figure 2.7a). To maintain a consistent nomenclature, these IgG loops are traditionally numbered according to the convention established by Kabat [110]. Areas of interest from the biotherapeutics perspective tend to be focused on the variable domains V_L and V_H of the Fab fragment, whose ability to vary is responsible for target binding, and the C_H2/C_H3 domains of the heavy chain, the Fc fragment, responsible for dimerizing the molecule and extending the antibody half -life.

The basic Y or T shape of the glycoprotein IgG was established early on by negative stain EM [111] (Figures 2.7a and 2.7b). These studies also established that IgGs were intrinsically flexible, something that often makes full-length IgG molecules fairly recalcitrant to crystallization. In contrast to the extensive information of the fragments of antibodies the information on full-length IgG structure is limited to a few native or slightly modified structures. That the flexibility in the hinges hinders crystallization was confirmed by early representative structures of Dob and Mcg, achieved by truncating the hinge regions between Fc and Fab [112, 113]. This flexibility can thwart crystallization efforts but is not insurmountable if the correct conditions or construct can be applied to initiate crystallization. Recent studies continued to demonstrate the flexible and nonsymmetric structures of full length IgG molecules [15] (Figure 2.7b).

2.6.1.1 IgG Fragments, Fabs, and Fc Domains.

Studies by Porter [6] established methods to generate independent fragments of antibodies separating the antigen binding (Fab or Fv) from the Fc fragments. These protocols involve cleavage of the IgG1 antibody by papain, which yields two Fabs and an Fc domain, which can be subsequently separated by a protein A or G column. Such methodologies have allowed the majority of the X-ray work to be targeted at these more accessible fragments.

Fabs being the antigen-binding portion of the antibody have attracted the majority of interest from the X-ray crystallographic and biotherapeutics community (Figure 2.8a). First solved in 1972 by Poljak et al. followed by atomic resolution structures and antigen complex structures in 1974 [10–12] (Figure 2.8a) these structures show the diversity of recombination of the variable regions, giving IgG the ability to bind a diverse array of molecular entities by variation of the six CDR regions. The Fab consists of four immunoglobulin domains with V and C regions from light and heavy chains both interacting with their corresponding domain. Between the V and C dimers, a linker region provides a variable elbow angle connecting variable and constant regions. In an exhaustive study of over 360 unique Fabs, Stanfield et al. [14] showed that the variation in the elbow angles could be partitioned between the λ and κ light-chain variants, whereby the λ chain showed much more variety of the elbow angle relative to the κ.

Fab structures in complex with their target epitopes are now very common and have been a primary source of protein–protein interaction information at the PDB for some years [114, 115]. Affinity is generally mediated by a subset of the six CDRs in heavy and light chains [116], often with the interactions being dominated by the CDR3 loops of the heavy and light chains.

The Fc domain (Figure 2.8b) is responsible for translating the interaction of the Fab region with antigen into a useful immune response, primarily by interaction with a variety of cell-surface Fc receptors [117]. From the biotherapeutic point of view, the Fc domain is of interest because of its ability to bind the FcRn receptor, which allows internalization and recycling of the IgG, which is the primary reason that some successful mAbs can have a dramatically improved half-life of up to a month, relative to many non-antibody biotherapeutics [118]. The structure of IgG receptor interactions has been solved for the rat IgG-Fc/FcRN receptor [119] and later the human IgG1 Fc/γRIII complex [120]. This work has greatly enabled structure-based protein design work to both enhance and abolish this interaction [118, 121, 122]; for a review see Reference 123. In addition Fc receptor interaction structures are also available for IgE [124–126] where affinity for the Fc with its receptor is extremely tight.

The Fc domain also contains two extracellular immunoglobulin domains, which are dimers of the C-terminus of the heavy chain but with different quaternary structures

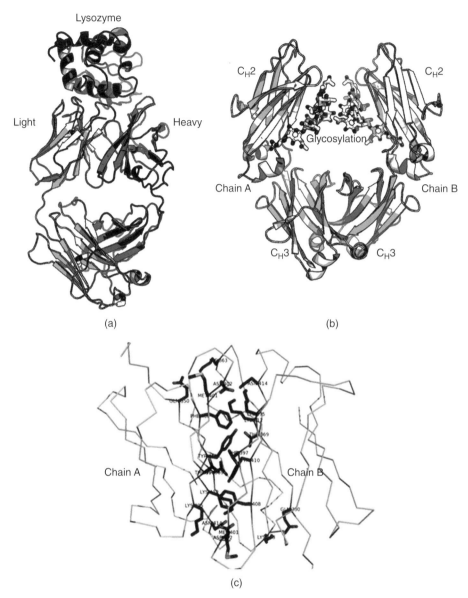

Figure 2.8. Structures of antibody, Fab, and fragment crystallizable (Fc) domains. (a) Ribbon representation of structure of Fab in complex with lysozyme; the antigen lysozyme is shaded black (PDB ID 3HFM [173]). (b) Structure of a typical Fc domain; conserved glycosylation is shown as stick representation (PDB ID 1FC1 [174]). (c) The C_H3 interface used in the design of asymmetric knob–hole interactions to generate heterodimeric monoclonal antibodies

relative to the Fab, the C_H3 domains forming a tight dimer while the C_H2 pair being completely separate and containing a glycosylation site whose presence can influence C_H2 conformation [127]. The N-termini of the Fc are fairly flexible regions which contain coordinated interchain disulfides prior to the heavy chain linker to the Fab.

2.6.1.2 Alternatives to Antibody Scaffolds. The structural information available for antibodies has allowed the biotechnologist to generate many different possible alternatives to the traditional biotherapeutic. The single-chain variable fragment (scFv) was initially engineered from conventional antibodies by combining V_L and V_H regions via the insertion of a 10–25 amino acid flexible linker (Figure 2.9a). These small variable regions can often be linked in various ways to produce multifunctional entities. Other searches for alternative scaffolds have extended into looking to antibodies generated by different species. The first of these, nonengineered proteins, commonly known as nanobodies or sdAbs, were discovered from the camelids (camels and llamas) [128–130] via structural studies [131, 132] (Figure 2.9b). These antibodies are equivalent to heavy chain-only IgGs whose three CDRs can generate comparable diversity to that of their two-chain counterparts. Due to the extended CDR H3, these nanobodies can often produce antibodies that can bind regions that conventional antibodies cannot (Figure 2.9b). Further study into the animal repertoire of antibody structures has also revealed that shark antibodies are also single domain entities, coined variable new antigen recognition (VNAR) domains (Figure 2.9c) [133, 134].

X-ray crystallography also plays a large role in the search for alternative scaffolds, distinct from traditional immunoglobulin scaffolds. Novel biotherapeutic molecules such as DARPins [135], Adnectins [136], affibodies, and anticalins [137] have been strongly influenced by the knowledge of their structures and these structural patterns are invariably used for selection of regions to optimize binding by techniques such as phage display [138] (Figure 2.9).

2.6.2 Antibody Design

In the future, structure-based design will most likely be commonplace in the design cycle of biologics. In the antibody field, structure-based affinity optimization, humanization, and enhancement of developable characteristics have been pursued for some time; for review see References 123, 139, and 140. Generally these structure-aided approaches utilize a combination of biological optimization such as phage display, combined with *in silico* techniques [141].

One of the most sought after molecular engineering feats in antibody design is to produce an easily generated heterodimeric heavy chain. Such a molecule would have the ability to incorporate two independent Fabs in one bifunctional antibody. From the many structural studies of Fc domains, characterization of the symmetry inherent in the C_H3 interface (Figure 2.7c) has established a number of strategies that can be employed to modulate the C_H3 interface, the simplest being the knobs-into-holes strategy of Ridgway et al. [142, 143] based on the original study of helix packing by Francis Crick [144]. Simply, a knob is formed at the interface by replacing a small residue with a larger one in one of the chains of the dimer and a hole via a complementary replacement of a

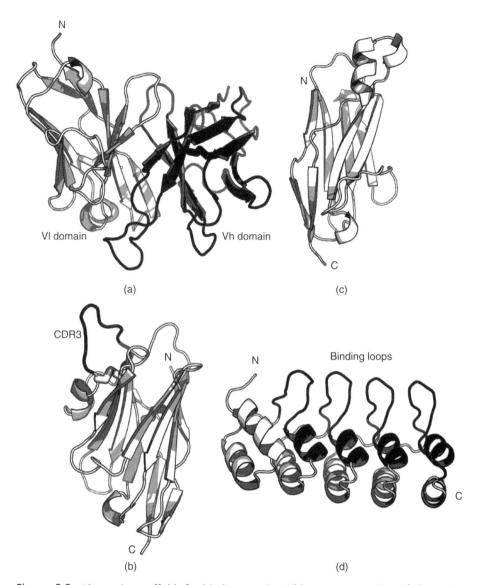

Figure 2.9. Alternative scaffolds for biotherapeutics. Ribbon representation of alternative scaffolds. (a) An engineering single-chain Fv antibody (PDB ID 2A9N [175]). (b) Camelid single domain antibody. Extended CDR3 shaded black (PDB ID 1MEL [131]). (c) Shark VNAR heavy chain variable domain (PDB ID 2I24 [134]). (d) Non-immunoglobulin-like DARPIN scaffold binding loops are formed from repeated loop/helix structure and are shaded black (PDB ID 2J8S [176])

larger residue for a smaller one in the opposing chain (Figure 2.7c). Variations of this rational technique have subsequently followed in the form of covalently linking the C_H3 domains via disulfides [145] or electrostatically matching via salt bridge formation; for a review see Carter [146].

2.6.3 Protein Receptor Interactions

Beyond Fab–antigen interactions, many other structures of protein complexes are available and are extremely useful as models for the understanding of protein–protein interactions. The biologics arena is mainly interested in the interaction between the biotherapeutic molecule and its cell-surface receptor and ways to modulate this activity. Thus insight into how a non-antibody biologic interacts with its target can be enormously beneficial in appreciating the interactions which modulate the biological activity of protein components. Information of this kind can be utilized in many different ways and will assist in the design of residues that can be modified to increase half-life via PEGylation or residues that can be mutated to enhance or detract from function. It can also aid in the design of Fc fusion constructs for the targeting protein, a biotherapeutic strategy that is extremely common for extending half-life.

Many structures of cell-surface receptors in complex with their cytokine ligands are now publicly available. Despite many of these complexes being made out of only a small number of homologous protein domains, the complexity of the quaternary structure of the complexes is astounding (Figure 2.10) [147]. Results such as the EPO/erythropoietin receptor complex [148], IL2 complex [149], GCSF/receptor complex, LIF, and CNTF all enhance our knowledge of cytokine receptor interactions. However, this diversity of form and structure currently tends to thwart our ability to reliably predict the structure of a complex without some prior knowledge from an X-ray structure which may be

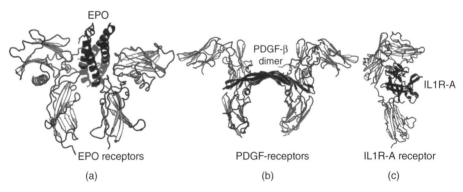

Figure 2.10. Ribbon diagrams of immunoglobulin domain receptors and their complex interactions with endogenous ligands. (a) Structure of the four-helix bundle erythropoietin (EPO), shaded black, dimerizing the erythropoietin receptor, shaded white (PDB ID 1CN4 [148]). (b) Structure of platelet-derived growth factor-β (PDGF-β), shaded black, dimerizing the PDGF-β receptor, shaded white (PDB ID 3MJG [177]). (c) Structure of interleukin-1 receptor antagonist (IL1R-A), shaded black, with its monomeric receptor shaded white (PDB 1IRA [170])

difficult to obtain. As such, many of the interactions of biotherapeutics with receptors still remain to be characterized at the atomic level.

2.7 FUTURE APPLICATIONS OF CRYSTAL STRUCTURES IN BIOTHERAPEUTICS

As the use of macromolecules for biomedical applications continues to expand, the use of structure beyond the typical applications described above also continues.

Structures are beginning to be used more extensively in the design of vaccine candidates. Preventative vaccines have become a standard proactive treatment for infectious diseases and techniques such as structural vaccinology [150] are being developed to aid the expansion of more potent vaccine candidates [151]. This approach is fairly similar to that taken for the combination of biotherapeutics with Fc domains to enhance half-life and generally works by using structures to concatenate and improve the accessibility of pathogenic antigen regions to the recipient. However, it is hoped that with the multitude of structural genomics studies on pathogenic organism proteins [152, 153] and their interactome, the ultimate goal of providing a structural basis for the determination of antigenicity of nonself-proteins will be realized. If possible, such an innovation would lead to a major advancement in the ability to design more potent initiators of the immune system. This area is beginning to be pursued within the bacterial and viral vaccine world and should provide complementary information to the results of reverse vaccinology whereby mice are exposed to large proteomic samples of bacterial proteins to find the most potent immunogens [154].

This chapter has been largely devoted to proteins as the vast majority of biologic-based therapeutics are proteins, but crystallographic applications are not limited to proteins. Nucleic acid-based molecules are also progressing as potential biologics and have been the subject of crystallographic investigation since the early 1960s. The ability of nucleic acid chains to match the capacity of proteins in both binding affinity and specificity and the potential ease of panning large libraries of nucleic acids for this functionality have made them attractive biotherapeutic candidates and are likely to be very important in the future if important issues such as bioavailability and half-life can be addressed. Structures of RNA aptamers in complex with proteins are beginning to emerge [155, 156].

2.7.1 Protein Engineering

2.7.1.1 Unnatural Amino Acids. Another emerging technique to build proteins with special properties is the utilization of nonnatural amino acid technologies [157] to change the properties of proteins completely. Here, unnatural amino acids with special properties are incorporated into proteins at structurally defined sites. The macromolecular structure is extremely useful for the design of both unnatural tRNA synthetases [158] and proteins incorporating such amino acids [159, 160]. Further these technologies with structure being utilized as a guide can be used to design half-life-improved proteins by the incorporation of combination sites for polyethylene glycol [161, 162] (Figure 2.11).

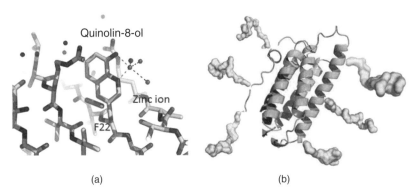

(a)   (b)

Figure 2.11. Modifying proteins with unnatural chemical entities. (a) Crystal structure of a protein incorporating a chemically functional zinc chelating unnatural amino acid (quinolin-8-01) (PDB ID 3FCA [160]). (b) Model of using structure to optimally position half-life extending polyethylene glycol (represented as filled blobs) to the surface of an interferon-α molecule

2.7.1.2 *Aggregation Propensity.*

Despite the high homology between scaffold constant domains of mAbs, unacceptable aggregation of the molecules can occur with relatively minor changes in the constant or CDR regions; such properties in biotherapeutic molecules are extremely undesirable. Structure can be invaluable in the analysis of these effects and algorithms have been developed to exploit high-resolution structures for the detection and prevention of aggregation. One such technique is spatial aggregation propensity (SAP) algorithm [163–165]. This shows great promise not only for the use of structure to identify regions of aggregation propensity, but also for the identification of protein-binding hotspots. The techniques work by performing a spatial and dynamic analysis of regions of hydrophobicity in a structure. In the molecular dynamics simulation the mean hydrophobicity of an atom is calculated from a sum of the hydrophobicity in spheres around the atom, thus building up a three-dimensional profile of regions prone to aggregation. As the general stickiness of a protein surface prone to aggregation may also have relation to a protein's ability to bind other proteins, the techniques may also be applied to estimate the binding surfaces of molecules involved in protein–protein interactions; such as the signaling of cytokines to their receptors. This sort of study can be extremely useful to provide indications of potential sites of protein–protein interaction when structural information on the complex is absent.

2.7.1.3 *Thermal Stability.*

The development cycle of biologics largely deals with the ability to mass produce an optimally bioactive molecule with the ability to withstand conditions beyond that of most proteins. This difference in formulation tends to be a major difference relative to traditional small-molecule drug discovery. The need to maintain the biomolecule in an aqueous state introduces a number of additional stability challenges for the biotherapeutic. Stress testing a protein in response to raised temperatures to simulate external conditions is often a way to test the degradation reactions. Utilizing structure to analyze aggregation propensity measures as previously described can also be applied at this point. In addition, enhancement of thermal stability with modern computational mutagenesis techniques may further enhance stability [166]. Such approaches can greatly improve the stability of molecules with only a few internal residue changes.

2.8 CONCLUSION

X-ray crystallography is an invaluable tool for the study and understanding of biologics. Modern parallel techniques for cloning, protein expression, and crystallization have revolutionized the field and are now allowing structure analysis to be incorporated into the design cycle of biotherapeutics. As molecular biology and expression technologies improve, the ability to rapidly modify a protein based on structure and to measure the consequences of these modifications biophysically and biochemically is fast becoming the norm in a similar mode to that traditionally used for small-molecule structure-based design. The ultimate goal of these endeavors is to provide de novo methods to predict the improvement of properties of the biotherapeutic such as potency, half-life, and stability. There is still a long way to go in this regard and computational speeds and predictive algorithms are ever advancing, but the orders of magnitude of complexity macromolecules possess relative to small molecules make quantitatively estimating the biophysical properties of biotherapeutics difficult. However, the inclusion of structural techniques such as X-ray crystallography at various points in the design cycle so that snapshots of improvement are captured can greatly facilitate the production of a biotherapeutic suitable for development.

This is an exciting time in the area of biotherapeutic X-ray crystallography. There are still many challenges remaining, from the production of large quantities of stable proteins and protein complexes to the application of structure to derive quantitative and predictive information for biotherapeutic design. While to fully utilize X-ray structures more progress in computation algorithms for biologic structure-based design needs to be made, there are few other techniques that have provided more information and enabled biotherapeutics more than X-ray crystallography, which is still progressing rapidly in step with biotherapeutic field as a whole.

ACKNOWLEDGMENTS

The author thanks Scott Lesley and Martin Siedel for continued support in the pursuit of structure. The author also thanks Peter Nollert and Beth Bacon from Emerald Bio, David Terrell from Art Robbins, Joesph Ferrara and Anita Yonick at Rigaku, and Chis Shaw and Andy Meyer at GNF Systems for help with diagrams.

All molecular figures were produced with the program PyMol [167].

REFERENCES

1. Sanger, F. and Tuppy, H. (1951) The amino-acid sequence in the phenylalanyl chain of insulin. 2. The investigation of peptides from enzymic hydrolysates. *Biochem. J.*, **49**, 481–490.

2. Sanger, F. and Tuppy, H. (1951) The amino-acid sequence in the phenylalanyl chain of insulin. I. The identification of lower peptides from partial hydrolysates. *Biochem. J.*, **49**, 463–481.

3. Keefer, L.M., Piron, M.A., and De Meyts, P. (1981) Human insulin prepared by recombinant DNA techniques and native human insulin interact identically with insulin receptors. *Proc. Natl Acad. Sci. U. S. A.*, **78**, 1391–1395.

4. Adams, M.J., et al. (1969) Structure of rhombohedral 2 zinc insulin crystals. *Nature*, **224**, 491–495.

5. Walsh, G. (2005) Therapeutic insulins and their large-scale manufacture. *Appl. Microbiol. Biotechnol.*, **67**, 151–159.

6. Porter, R.R. (1958) Separation and isolation of fractions of rabbit gamma-globulin containing the antibody and antigenic combining sites. *Nature*, **182**, 670–671.

7. Goldstein, D.J., Humphrey, R.L., and Poljak, R.J. (1968) Human Fc fragment: crystallographic evidence for two equivalent subunits. *J. Mol. Biol.*, **35**, 247–249.

8. Deisenhofer, J., Colman, P.M., Epp, O., and Huber, R. (1976) Crystallographic structural studies of a human Fc fragment. II. A complete model based on a Fourier map at 3.5 A resolution. *Hoppe Seylers Z. Physiol. Chem.*, **357**, 1421–1434.

9. Deisenhofer, J., et al. (1976) Crystallographic structural studies of a human Fc-fragment. I. An electron-density map at 4 A resolution and a partial model. *Hoppe Seylers Z. Physiol. Chem.*, **357**, 435–445.

10. Amzel, L.M., et al. (1974) The three dimensional structure of a combining region-ligand complex of immunoglobulin NEW at 3.5-A resolution. *Proc. Natl Acad. Sci. U. S. A.*, **71**, 1427–1430.

11. Poljak, R.J., et al. (1974) The three-dimensional structure of the fab' fragment of a human myeloma immunoglobulin at 2.0-angstrom resolution. *Proc. Natl Acad. Sci. U. S. A.*, **71**, 3440–3444.

12. Segal, D.M., et al. (1974) The three-dimensional structure of a phosphorylcholine-binding mouse immunoglobulin Fab and the nature of the antigen binding site. *Proc. Natl Acad. Sci. U. S. A.*, **71**, 4298–4302.

13. Kohler, G. and Milstein, C. (1975) Continuous cultures of fused cells secreting antibody of predefined specificity. *Nature*, **256**, 495–497.

14. Stanfield, R.L., Zemla, A., Wilson, I.A., and Rupp, B. (2006) Antibody elbow angles are influenced by their light chain class. *J. Mol. Biol.*, **357**, 1566–1574.

15. Saphire, E.O., et al. (2003) Crystal structure of an intact human IgG: antibody asymmetry, flexibility, and a guide for HIV-1 vaccine design. *Adv. Exp. Med. Biol.*, **535**, 55–66.

16. Klock, H.E., Koesema, E.J., Knuth, M.W., and Lesley, S.A. (2008) Combining the polymerase incomplete primer extension method for cloning and mutagenesis with microscreening to accelerate structural genomics efforts. *Proteins*, **71**, 982–994.

17. Coggill, P., Finn, R.D., and Bateman, A. (2008) Identifying protein domains with the Pfam database. *Curr. Protoc. Bioinform.* Chapter 2, Unit 2.5.

18. Finn, R.D., et al. (2010) The Pfam protein families database. *Nucleic Acids Res.*, **38**, D211–D222.

19. Hunter, S., et al. (2009) InterPro: the integrative protein signature database. *Nucleic Acids Res.*, **37**, D211–D215.

20. Mulder, N.J. and Apweiler, R. (2008) The InterPro database and tools for protein domain analysis. *Curr. Protoc. Bioinform.*, Chapter 2, Unit 2.7.

21. Marchler-Bauer, A., et al. (2011) CDD: a Conserved Domain Database for the functional annotation of proteins. *Nucleic Acids Res.*, **39**, D225–D229.

22. Chi, P.H., Pang, B., Korkin, D., and Shyu, C.R. (2009) Efficient SCOP-fold classification and retrieval using index-based protein substructure alignments. *Bioinformatics*, **25**, 2559–2565.

23. Cuff, A., et al. (2009) The CATH hierarchy revisited-structural divergence in domain super-families and the continuity of fold space. *Structure*, **17**, 1051–1062.

24. Altschul, S.F., et al. (1990) Basic local alignment search tool. *J. Mol. Biol.*, **215**, 403–410.

25. Gasteiger, E., et al. (2003) ExPASy: the proteomics server for in-depth protein knowledge and analysis. *Nucleic Acids Res.*, **31**, 3784–3788.

26. Julenius, K., Molgaard, A., Gupta, R., and Brunak, S. (2005) Prediction, conservation analysis, and structural characterization of mammalian mucin-type O-glycosylation sites. *Glycobiology*, **15**, 153–164.

27. Chang, V.T., et al. (2007) Glycoprotein structural genomics: solving the glycosylation problem. *Structure*, **15**, 267–273. doi:10.1016/j.str.2007.01.011

28. Kyte, J. and Doolittle, R.F. (1982) A simple method for displaying the hydropathic character of a protein. *J. Mol. Biol.*, **157**, 105–132.

29. Tusnady, G.E. and Simon, I. (2010) Topology prediction of helical transmembrane proteins: how far have we reached? *Curr. Protein Peptide Sci.*, **11**, 550–561.

30. Rost, B. (2001) Review: protein secondary structure prediction continues to rise. *J. Struct. Biol.*, **134**, 204–218. doi:10.1006/jsbi.2001.4336

31. Wooten, J.C. (1994) Non-globular domains in protein sequences: automated segmentation using complexity measures. *Comput. Chem.*, **18**, 269–285.

32. Linding, R., et al. (2003) Protein disorder prediction: implications for structural proteomics. *Structure*, **11**, 1453–1459.

33. Ward, J.J., et al. (2004) The DISOPRED server for the prediction of protein disorder. *Bioinformatics*, **20**, 2138–2139.

34. Ward, J.J., et al. (2004) Prediction and functional analysis of native disorder in proteins from the three kingdoms of life. *J. Mol. Biol.*, **337**, 635–645.

35. Pantazatos, D., et al. (2004) Rapid refinement of crystallographic protein construct definition employing enhanced hydrogen/deuterium exchange MS. *Proc. Natl Acad. Sci. U. S. A.*, **101**, 751–756.

36. Spraggon, G., et al. (2004) On the use of DXMS to produce more crystallizable proteins–structures of the Thermotoga maritima proteins TM0160 and TM1171. *Protein Sci.*, **13**, 3187–3199.

37. Slabinski, L., et al. (2007) XtalPred: a web server for prediction of protein crystallizability. *Bioinformatics*, **23**, 3403–3405.

38. Jackson, D.A., Symons, R.H., and Berg, P. (1972) Biochemical method for inserting new genetic information into DNA of Simian Virus 40: circular SV40 DNA molecules containing lambda phage genes and the galactose operon of Escherichia coli. *Proc. Natl Acad. Sci. U. S. A.*, **69**, 2904–2909.

39. Cohen, S.N., Chang, A.C., Boyer, H.W., and Helling, R.B. (1973) Construction of biologically functional bacterial plasmids in vitro. *Proc. Natl Acad. Sci. U. S. A.*, **70**, 3240–3244.

40. Klock, H.E. and Lesley, S.A. (2009) The Polymerase Incomplete Primer Extension (PIPE) method applied to high-throughput cloning and site-directed mutagenesis. *Methods Mol. Biol. High Throughput Protein Express. Purif.*, **498**, 91–103.

41. Campion, E.M., Loughran, S.T., and Walls, D. (2011) Protein quantitation and analysis of purity. *Methods Mol. Biol.*, **681**, 229–258.

42. Loughran, S.T. and Walls, D. (2011) Purification of poly-histidine-tagged proteins. *Methods Mol. Biol.*, **681**, 311–335.

43. Walls, D. and Loughran, S.T. (2011) Tagging recombinant proteins to enhance solubility and aid purification. *Methods Mol. Biol.*, **681**, 151–175.

44. Carson, M., et al. (2007) His-tag impact on structure. *Acta Crystallogr. D Biol. Crystallogr.*, **63**, 295–301.

45. Smyth, D.R., et al. (2003) Crystal structures of fusion proteins with large-affinity tags. *Protein Sci.*, **12**, 1313–1322.

46. Chapman, H.N., et al. (2011) Femtosecond X-ray protein nanocrystallography. *Nature*, **470**, 73–77. doi:10.1038/nature09750

47. Bernal, J.D. and Crowfoot, D. (1934) X-ray photographs of crystalline pepsin. *Nature*, **133**, 794–795.

48. Federov, D.S. (1891) Symmetry of regular systems of figures. *Zap. Mineral. Obch.*, **28**, 1–146.

49. Schönflies, A.M. (1891) Kristallsysteme und kristallstruktur, Druck und verlag von B.G. Teubner, Liepzig, Germany.

50. Barlow, W. (1894) Über die geometrischen Eigenschaften starrer Strukturen und ihre Anwendung auf Kristalle. *Z. Kristallogr.*, **23**, 1–63.

51. Hahn, T. (2006) *International Tables for Crystallography. Space-Group Symmetry A.* Springer, Dordrecht, The Netherlands.

52. Sumner, J.B. (1926) The isolation and crystallization of the enzyme urease. Preliminary paper. *J. Biol. Chem.*, **69**, 435–441.

53. Haas, C. and Drenth, J. (1999) Understanding protein crystallization on the basis of phase diagrams. *J. Cryst. Growth*, **196**, 388–394.

54. Jancarik, J. and Kim, S.-H. (1991) Sparse matrix sampling: a screening method for crystallization of proteins. *J. Appl. Crystallogr.*, **24**, 409–411.

55. Page, R., et al. (2003) Shotgun crystallization strategy for structural genomics: an optimized two-tiered crystallization screen against the Thermotoga maritima proteome. *Acta Crystallogr. D Biol. Crystallogr.*, **59**, 1028–1037.

56. Santarsiero, B.D., et al. (2002) An approach to rapid protein crystallization using nanodroplets. *J. Appl. Crystallogr.*, **35**, 278–281.

57. Luft, J.R., et al. (2003) A deliberate approach to screening for initial crystallization conditions of biological macromolecules. *J. Struct. Biol.*, **142**, 170–179.

58. Hansen, C. and Quake, S.R. (2003) Microfluidics in structural biology: smaller, faster em leader better. *Curr. Opin. Struct. Biol.*, **13**, 538–544.

59. Gerdts, C.J., et al. (2010) Nanovolume optimization of protein crystal growth using the microcapillary protein crystallization system. *J. Appl. Crystallogr.*, **43**, 1078–1083.

60. Helliwell, J.R. (1998) Synchrotron radiation facilities. *Nat. Struct. Biol.*, **6**, 614–617.

61. Ravelli, R.B. and Garman, E.F. (2006) Radiation damage in macromolecular cryocrystallography. *Curr. Opin. Struct. Biol.*, **16**, 624–629. doi:10.1016/j.sbi.2006.08.001

62. Garman, E.F. and Mitchell, E.P. (1996) Glycerol concentrations required for cryoprotection of 50 typical protein crystallization solutions. *J. Appl. Crystallogr.*, **29**, 584–587.

63. Dauter, Z. (1999) Data-collection strategies. *Acta Crystallogr. D Biol. Crystallogr.*, **55**, 1703–1717.

64. Nave, C. and Garman, E.F. (2005) Towards an understanding of radiation damage in cryocooled macromolecular crystals. *J. Synchrotron Radiat.*, **12**, 257–260.

65. Garman, E.F. and McSweeney, S.M. (2007) Progress in research into radiation damage in cryo-cooled macromolecular crystals. *J. Synchrotron Radiat.*, **14**, 1–3.

66. Ercan, A., Tate, M.W., and Gruner, S.M. (2006) Analog pixel array detectors. *J. Synchrotron Radiat.*, **13**, 110–119.

67. Patterson, A.L. (1935) A direct method for the determination of the components of interatomic distances in crystals. *Z. Krist.* **A90**, 517–542.

68. Sheldrick, G.M., et al. (1993) The application of direct methods and Patterson interpretation to high-resolution native protein data. *Acta Crystallogr. D Biol. Crystallogr.*, **49**, 18–23.

69. Uson, I. and Sheldrick, G.M. (1999) Advances in direct methods for protein crystallography. *Curr. Opin. Struct. Biol.*, **9**, 643–648.

70. Hoppe, W. (1957) Die 'Faltmolelciilmethode' – eine neue Merhode zur Bestimmung der Kristallstruktur bei ganz oder teilweise bekannter Molekiilstruktur. *Acta. Crystallogr.*, **10**, 750–751.

71. Rossmann, M.G. and Blow, D.M. (1962) The detection of sub-units within the crystallographic asymmetric unit. *Acta Crystallogr.* **A15**, 24–51.

72. McCoy, A.J., et al. (2007) Phaser crystallographic software. *J. Appl. Crystallogr.*, **40**, 658–674.

73. Read, R.J. (2001) Pushing the boundaries of molecular replacement with maximum likelihood. *Acta Crystallogr. D Biol. Crystallogr.*, **57**, 1373–1382.

74. Green, D.W., Ingram, M., and Perutz, M.F. (1954) The structure of haemoglobin: IV. Sign determination by the isomorphous replacement method. *Proc. R. Soc. Lond. Ser. A*, **225**, 287–307.

75. Hendrickson, W.A. and Ogata, C. (1997) Phase determination from multiwavelength anomalous diffraction measurements. *Methods Enzymol.*, **297**, 494–523.

76. Hendrickson, W.A., Horton, J.R., and LeMaster, D.M. (1990) Selenomethionyl proteins produced for analysis by multiwavelength anomalous diffraction (MAD): a vehicle for direct determination of three-dimensional structure. *EMBO J.*, **9**, 1665–1672.

77. Terwilliger, T.C. (1999) σ^2_R a reciprocal-space measure of the quality of macromolecular electron-density maps. *Acta Crystallogr.*, **D55**, 1174–1178.

78. Read, R.J. (1986) Improved Fourier coefficients for maps using phases from partial structures with errors. *Acta Crystallogr.*, **A42**, 140–149.

79. Morris, R.J., Perrakis, A., and Lamzin, V.S. (2003) ARP/wARP and automatic interpretation of protein electron density maps. *Methods Enzymol.*, **374**, 229–244.

80. Perrakis, A., Harkiolaki, M., Wilson, K.S., and Lamzin, V.S. (2001) ARP/wARP and molecular replacement. *Acta Crystallogr. D Biol. Crystallogr.*, **57**, 1445–1450.

81. Terwilliger, T.C. (2001) Maximum-likelihood density modification using pattern recognition of structural motifs. *Acta Crystallogr.*, **D57**, 1755–1762.

82. Engh, R.A. and Huber, R. (1991) Accurate bond and angle parameters for X-ray protein structure refinement. *Acta Crystallogr. A*, **A47**, 392–400.

83. Jones, T.A., Zou, J.Y., Cowan, S.W., and Kjeldgaard, M. (1991) Improved methods for building protein models in electron density maps and the location of errors in these models. *Acta Crystallogr. A*, **47**, 110–119.

84. McRee, D.E. (1999) XtalView/Xfit- a versatile program for manipulating atomic coordinates and electron density. *J. Struct. Biol.*, **125**, 156–165.

85. Emsley, P. and Cowtan, K. (2004) Coot: model-building tools for molecular graphics. *Acta Crystallogr.*, **D60**, 2126–2132.

86. Murshudov, G.N., Vagin, A.A., and Dodson, E.J. (1997) Refinement of macromolecular structures by the maximum-likelihood method. *Acta Crystallogr.*, **D53**, 240–255.

87. Adams, P.D., et al. (2004) Recent developments in the PHENIX software for automated crystallographic structure determination. *J. Synchrotron Radiat.*, **11**, 53–55.

88. Blanc, E., et al. (2004) Refinement of severely incomplete structures with maximum likelihood in BUSTER-TNT. *Acta Crystallogr. D Biol. Crystallogr.*, **60**, 2210–2221.

89. Roberts, A.L. and Brunger, A.T. (1995) Phase improvement by cross-validated density modification. *Acta Crystallogr. D Biol. Crystallogr.*, **51**, 990–1002.

90. Ramachandran, G.N., Ramakrishnan, C., and Sasisekharan, V. (1963) Stereochemistry of polypeptide chain configurations. *J. Mol. Biol.*, **7**, 95–99.

91. Lakowski, R.A., MacArthur, M.W., Moss, D.S., and Thornton, J.M. (1993) PROCHECK: a program to check the stereochemical quality of protein structures. *J. Appl. Crystallogr.*, **26**, 283–291.

92. Winn, M.D., Isupov, M.N., and Murshudov, G.N. (2001) Use of TLS parameters to model anisotropic displacements in macromolecular refinement. *Acta Crystallogr. D Biol. Crystallogr.*, **57**, 122–133.

93. Berman, H.M., et al. (2000) The Protein Data Bank. *Nucleic Acids Res.*, **28**, 235–242.

94. Lo Conte, L., et al. (2000) SCOP: a structural classification of proteins database. *Nucleic Acids Res.*, **28**, 257–259.

95. Orengo, C.A., Pearl, F.M., and Thornton, J.M. (2003) The CATH domain structure database. *Methods Biochem. Anal.*, **44**, 249–271.

96. Pirovano, W. and Heringa, J. (2010) Protein secondary structure prediction. *Methods Mol. Biol.*, **609**, 327–348.

97. Kabsch, W. and Sander, C. (1983) Dictionary of protein secondary structure: pattern recognition of hydrogen-bonded and geometrical features. *Biopolymers*, **22**, 2577–2637.

98. Richards, F.M. and Kundrot, C.E. (1988) Identification of structural motifs from protein coordinate data: secondary structure and first-level supersecondary structure. *Proteins*, **3**, 71–84.

99. Martin, J., et al. (2005) Protein secondary structure assignment revisited: a detailed analysis of different assignment methods. *BMC Struct. Biol.*, **5**, 17.

100. Sibanda, B.L., Blundell, T.L., and Thornton, J.M. (1989) Conformation of β-hairpins in protein structures: a systematic classification with applications to modelling by homology, electron density fitting and protein engineering. *J. Mol. Biol.*, **206**, 759–777.

101. Bork, P. (1991) Shuffled domains in extracellular proteins. *FEBS Lett.*, **286**, 47–54.

102. Holm, L., Kaariainen, S., Rosenstrom, P., and Schenkel, A. (2008) Searching protein structure databases with DaliLite v.3. *Bioinformatics*, **24**, 2780–2781.

103. Holm, L., Kaariainen, S., Wilton, C., and Plewczynski, D. (2006) Using Dali for structural comparison of proteins. *Curr. Protoc Bioinform.* Chapter 5, Unit 5.5.

104. Kobe, B., et al. (2008) Crystallography and protein-protein interactions: biological interfaces and crystal contacts. *Biochem. Soc. Trans.*, **36**, 1438–1441.

105. Bordner, A.J. and Gorin, A.A. (2008) Comprehensive inventory of protein complexes in the Protein Data Bank from consistent classification of interfaces. *BMC Bioinform.*, **9**, 234.

106. Valdar, W.S. and Thornton, J.M. (2001) Conservation helps to identify biologically relevant crystal contacts. *J. Mol. Biol.*, **313**, 399–416.

107. Krissinel, E. and Henrick, K. (2007) Inference of macromolecular assemblies from crystalline state. *J. Mol. Biol.*, **372**, 774–797.

108. Ponstingl, H., Henrick, K., and Thornton, J.M. (2000) Discriminating between homodimeric and monomeric proteins in the crystalline state. *Proteins*, **41**, 47–57.

109. Bork, P., Holm, L., and Sander, C. (1994) The immunoglobulin fold. Structural classification, sequence patterns and common core. *J. Mol. Biol.*, **242**, 309–320.

110. Wu, T.T. and Kabat, E.A. (1970) An analysis of the sequences of the variable regions of Bence Jones proteins and myeloma light chains and their implications for antibody complementarity. *J. Exp. Med.*, **132**, 211–250.

111. Valentine, R.C. and Green, N.M. (1967) Electron microscopy of an antibody-hapten complex. *J. Mol. Biol.*, **27**, 615–617.

112. Fett, J.W., Deutsch, H.F., and Smithies, O. (1973) Hinge-region deletion localized in the IgG-globulin Mcg. *Immunochemistry*, **10**, 115–118.

113. Steiner, L.A. and Lopes, A.D. (1979) The crystallizable human myeloma protein Dob has a hinge-region deletion. *Biochemistry*, **18**, 4054–4067.

114. Davies, D.R., Sheriff, S., and Padlan, E.A. (1988) Antibody-antigen complexes. *J. Biol. Chem.*, **263**, 10541–10544.

115. Davies, D.R. and Cohen, G.H. (1996) Interactions of protein antigens with antibodies. *Proc. Natl Acad. Sci. U. S. A.*, **93**, 7–12.

116. Tong, J.C., et al. (2008) BEID: database for sequence-structure-function information on antigen-antibody interactions. *Bioinformation*, **3**, 58–60.

117. Raghavan, M. and Bjorkman, P.J. (1996) Fc receptors and their interactions with immunoglobulins. *Annu. Rev. Cell Dev. Biol.*, **12**, 181–220.

118. Roopenian, D.C. and Akilesh, S. (2007) FcRn: the neonatal Fc receptor comes of age. *Nat. Rev. Immunol.*, **7**, 715–725.

119. Martin, W.L., West, A.P., Jr., Gan, L., and Bjorkman, P.J. (2001) Crystal structure at 2.8 A of an FcRn/heterodimeric Fc complex: mechanism of pH-dependent binding. *Mol. Cell*, **7**, 867–877.

120. Sondermann, P., Huber, R., Oosthuizen, V., and Jacob, U. (2000) The 3.2-A crystal structure of the human IgG1 Fc fragment-Fc gammaRIII complex. *Nature*, **406**, 267–273.

121. Sondermann, P. and Oosthuizen, V. (2002) The structure of Fc receptor/Ig complexes: considerations on stoichiometry and potential inhibitors. *Immunol. Lett.*, **82**, 51–56.

122. Sondermann, P., Kaiser, J., and Jacob, U. (2001) Molecular basis for immune complex recognition: a comparison of Fc-receptor structures. *J. Mol. Biol.*, **309**, 737–749.

123. Presta, L.G. (2008) Molecular engineering and design of therapeutic antibodies. *Curr. Opin. Immunol.*, **20**, 460–470.

124. Garman, S.C., et al. (2000) Structure of the Fc fragment of human IgE bound to its high-affinity receptor Fc epsilonRI alpha. *Nature*, **406**, 259–266.

125. Henry, A.J., et al. (2000) Conformation of the isolated cepsilon3 domain of IgE and its complex with the high-affinity receptor, FcepsilonRI. *Biochemistry*, **39**, 7406–7413.

126. Holdom, M.D., et al. (2011) Conformational changes in IgE contribute to its uniquely slow dissociation rate from receptor FcvarepsilonRI. *Nat. Struct. Mol. Biol.*, **18**, 571–576.

127. Krapp, S., et al. (2003) Structural analysis of human IgG-Fc glycoforms reveals a correlation between glycosylation and structural integrity. *J. Mol. Biol.*, **325**, 979–989.

128. Harmsen, M.M., et al. (2000) Llama heavy-chain V regions consist of at least four distinct subfamilies revealing novel sequence features. *Mol. Immunol.*, **37**, 579–590.

129. Spinelli, S., et al. (2000) Camelid heavy-chain variable domains provide efficient combining sites to haptens. *Biochemistry*, **39**, 1217–1222.

130. Muyldermans, S. and Lauwereys, M. (1999) Unique single-domain antigen binding fragments derived from naturally occurring camel heavy-chain antibodies. *J. Mol. Recognit.*, **12**, 131–140.

131. Desmyter, A., et al. (1996) Crystal structure of a camel single-domain VH antibody fragment in complex with lysozyme. *Nat. Struct. Biol.*, **3**, 803–811.

132. Woolven, B.P., Frenken, L.G., van der Logt, P., and Nicholls, P.J. (1999) The structure of the llama heavy chain constant genes reveals a mechanism for heavy-chain antibody formation. *Immunogenetics*, **50**, 98–101.

133. Stanfield, R.L., et al. (2007) Maturation of shark single-domain (IgNAR) antibodies: evidence for induced-fit binding. *J. Mol. Biol.*, **367**, 358–372.

134. Stanfield, R.L., Dooley, H., Flajnik, M.F., and Wilson, I.A. (2004) Crystal structure of a shark single-domain antibody V region in complex with lysozyme. *Science*, **305**, 1770–1773.

135. Stumpp, M.T. and Amstutz, P. (2007) DARPins: a true alternative to antibodies. *Curr. Opin. Drug Discov. Dev.*, **10**, 153–159.

136. Lipovsek, D. (2011) Adnectins: engineered target-binding protein therapeutics. *Protein Eng. Des. Sel.*, **24**, 3–9.

137. Skerra, A. (2007) Anticalins as alternative binding proteins for therapeutic use. *Curr. Opin. Mol. Ther.*, **9**, 336–344.

138. Umetsu, M., et al. (2010) Protein-protein interactions and selection: generation of molecule-binding proteins on the basis of tertiary structural information. *FEBS J.*, **277**, 2006–2014.

139. Caravella, J.A., Wang, D., Glaser, S.M., and Lugovskoy, A. (2010) Structure-guided design of antibodies. *Curr. Comput. Aided Drug Des.*, **6**, 128–138.

140. Presta, L.G., et al. (2002) Engineering therapeutic antibodies for improved function. *Biochem. Soc. Trans.*, **30**, 487–490.

141. Barderas, R., et al. (2008) Affinity maturation of antibodies assisted by in silico modeling. *Proc. Natl Acad. Sci. U. S. A.*, **105**, 9029–9034.

142. Ridgway, J.B., Presta, L.G., and Carter, P. (1996) 'Knobs-into-holes' engineering of antibody CH3 domains for heavy chain heterodimerization. *Protein Eng.*, **9**, 617–621.

143. Zhu, Z., Presta, L.G., Zapata, G., and Carter, P. (1997) Remodeling domain interfaces to enhance heterodimer formation. *Protein Sci.*, **6**, 781–788.

144. Crick, F.H.C. (1953) The packing of α-helices: simple coiled-coils. *Acta Crystallogr. A*, **6**, 689–697.

145. Merchant, A.M., et al. (1998) An efficient route to human bispecific IgG. *Nat. Biotechnol.*, **16**, 677–681.

146. Carter, P. (2001) Bispecific human IgG by design. *J. Immunol. Methods*, **248**, 7–15.

147. Wang, X., Lupardus, P., Laporte, S.L., and Garcia, K.C. (2009) Structural biology of shared cytokine receptors. *Annu. Rev. Immunol.*, **27**, 29–60.

148. Syed, R.S., et al. (1998) Efficiency of signalling through cytokine receptors depends critically on receptor orientation. *Nature*, **395**, 511–516.

149. Wang, X., Rickert, M., and Garcia, K.C. (2005) Structure of the quaternary complex of interleukin-2 with its alpha, beta, and gammac receptors. *Science*, **310**, 1159–1163.

150. Rinaudo, C.D., Telford, J.L., Rappuoli, R., and Seib, K.L. (2009) Vaccinology in the genome era. *J. Clin. Invest.*, **119**, 2515–2525.

151. Nuccitelli, A., et al. (2011) Structure-based approach to rationally design a chimeric protein for an effective vaccine against Group B Streptococcus infections. *Proc. Natl Acad. Sci. U. S. A.*, **108**, 10278–10283.

152. Verlinde, C.L., et al. (2009) Fragment-based cocktail crystallography by the medical structural genomics of pathogenic protozoa consortium. *Curr. Top. Med. Chem.*, **9**, 1678–1687.

153. Fan, E., et al. (2008) Structural genomics of pathogenic protozoa: an overview. *Methods Mol. Biol.*, **426**, 497–513.

154. Rappuoli, R. (2000) Reverse vaccinology. *Curr. Opin. Microbiol.*, **3**, 445–450.

155. Nomura, Y., et al. (2010) Conformational plasticity of RNA for target recognition as revealed by the 2.15 A crystal structure of a human IgG-aptamer complex. *Nucleic Acids Res.*, **38**, 7822–7829.

156. Long, S.B., Long, M.B., White, R.R., and Sullenger, B.A. (2008) Crystal structure of an RNA aptamer bound to thrombin. *RNA*, **14**, 2504–2512.

157. Wang, L. and Schultz, P.G. (2002) Expanding the genetic code. *Chem. Commun. (Camb.)*, 1–11.

158. Turner, J.M., Graziano, J., Spraggon, G., and Schultz, P.G. (2006) Structural plasticity of an aminoacyl-tRNA synthetase active site. *Proc. Natl Acad. Sci. U. S. A.*, **103**, 6483–6488.

159. Xie, J., et al. (2004) The site-specific incorporation of p-iodo-L-phenylalanine into proteins for structure determination. *Nat. Biotechnol.*, **22**, 1297–1301.

160. Lee, H.S., Spraggon, G., Schultz, P.G., and Wang, F. (2009) Genetic incorporation of a metal-ion chelating amino acid into proteins as a biophysical probe. *J. Am. Chem. Soc.*, **131**, 2481–2483.

161. Cho, H., et al. (2011) Optimized clinical performance of growth hormone with an expanded genetic code. *Proc. Natl Acad. Sci. U. S. A.*, **108**, 9060–9065.

162. Deiters, A., et al. (2004) Site-specific PEGylation of proteins containing unnatural amino acids. *Bioorg. Med. Chem. Lett.*, **14**, 5743–5745.

163. Chennamsetty, N., et al. (2009) Design of therapeutic proteins with enhanced stability. *Proc. Natl Acad. Sci. U. S. A.*, **106**, 11937–11942.

164. Voynov, V., et al. (2009) Predictive tools for stabilization of therapeutic proteins. *mAbs*, **1**, 580–582.

165. Chennamsetty, N., et al. (2009) Aggregation-prone motifs in human immunoglobulin G. *J. Mol. Biol.*, **391**, 404–413.

166. Korkegian, A., Black, M.E., Baker, D., and Stoddard, B.L. (2005) Computational thermostabilization of an enzyme. *Science*, **308**, 857–860.

167. DeLano, W. *The PyMOL Molecular Graphics System*, Version 1.2r3pre, Schrödinger, LLC.

168. Ruzheinikov, S.N., et al. (2003) High-resolution crystal structure of the Fab-fragments of a family of mouse catalytic antibodies with esterase activity. *J. Mol. Biol.*, **332**, 423–435.

169. Arkin, M.R., et al. (2003) Binding of small molecules to an adaptive protein-protein interface. *Proc. Natl Acad. Sci. U. S. A.*, **100**, 1603–1608. doi:10.1073/pnas.252756299

170. Schreuder, H., et al. (1997) A new cytokine-receptor binding mode revealed by the crystal structure of the IL-1 receptor with an antagonist. *Nature*, **386**, 194–200.

171. Vajdos, F.F., et al. (2001) Crystal structure of human insulin-like growth factor-1: detergent binding inhibits binding protein interactions. *Biochemistry*, **40**, 11022–11029.

172. Harris, L.J., Skaletsky, E., and McPherson, A. (1998) Crystallographic structure of an intact IgG1 monoclonal antibody. *J. Mol. Biol.*, **275**, 861–872.

173. Padlan, E.A., et al. (1989) Structure of an antibody-antigen complex: crystal structure of the HyHEL-10 Fab-lysozyme complex. *Proc. Natl Acad. Sci. U. S. A.*, **86**, 5938–5942.

174. Deisenhofer, J. (1981) Crystallographic refinement and atomic models of a human Fc fragment and its complex with fragment B of protein A from Staphylococcus aureus at 2.9- and 2.8-A resolution. *Biochemistry*, **20**, 2361–2370.

175. Honegger, A., Spinelli, S., Cambillau, C., and Pluckthun, A. (2005) A mutation designed to alter crystal packing permits structural analysis of a tight-binding fluorescein-scFv complex. *Protein Sci.*, **14**, 2537–2549.

176. Sennhauser, G., et al. (2007) Drug export pathway of multidrug exporter AcrB revealed by DARPin inhibitors. *PLoS Biol.*, **5**, e7.

177. Shim, A.H., et al. (2010) Structures of a platelet-derived growth factor/propeptide complex and a platelet-derived growth factor/receptor complex. *Proc. Natl Acad. Sci. U. S. A.*, **107**, 11307–11312.

3

SOLUBILITY AND EARLY ASSESSMENT OF STABILITY FOR PROTEIN THERAPEUTICS

Sheng-Jiun Wu[1], Gary L. Gilliland[1], and Yiqing Feng[2]

[1]*Janssen R&D, LLC, Spring House, PA, USA*
[2]*Lilly Research Laboratory, Indianapolis, IN, USA*

Biophysical Methods for Biotherapeutics: Discovery and Development Applications, First Edition. Edited by Tapan K. Das.
© 2014 John Wiley & Sons, Inc. Published 2014 by John Wiley & Sons, Inc.

3.1   INTRODUCTION

Biotherapeutics or biologics are complex molecules that have been designed and in some cases engineered to modulate biological processes that lead to therapeutic benefits. Biotherapeutics are comprised of a diverse set of biomolecules, of which antibodies represent the largest class. Thus, this chapter will focus primarily on antibodies, but the principles and methods outlined and described here, with a few exceptions, have general applicability to all classes of biotherapeutics.

Solubility and stability are critical factors that can dictate success or failure in the development of therapeutic antibodies. These properties as well as all the functional and other biophysical properties of antibodies depend on amino acid sequence, posttranslational modifications, and the solution environment. In current discovery programs, lead molecules are identified by optimizing not only their therapeutic potential, but also their biophysical and chemical properties. Once a lead is identified, formulation development establishes and optimizes conditions that maximize both solubility and stability.

The importance of high solubility of an antibody has its foundation in dosing requirements. Because biotherapeutics are administered as solutions, dosing depends on solubility. For example, many newer mAb therapies rely primarily on subcutaneous injection for convenience and compliance. Subcutaneous administration places high demands on a biotherapeutic's biophysical properties due to the constraints of dosage volume (preferred at <1.5 mL) and high doses (>10 mg/kg) [1]. Subcutaneously delivered mAbs are often formulated at very high concentrations under conditions that maintain low viscosity, chemical integrity, and a monomeric state. Thus, antibody solubility is important to the success of formulation development.

The stability of antibodies is very important for their manufacturing, long-term storage, delivery, and efficacy. The physical and chemical stabilities of therapeutic proteins have fundamental implications in the development process and therapeutic applications. Highly stable proteins will likely encounter less hurdles in the manufacturing processes. The ability of these proteins to remain in the functional native state under storage conditions without aggregation or chemical alteration over a long period of time is crucial for successful therapeutic applications [2].

In the sections that follow, experimental methods for directly and indirectly assessing biologic solubility and stability are described. Also, useful computational methods for predicting the aggregation potential as well as identifying amino acid sequence regions that may be problematic are presented. A number of general approaches are outlined for enhancing biologic solubility and stability that includes pegylation, glycosylation, site-directed mutagenesis, and formulation optimization. This is followed by a discussion of the trend toward developing rapid to high-throughput screening (HTS) approaches for assessing solubility and stability.

3.2 MEASURING PROTEIN SOLUBILITY

Protein solubility is an important prerequisite for structural and biophysical studies. The solubility of a protein varies with the solution conditions. Factors determining solubility include pH, ionic strength, temperature, and co-solutes [3–5]. The intrinsic properties of proteins such as size, hydrophobicity, electrostatics, and charge distribution play important roles in protein solubility [6–10]. Antibody solubility measurements in discovery and development have different goals. In lead identification, solubility determination is used to rank-order hits, to flag proteins with potential manufacturing liabilities. The major focus of solubility assays in development is to improve formulation and minimize aggregation. As mentioned above, many newer mAb therapies that require subcutaneous injection are often formulated at very high concentrations of >100 mg/mL [1]. Therefore, the methods used are similar but not identical in discovery and development.

The solubility of an antibody is often defined as the antibody concentration where the chemical potential of the antibody in solution is greater than that for the solid phase, crystalline precipitate, or amorphous precipitate. Thus, there are two types of solubility measurements, amorphous solubility and crystalline or thermodynamic solubility [5]. Although solubility is a thermodynamic characteristic of an antibody, for biochemical and ranking purposes, amorphous solubility would be more useful and relevant in discovery and early development stages. There are various techniques for determining protein solubility including filtration, precipitation, and the determination of second virial coefficient (B_{22}) using light scattering or self-interaction chromatography (SIC).

3.2.1 Direct Measurement of Solubility: Concentration to Precipitation

Operationally, solubility for proteins could be defined as the maximum protein concentration that can be achieved before the observance of amorphous precipitate. Thus, the term solubility is used here to describe amorphous solubility rather than crystalline or thermodynamic solubility. The techniques used to concentrate proteins in solution were compared and well summarized, with their advantages and limitations, in a review article by Shire et al. [11]. Among the commonly used methods, ultrafiltration is perhaps the most direct method used to evaluate protein solubility in a given solution [12–16]. In this method, the protein solution is concentrated using a membrane with an appropriate molecular weight cutoff that retains the protein as the solution volume is decreased until an amorphous precipitate is observed. After precipitation, the solution can be filtered and the protein concentration determined by measuring the absorbance at 280 nm. One advantage of this technique is that solubility can be determined in a pharmaceutically relevant solution. However, this method suffers from the fact that large quantities of highly soluble proteins are required to reach the concentration at which precipitation occurs. Thus, the requirement for multi-milligram quantities of purified proteins makes the use of the method impractical in the early stages of discovery.

Alternatively, determination of protein solubility can be achieved through the presence of increasing concentrations of precipitant, usually ammonium sulfate or polyethylene glycol (PEG). The data generated from such experiments are then extrapolated to zero precipitant concentration to determine the solubility [5, 17–19]. Ammonium sulfate precipitation requires relatively small amounts of protein (10 mg or less)—even with a highly soluble protein. This method can give quick and relatively accurate solubility values of variants of the same protein, and it can be performed at different pH values. Protein solubility is highly pH dependent and solubility is minimal near the isoelectric point (pI) of a protein. In addition, there are other techniques that have been described for estimating or directly measuring protein solubility using dialysis, evaporation, and lyophilization methods [11]. These methods are, however, not always amenable to pharmaceutically relevant solutions.

3.2.2 Indirect Assessment of Solubility: The Second Virial Coefficient (B$_{22}$) and Self-Interaction Chromatography

The osmotic second virial coefficient (B$_{22}$) provides a functional measure of a protein's solubility. B$_{22}$ is related to thermodynamic properties contributed by hydrophobic effects, van der Waals interactions, hydration forces, and electrostatic interactions providing a measure of the aggregation propensity of the molecule [20]. The B$_{22}$ value reflects both the magnitude and sign of the self-interaction between two protein molecules in solution. Briefly, a positive B$_{22}$ value indicates repulsive interactions between protein molecules while negative values represent attractive interactions [21]. Protein solutions with large negative B$_{22}$ values indicate a tendency for aggregation and precipitation. Another application of B$_{22}$ is in the prediction of conditions appropriate for protein crystallization (the crystallization slot) [22]. For crystallization, the ideal B$_{22}$ values are determined to be in the range between -1 and -8×10^{-4} mol mL/g^2. In contrast, for therapeutic applications, where solubility is being assessed, it is important for the protein to have B$_{22}$ values ranging from slightly negative to positive values in pharmaceutically relevant buffers.

The B$_{22}$ coefficient can be derived by a number of methods [23]. The two most commonly used approaches use static light scattering (SLS) measurements or SIC [22, 24–26]. The SLS method is generally the most reliable. However, the measurements are time consuming and require a large amount of recombinant protein and the use of specialized instrumentation. In contrast SIC is a simple method that uses conventional chromatography equipment and can be adapted to greatly reduce the amount of protein required. For the SIC measurements, the protein of interest is immobilized onto a chromatographic support, and the same protein is flowed over the column in an appropriate buffer. The B$_{22}$ value can then be calculated from the chromatographic retention factor, k', obtained from the observed retention time [27]. Proteins with negative B$_{22}$ values are retained longer on the protein-coupled column than on the column matrix due to their attraction to the proteins in the stationary phase. Proteins that repel each other and pass rapidly through the column lead to more positive B$_{22}$ values.

Recently a correlation between B$_{22}$ values and antibody solubility has been demonstrated [13,14]. B$_{22}$ values were determined for the antibody CNTO607 and its solubility-enhanced variants in PBS buffer using SLS, and the solubility of the antibody was determined by the ultrafiltration method. Figure 3.1 shows that the B$_{22}$ values become

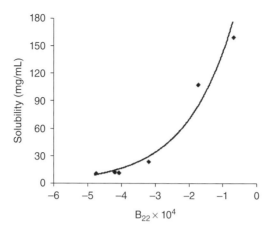

Figure 3.1. A plot of the B_{22} value versus antibody solubility of CNTO607 and its solubility-enhanced variants. Wild-type CNTO607 antibody and variants are all in PBS buffer, pH 7.4.

more positive as the solubility of CNTO607 variant increases (Wu et al. unpublished data). These data suggest that a B_{22} value of greater than -2×10^{-4} is necessary to achieve the solubility of >100 mg/mL. The results obtained demonstrate that the B_{22} values can be used as a predictor for evaluating the relative solubility of biotherapeutic proteins. A recent report on an antibody formulation study found that the most promising buffer condition by B_{22} was confirmed in a stress study conducted for 12 weeks at 40°C, suggesting SIC is a promising screening tool for physical characterization of protein formulations [28].

3.3 ASSESSMENT OF PROTEIN STABILITY

The stability of a biologic has a broad meaning in the context of its physical and chemical properties, and it relates directly to its solution behavior and biological properties. As mentioned in the introduction, long-term stability of therapeutics during manufacturing, storage, distribution, and administration is essential for its success. It is therefore important to characterize the stability of therapeutic candidates during discovery and given similar biological activity profiles use it as one of the criteria for selecting candidates that move into product development.

3.3.1 Thermal Stability

The stability of a therapeutic protein is determined by intrinsic properties as well as external factors. Thermal stability refers to how well a protein can withstand increasing temperature before unfolding, and it is characterized by its melting temperature (T_m). Thermal instability presumably leads to fully or partially unfolded states or folding intermediates of the molecules that can interact with each other leading to protein aggregation [29]. Therefore, temperature stability is arguably one of the most critical external factors for a therapeutic candidate as it progresses through the development and commercialization processes.

The melting temperature of a protein can be measured by a number of methods, including differential scanning calorimetry (DSC), circular dichroism (CD), fluorescence, and activity assays. For multidomain proteins such as monoclonal antibodies, the thermal unfolding profile is more complex and often requires careful deconvolution of the thermogram or spectra [30–33]. For antibodies, the melting temperature of the Fab domains, which is often characterized by the transition with the largest enthalpy in the DSC thermogram, can vary significantly even for those with closely related sequences. In contrast, the sequences of the Fc constant domains are highly homologous and iso-type dependent so their unfolding transition temperatures are similar for antibodies of the same isotype. Thus, the primary focus of the thermal stability characterization of antibody therapeutic candidates is on the Fab thermal transition.

The DSC thermograms of intravenous immunoglobulin G (IVIG) derived from plasma fractionation shows a broad melting profile between 60°C and 90°C [34]. This is not unexpected since it is the sum of the unfolding of molecules from all subclasses of human IgG. Most of the thermal melting profiles of monoclonal antibodies reported in the literature fall within the same range.

The T_m measurements of a series of related antibodies demonstrate that a few amino acid differences can result in a range of Fab T_m values. In one report the values range from 62°C to 69°C (Table 3.1; [13]). It is worthwhile to note that within this range, there is no correlation between the melting temperature and solubility. Another recent report on engineering antibody solubility, which employed primarily isotype change and aglycosylation, also showed a lack of correlation between T_m and solubility (Table 3.2; [35]). These results suggest that once the T_m is within the range of the naturally occur-ring IgG antibodies, its role in determining solubility is limited. Garber and Demarest studied the thermal stability and the amino acid sequences of a series of 17 mAbs with the Fab T_m values ranging from 57.2°C to 81.6°C [33]. Their study suggested that

TABLE 3.1. Biophysical characterization and binding affinities of anti-IL-13 IgG1 and variants

MAb ID	K_D (pM)	Solubility (mg/mL)	ΔT_r (min)	T_m (°C)
CNTO607	18.4 ± 1.9	13.3	0.0	63.7
Mab II	23.3 ± 5.6	29.1	−0.3	ND
Mab III	70.8 ± 0.9	25.4	−1.5	ND
Mab IV	29.1 ± 0.8	12.4	−1.6	ND
Mab V	7.6 ± 0.5	29.2	−2.8	69.4
Mab VI	>45,000	>164	−6.5	64.9
Mab VII	22.2 ± 6.6	>110	ND	61.9

Source: Adapted from Reference 13.
The equilibrium dissociation constant K_D is calculated from the ratio of k_{off}/k_{on} using a BIAcore analysis. Data were obtained from a series of protein concentrations. The retention time, T_r, of antibodies was obtained in hydrophobic interaction chromatography (HIC). $\Delta T_r = T_r$ (variant) − T_r (Mab I). ND, not determined. T_m was determined by DSC on Fv transition.

TABLE 3.2. Impact of Li33 IgG frameworks on solubility

Li33 isotype	Solubility[a] (mg/mL)	LINGO-1 binding, EC50 (nM)	Stability (°C) TM1	Stability (°C) TM2	SEC (% monomer)
IgG1wt pH 7.0	0.9	0.12	69	76	99
IgG1agly pH 7.0	0.3	0.12	60	77	>99
IgG4wt pH 7.0	>30	0.35	64	72	98
IgG4agly pH 7.0	0.3	0.37	56	73	95
IgG2wt pH 7.0	>50	0.23	69	76	96
IgG2agly pH 7.0	0.2	0.26	59	76	98
IgG2-V234A/G237A pH 7.0	5.6	0.19	69	76	95
IgG1 Fab2 pH 7.0	0.3	0.10	–	77	98
IgG2 Fab2 pH 7.0	>50	0.39	–	77	98
IgG1 Fab pH 7.0	>50	0.68	–	76	95
IgG1agly reduced pH 7.0	>40	0.12	55	75	98
IgG1wt reduced pH 7.0	>50	0.15	63	75	98
IgG1wt pH 6.5	1.7	0.10	69	77	
IgG1wt pH 6.0	2.4	0.10	69	78	
IgG1wt pH 5.5	30	0.16	66	81	
IgG1wt pH 5.0	>50	0.45	66	81	
IgG1wt pH 4.5	>50	2.1	62	82	
IgG1wt pH 4.0	>50	16	54	78	
IgG1wt pH 3.5	>50	34	46/66	74	
IgG1wt pH 3.0	>50	>100	34/52	72	

Source: Reproduced with permission from Reference 35.
Li33 Mabs were characterized for the limit in their solubility at 4°C, apparent binding affinity for LINGO-1, thermal stability, and aggregation state by SEC (samples loaded at 300 μg/mL). Samples at pH 7.0 were prepared in 20 mM sodium phosphate, 150 mM NaCl. Samples at other pHs were prepared in 10 mM sodium citrate, 150 mM NaCl.
[a]The protocol for assessing solubility was described in the Reference 35 (Solubility studies in the Methods section).

Fab instability may be related to unusually long or short Complementarity determining region (CDR) loops and/or high levels of infrequently used amino acid residues in the CDRs that may also contribute to poor expression. Thus, at the discovery research stage it is important to ensure that the T_m of therapeutic candidates is in the normally observed range. An abnormally low T_m can be an indication of improper folding of the molecule, which ultimately may lead to problems in the development phase.

3.3.2 Aggregation

Therapeutic proteins may aggregate under a variety of conditions. Of particular concern are the classes of aggregate that have the potential to enhance immune responses and cause adverse clinical effects or aggregates that may compromise the safety and efficacy of the drug product [36]. A protein with an elevated T_m does not necessarily possess a lower tendency to aggregate. Very often, it is the actual solution conditions and environment that control protein aggregation. It is important to recognize that the stability of a therapeutic protein is influenced by a variety of external factors such as the fermentation and purification processes, formulation and storage conditions, as well as packaging materials [37, 38]. For example, tungsten pins used in the supplier's syringe barrel-forming process have been associated with protein aggregation in prefilled syringes [39]. Hence, caution should be used when assessing aggregation early in the discovery stage where only a small portion of the development space can be explored.

The analytical challenge to detect protein aggregation early in the discovery process is similar to that in development. The level of protein aggregation measured in samples can differ significantly depending on the analytical methods employed. The majority of analytical methods, such as size-exclusion chromatography, involve a dilution of the samples which may prevent the detection of reversible aggregation. The commonly used analytical techniques and emerging technologies for aggregate detection, characterization, and quantification have been described in several recent review articles (e.g., references 37 and 40). It is well recognized that for the analysis of protein aggregation, no single analytical method exists to cover the entire size range or type of aggregates (Figure 3.2). Every analytical method has its specific advantages and limitations, limits of detection, and the potential for creating artifacts through sample preparation by inducing or destroying aggregates. Therefore, orthogonal analytical methods must be used in order to adequately detect aggregation and to cover the particle size range that can vary from a few nanometers to large visible particles.

High-concentration formulations, driven by the demand for more convenient administration methods such as subcutaneous injection using prefilled syringes, place urgency on the development of methods to study proteins directly at high concentrations. The incidence of intermolecular interaction increases dramatically at elevated concentrations. The most common and challenging issues encountered in developing these administration methods are protein aggregation and high viscosity [41, 42]. The high-concentration solution properties deviate significantly from the ideal solution behaviors observed at lower concentrations due to the crowding of the protein molecules as the concentration increases. Extrapolation of stability indicating parameters obtained from dilute studies should not be assumed as representative of those for high-concentration solutions. Adaptation of a number of biophysical methods including fluorescence, UV absorbance, CD, FTIR, and DSC that permit direct study of biotherapeutics at high concentrations has been reported. These approaches require minor or no modifications to standard instrumentation [43]. In addition to these techniques, analyzing protein aggregation in highly concentrated samples by fluorescence microscopy after staining with the hydrophobic probe Nile Red has recently been described [44]. While the value that these techniques offer for formulation development is evident, it remains to be defined how they can be used effectively in early assessment of stability given the limited amounts of candidate proteins available for biophysical property screening during the discovery studies.

3.3.3 Chemical Modifications

Biotherapeutics are complex molecules that are susceptible to chemical alteration and degradation. The inherent heterogeneity and instability of monoclonal antibodies have

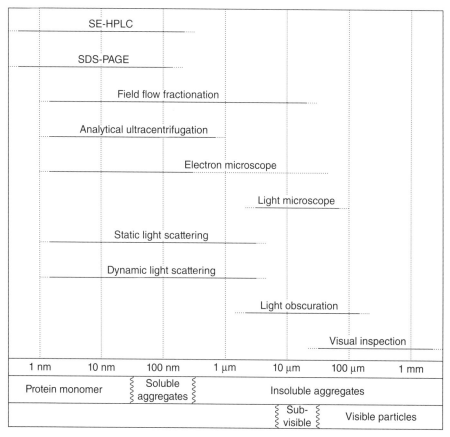

Figure 3.2. Schematic representation of the approximate range of detectable protein sizes (diameter) of various analytical methods. Reproduced with permission from Reference 37.

been thoroughly described in a number of reports (e.g., References 45 and 46) as well as in other chapters of this book. A variety of chemical modification and degradation pathways have been observed. These include disulfide bond formation and exchange, oxidation, deamidation, isomerization, hydrolysis, N-terminal pyroglutamate formation, C-terminal clipping, and glycation. These chemical changes can potentially affect activity as well as the solution behavior of the proteins, in addition to impacting the consistency of the drug product quality [47]. For example, the isomerization of residues in the CDR regions of an anti-human IgE antibody and anti-HER2 antibody has been shown to result in significantly reduced binding affinity to the antigen or biological activity [48, 49]. Although all IgG antibodies share a high level of sequence and structural homology, each antibody presents unique and sometimes unpredictable chemical instability challenges. Many of the chemical modification and degradation pathways are strongly influenced by the protein formulation conditions, making it difficult to determine the exact risk at the discovery stage when candidates are selected. For example, the rate of deamidation is much reduced near or below neutral pH, and the rate of oxidation is also influenced by the formulation pH due to changes in the oxidation potential of the reactants, the binding affinity of catalytic metal ions, ionizable amino acids, and the stability of oxidation intermediates. Nevertheless, when a large number of candidates are available to choose from, it is beneficial to select candidates that are less susceptible to such modifications or to replace or eliminate susceptible residues in the top candidates, especially within the CDRs. Characterization of chemical modifications is often performed using a combination of liquid chromatography and electrophoresis or mass spectrometry. Because chemical modification or degradation processes typically take a long time to manifest themselves, accelerated or stress conditions are often introduced to facilitate the determination of susceptibility. With the potential hot spots of chemical alteration identified from sequences and structural models, stress studies followed by protease digestion combined with LC/MS provide a powerful approach for assessing the chemical liabilities in therapeutic candidates.

Various methods have been used to accelerate protein degradation such as oxidation, isomerization, and deamidation. Cacia and coworkers [48] accelerated the isomerization of an Asp in the CDR of an antibody by incubating the antibody sample at pH 5.2 and 37°C, while Wakankar and colleagues [50] stressed the antibody samples at 50°C at pH 5–8. Chemical stress induced by using *tert*-butylhydroperoxide (tBHP) was used to investigate methionine oxidation in a monoclonal antibody [46]. In addition, the effects of photo-induced degradation in biopharmaceuticals have been reported [51]. The residues that undergo primary photooxidation were tryptophan, tyrosine, phenylalanine, and cysteine/cystine. Risk assessment using such stress conditions provides guidance to candidate selection and candidate engineering.

3.4   COMPUTATIONAL PREDICTIONS

In antibody discovery, the amino acid sequence of the antibody is available providing input for computational approaches that can aid in the interpretation of the biophysical properties. These approaches include prediction of aggregation potential through the identification of specific motifs, and identification of potential posttranslational modification liabilities (N-linked glycosylation, deamidation, oxidation, etc., sites). The computational approaches become even more powerful with the availability of the three-dimensional structure of the antibody and its binding partner. If structural data are not available reasonably accurate models can be built for the antibody. Modeling of the antibody–antigen complex is much more problematic. If the structure of the antigen is unknown, it would involve homology modeling followed by computational docking approaches. Regardless, the structure provides context for the amino acid sequence and is essential for assessing risks and identifying potential solutions when liabilities are discovered.

With respect to antibody modeling, current available methods can provide a reasonably accurate model of the Fab structure for use in interpreting the experimental results from a biophysical property assessment. The quality of such models was assessed in a recent study that compared four different antibody modeling software packages [52]. In this study the sequences of the V-regions of nine different antibodies for which structures have been determined were used as input to the software packages. The models generated in the study were compared with the X-ray structures. Not surprisingly, all of the models agreed fairly well with the structures (~1.2 Å average rmsd). The average rmsd value for the hypervariable loops (L1, L2, L3, H1, and H2) were close to 1.0 Å. The largest deviations were found for H3 (~3.0 Å average rmsd). A number of strengths and weaknesses were identified for each package. Regardless, this study indicates that antibody modeling is at a fairly mature state, and the models are a reasonable representation of the three-dimensional structure of the antibody variable region.

3.4.1 Identifying Aggregation Promoting Regions

A variety of computational approaches have been developed for identifying regions of protein sequences that have a propensity for participating in aggregation. These methods have been recently reviewed [53]. These include a number of prediction methods that use only the amino acid sequence of the biotherapeutic as input, and output short (five to nine residues) aggregation-prone regions that could potentially form amyloid-like fibrils. Other approaches have been reported that rely on three-dimensional profiling, pattern matching, and/or molecular simulations. Of these reports only four methods have been applied and experimentally validated in improving the developability of biotherapeutics [53]. These include TANGO [54], PAGE [55], Zyggregator [56], and spatial aggregation propensity (SAP) [57].

Many software packages have been developed to identify aggregation-prone regions that contribute to amyloid fibril formation and a number have been extended to the more general structured globular proteins. TANGO, PAGE, and Zyggregator are examples that use such an approach. TANGO is based on a statistical mechanics algorithm, which takes into account competing conformations of a peptide, β-sheet, β-turn, α-helix, the folded state, and β-aggregates [54]. TANGO also considers protein stability and physicochemical parameters in the calculation of a partition function used to predict the β-aggregation potential. The second package, PAGE, also takes into account the physicochemical properties of the protein, but it employs a computational approach that analyzes the amino acid sequence as well as conformational space through molecular dynamics sampling [55]. The third approach, Zyggregator, uses a protein's or peptide's amino acid sequence to calculate its intrinsic aggregation propensity and its propensity for folding to determine a unique aggregation propensity profile [56]. All of these methods were independently assessed by comparing the prediction methods with well-studied peptides and proteins that have experimentally determined aggregation propensities.

The algorithms employed in the Zyggregator software package was used in the protein engineering of human calcitonin to produce a bioactive aggregation-resistant biopeptide [58]. Calcitonin is a 32-amino acid residue peptide hormone involved in the regulation of calcium. It is used to treat a number of diseases such as osteoporosis and Paget's disease. Unfortunately human calcitonin has solubility issues that can lead to the formation of amyloid fibrils. Because of this salmon calcitonin (50% sequence identity to human) is used as a clinical alternate, since it has improved solubility properties over the human peptide. However, it can lead to the obvious problem of immunogenicity.

The result of the protein engineering effort was three human calcitonin variants with improved solubility properties and with suitable physiological activity.

The human calcitonin engineering approach was straightforward in that residues known to contribute to physiological function were avoided, α-helical stabilizing amino acid substitutions were introduced in the region of highest helical propensity, and the maximum number of changes was restricted to 20% of the total number of residues of the human sequence [58]. The computational predictions of aggregation propensity were made on >600 peptide variants as an initial screen for selecting possible variants that would improve solubility. Screening involved variants with multiple amino acid substitutions. It was observed that as few as four changes could be introduced to improve the human calcitonin aggregation propensity to that observed for salmon calcitonin. The study illustrates that a predictive computational approach can direct therapeutic engineering efforts to potential lead molecules with appropriate physicochemical properties.

TANGO and PAGE were recently employed in a study that analyzed the amino acid sequences of the majority of commercial antibody therapeutics to identify regions of aggregation propensity [59]. The two different methods were employed to avoid bias from parameterization and the training sets. The software predicted two to eight aggregation-prone regions in heavy or light chains. Interestingly, these regions were distributed throughout all antibody domains with the exception of the CH1 domains, and amino acid residues in the CDR regions and adjacent framework residues were most frequently predicted in the variable domains. Unfortunately, this study has not been experimentally verified. The authors did indicate that they used this approach to identify a single-point mutation in an aggregation-prone region in the FR2-L2 region of an IgG2 mAb that improved the solubility based on experimental biophysical measurements.

A related study applied the predictive tools TANGO and PAGE to identify aggregation-prone regions in Fab sequences for which there are known structures [60]. The study was carried out on Fab sequences from 24 Fab–antigen complexes (the resolution is 2.5 Å or better) that have been deposited in the Protein Data Bank [61]. The aggregation-prone regions were in general well correlated with the antigen recognition features (CDRs) of the antibodies. Because of this correlation, aggregation can lead to loss of function. The aggregation-prone regions were found primarily in the H2 CDRs (45%). Surprisingly, the H3 CDRs had only 7% of the aggregation-prone regions. These regions often contained Tyr and Trp residues that appear to be key residues in antibody–antigen interaction surfaces. It was suggested that it may be difficult to alter these residues because of an impact on binding affinity. It was recommended that adjacent residues in the aggregation-prone regions should be altered first.

In a different approach, the SAP method [57] identifies exposed hydrophobic patches on the surface of a protein that may lead to unwanted protein–protein interactions that increase aggregation and reduce solubility. This method uses molecular simulations of the three-dimensional structure to produce an ensemble of structures resulting from the dynamic fluctuations of the molecule that are then used collectively in the identification of hydrophobic patches. The method was applied to two therapeutic antibodies [57]. The three-dimensional structure of one of the antibodies was assembled from the structures of the Fab and the Fc fragments. For the second structure, a homology model was built for the Fab and it was used alone in the calculations. For both molecules, aggregation-prone regions were determined and ranked. Mutants that reduced hydrophobicity were proposed, produced and tested for reduced aggregation behavior. From the variant pool, mutations were found that did improve the stability and solubility for both therapeutic antibodies.

3.4.2 Interaction Hot Spots

Specific protein–protein interactions play a critical role in countless biological processes. The analyses of such interactions have determined that amino acid residues making up the complementary surfaces vary in terms of their contribution to the overall binding affinity. The subset of the residues that contribute most significantly to the strength of binding are termed a "hot spot." Traditionally these residues have been identified using alanine-scanning mutagenesis [62]. The aggregation-prone regions defined in the previous section can be considered hot spots, since the identified regions in a molecule in many cases bind through specific interactions with other molecules creating an aggregate.

Protein–protein hot spot interactions, including those involved in aggregation, can be characterized by van der Waals, hydrogen bond, electrostatic, and hydrophobic interactions. The strength of the interaction is the result of the specific interactions just mentioned as well as the size (>600 Å^2) and complementarity of the two surfaces in contact. From a structural analysis of 23 protein–protein complexes and alanine mutational analysis data, hot spots were characterized as having a central region with the strongest interactions that were inaccessible to water molecules and that are surrounded by residues with weaker binding interactions that directly interact with the solvent [63]. This O-ring-like structure is the central tenet of what is now known as the O-ring theory. The sequestered central amino acid residues, which are frequently tryptophan, arginine, and tyrosine, can form hydrogen bonds or electrostatic interactions that are strengthened in the absence of water molecules.

A computational analysis of the detailed interactions between two protein molecules requires three-dimensional structural data for the complex. This is best accomplished using experimentally determined structural data, but modeling approaches can be used especially if the individual structures of the two proteins are known. All such procedures are computationally intense and provide a ranking of the contributions of the amino acid residues to the energetics of the interaction which can then be validated experimentally (for a review see Reference 64). Unfortunately, structural data are rarely available in the study of aggregation of a potential biotherapeutic molecule. However, if the crystal structure of a therapeutic or therapeutic candidate is available, some clues for which residues are involved in the interaction (hot spots) may be ascertained [13, 65]. The task is considerably easier for antibodies, in which the V-region, especially the CDRs, are the focus of scrutiny since the rest of the sequences are strongly conserved and should remain so for therapeutics in order to avoid immunogenicity.

3.5 ENHANCE THE SOLUBILITY OF BIOTHERAPEUTICS

To overcome the challenge of solubility-limited proteins, whether cytokines or enzymes or monoclonal antibodies, a number of strategies have been employed to increase protein solubility [6, 9, 13, 66, 67]. A successful case involves fusing peptides to protein targets to enhance solubility [68,69]. The acidic tail of synuclein (ATS) peptide was fused to three therapeutic proteins, human growth hormone, granulocyte colony-stimulating factor (G-CSF), and human leptin, at either the N- or the C-terminus. The results demonstrated that the ATS peptide improved the solubility and stability of these proteins. While such an approach is helpful with research proteins, it is usually not desirable to introduce additional sequences into native proteins for therapeutics. Therefore, four commonly used strategies for increasing protein solubility, namely site-directed mutagenesis, pegylation, glycosylation, and formulation optimization, will be discussed in this section.

3.5.1   Site-Directed Mutagenesis

In several cases reducing surface hydrophobicity by substitution with hydrophilic residues has proven to be an effective strategy for improving solubility [7, 70, 71]. This approach is most successful when the three-dimensional structure of the protein is known. This allows a judicious selection of surface residues to vary. Without the structure the investigator increases the risk for introducing mutations that could destabilize or even unfold the protein. Hydrophobic interactions can be major contributors to protein aggregation, especially at elevated protein concentrations [7]. Successful examples of the use of this approach include members of the ankyrin repeat proteins and the subtilisin family of proteases. In both cases, the replacement of surface-exposed hydrophobic residues with arginine significantly improved the protein solubility at physiological pH [7, 72]. A summary of the data from solubility-changing mutations by Trevino and coworkers indicates that three hydrophilic residues, Asp, Glu, and Ser, contribute more significantly to improving protein solubility than the other hydrophilic amino acids [5, 9]. They suggest the following approach to improve protein solubility: when the structure of the protein is known, replace surface hydrophobic residues with Asp, Glu, or Ser; when the protein structure is not available, mutate the putative surface-exposed residues, Asn, Gln, or Thr to Asp, Glu, or Ser.

A structure-based engineering approach that combines three-dimensional structure and mutagenesis data provides an effective route for enhancing protein solubility. In an anti-IL13 antibody engineering case study, the authors identified an aggregation "hot spot" in heavy-chain CDR3 (H-CDR3) that contains three aromatic residues ($_{99}$FHW$_{100a}$) [13, 65]. These three residues are located near the middle of the hydrophobic ridge occupying the center of the antigen recognition site in the three-dimensional structure of the Fab–IL-13 complex (Figure 3.3). The potential contribution of these residues to antibody aggregation was confirmed by mutating them all to Ala residues. The triple Ala mutant was found to be highly soluble in PBS buffer (>160 mg/mL, compared to 13 mg/mL for wild type) and is significantly less hydrophobic as indicated by hydrophobic interaction chromatography. This result pinpointed the aromatic triad in the H-CDR3 sequence as a "hot spot" for aggregation [13].

3.5.2   Pegylation

Pegylation [73, 74] and glyco-engineering [75, 76] have been applied to improve therapeutic protein properties such as molecular stability, solubility, increased *in vivo* activity, and reduced immunogenicity. PEG conjugation to drug molecules, that is, proteins,

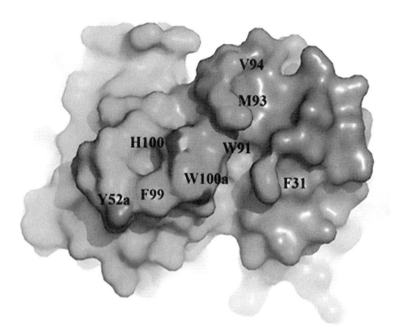

Figure 3.3. Antigen recognition surface of CNTO607 and its interaction with IL-13. The hydrophobic ridge in the CNTO607 antigen-binding site. VL and VH are colored cyan and magenta. The hydrophobic ridge is colored green. N, S, and O atoms in the ridge are colored blue, yellow, and red. Adapted from Reference 13. (See color insert)

peptides, oligonucleotides, antibody fragments, and small organic molecules, is also commonly used to prolong a biotherapeutic's half-life *in vivo*.

An excellent illustration of the use of pegylation to improve the biophysical properties of a biotherapeutic is the modification of interferon-β-1b (IFN-β-1b) that resulted in an increase in both solubility and stability [77]. When native IFN-β-1b was formulated in aqueous solution without a surfactant, the protein rapidly precipitated within 7 days at neutral pH, whereas the 40-kDa pegylated IFN-β-1b remained soluble for the same duration. The chromatographic analysis of the 40-kDa pegylated IFN-β-1b (0.1 mg/mL), both reverse-phase and size-exclusion HPLCs, showed no formation of soluble aggregates during a month-long storage at 4°C at pH 3.7–4.0. However, storage at higher pH, 5.0–8.5, resulted in increasing amounts of soluble aggregate. A trend of improved solubility and stability was observed when IFN-β-1b was conjugated with increasingly higher molecular weight polymers. Independent biochemical and bioactivity analyses demonstrated that the functional activities of the engineered IFN-β-1b conjugates were maintained.

In a second example, this approach was used to reduce the aggregation propensity of G-CSF. N-terminal conjugation to a 20-kDa PEG moiety prevented the precipitation of G-CSF by rendering the protein highly soluble and slowing the aggregation rate relative to the naked G-CSF [78]. These results suggest that the improved solubility was mediated by the presence of the hydrophilic PEG moiety, while biological activity measurements demonstrated that the 20-kDa PEG-G-CSF retained greater activity than naked G-CSF.

In summary, pegylation of biologically active proteins greatly increases aqueous solubility while still allowing relevant protein–protein interactions to occur. In addition, due to an increase in the overall M_w and effective hydrodynamic diameter, the conjugates exhibit a longer circulating half-life, leading to greater bioavailability, and reduced dosing frequency.

3.5.3 Glycosylation

Glyco-engineering has been used to improve the biophysical properties of proteins. For example, adding an N-linked glycosylation consensus sequence into a protein leading to a new glycoform of the molecule can be used to improve the biophysical properties of a therapeutic. Glyco-engineering has also been shown to increase the half-life and prolong the activity of protein therapeutics, thus allowing reduced frequency of administration [75, 76]. In particular, sialic acid-containing oligosaccharides are highly hydrophilic structures and increase protein solubility by shielding hydrophobic residues [79]. In general, the presence of covalently bound carbohydrate enhances molecular solubility and stability and improves resistance to protease digestion.

N-linked carbohydrates are usually attached to the Asn located in the Asn-X-Ser/Thr consensus sequence where X can be any amino acid except Pro [80]. Glyco-engineering introduces new N-linked glycosylation sites into target positions in the peptide backbone to produce proteins with increased sialic acid-containing carbohydrates when expressed in mammalian cells, in order to increase the biological activity and duration of action. For example, aglycosylated erythropoietin (EPO) is readily denatured by guanidine-HCl, heat, and acidic pH, while glycosylated EPO shows reversible folding and remains soluble at acidic pH [81,82]. Studies of human EPO glycosylation analogs showed a direct and positive relationship between the number of glycosylation sites and *in vivo* activity [83]. This strategy has been validated clinically and applied to other therapeutic proteins.

In another example, a glyco-engineering approach based on the three-dimensional structural data of a Fab was successfully employed to enhance the solubility of a monoclonal antibody [13]. An aggregation "hot spot" in H-CDR3 of an anti-IL-13 antibody, CNTO607, was identified using the three-dimensional structure and was found to be essential for high affinity binding to IL-13. Based on the spatial proximity and germline sequence, an N-linked glycosylation site in H-CDR2 was introduced, anticipating that the carbohydrate moiety would shield the aggregation "hot spot" in H-CDR3 without interfering with antigen binding. This glyco-engineered variant showed greatly improved solubility (from 13 to >110 mg/mL) and bound IL-13 with an affinity similar to that of CNTO607. In this case, the introduction of a single carbohydrate moiety in the H-CDR2 loop served to inhibit self-interactions. Due to the single-point attachment, the carbohydrate moiety has sufficient flexibility so that it does not interfere with antigen binding. In another example, Pepinsky et al. reported a successful case that incorporated N-linked glycosylation sites into the anti-LINGO-1 mAb Li33 IgG1 CH1 framework at positions where the added glycan could disrupt aggregation through steric effects [35]. With these changes they were able to increase the solubility of this antibody from 0.3 to >50 mg/mL. Thus, glyco-engineering a highly soluble and potent antibody by adding a consensus N-linked glycosylation site may pave the way toward high-concentration formulations.

3.5.4 Formulation Optimization

The success of biotherapeutics is a balance between activity and molecular properties. Given good activity, it is possible to manage the behavior of protein solutions through formulation development for therapeutic applications. The major purpose of formulation is to find conditions that prevent aggregation and degradation. This process requires identifying stabilizing pH range, buffering agents, and excipients and optimizing the concentrations of the excipients. This is an empirical process that must be repeated for each protein. As described earlier, many current monoclonal antibody therapies rely on subcutaneous administration at high concentration (>100 mg/mL). The concentration-dependent aggregation is considered the greatest challenge of protein formulation. Since protein aggregates generally reduce activity and increase immunogenicity [84], the aggregate level in commercial IV antibody products is generally limited to less than 5% [85]. In addition, high viscosity at high protein concentration poses a new challenge for the manufacturing process and the injection device.

Generally the aggregation in protein solutions is reduced by adjustment of pH and addition of excipients (see Reference 86 and references therein). These excipients also improve the stability of protein samples at high concentration and prevent degradation and precipitation over time. The use of excipients has been a common practice by protein formulation scientists as well as by structural biologists. Adding amino acids is one way to enhance solubility and stability of proteins. Golovanov et al. [87] proposed a simple and effective way of increasing protein solubility and long-term stability by adding the charged L-amino acids Arg and Glu as excipients for different protein solutions. By adding 50 mM of Arg and Glu, they were able to increase the solubility of several proteins by 2.5 to ~9 times. One simple explanation for this effect is that the charges of Arg and Glu interact with oppositely charged groups on protein surface while the aliphatic side chains of these residues interact with adjacent exposed hydrophobic surfaces of the protein. This reduces the potential for both electrostatic and hydrophobic self-interactions of the protein therapeutic. Indeed the underlying mechanism is likely to be far more complicated, as Arg and Glu may interact with water and ions in the buffer as well. Cyclodextrins and their derivatives belong to another class of commonly used excipients to enhance the solubility and stability of proteins [88]. The inclusion of 2-hydroxypropyl-β-cyclodextrin rendered interleukin-2 soluble upon lyophilization [89]. Recently, hydroxypropyl-β-cyclodextrin was reported to help reduce protein aggregation induced by shaking [90].

Formulation excipients are also effective in maintaining protein stability. Sugars are most commonly used among the excipients [86]. A variety of carbohydrates or polyol compounds such as sucrose, trehalose, and mannitol have been shown to provide this stabilizing function [91]. Some antibodies are unstable in liquid formulations. For such cases, lyophilized formulation can be an alternative approach for maintaining antibody stability. Addition of sorbitol to a sucrose or trehalose formulation resulted in the enhancement of stability of a lyophilized antibody [92]. Stabilizing is achieved through the inclusion of a carbohydrate excipient in sufficient quantities to compensate for the loss of water molecules that normally interact with polar amino acid residues on the protein surface [93]. Without going through a full-scale formulation development process, a few excipient conditions can be explored in the discovery stage to determine whether a therapeutic candidate is suitable to move forward.

Highly concentrated protein solutions can be very viscous, which is problematic for handling. Different approaches have been shown to reduce viscosity and the effectiveness appears to be protein specific. In one case, the viscosity of an antibody solution was reduced by the inclusion of a moderate amount of NaCl [94]. However, the viscosity of another antibody solution is increased in the presence of NaCl [95]. This reflects the fact that each protein is governed by a unique set of interactions that dictate their solution behavior and underscores the necessity of formulation optimization for individual protein.

For potent proteins that are too difficult to maintain in solution, lyophilization followed by reconstitution offers another venue to proceed with biotherapeutics [93]. It is of note that crystallization has proven to be effective in the development of protein formulation [86, 96, 97]. Shenoy et al. compared the stability of two proteins, glucose oxidase and lipase, and demonstrated that the crystalline state preserves protein structure and activity better than the amorphous state.

3.6 DEVELOPMENT OF RAPID METHODS TO IDENTIFY SOLUBLE AND STABLE BIOTHERAPEUTICS

Most of the biophysical or analytical methods mentioned above suffer from relatively low throughput primarily because they require substantial amounts of proteins. Thus, the use of such methods in the early discovery stage contributes to increased costs and extended timelines. During this phase it is common to have a large number of therapeutic candidates to characterize that have been produced on a small scale of a few milligrams or less. Thus, the biotherapeutic discovery phase could benefit significantly from the development of HTS methods that require relatively small quantities of purified protein for ranking potential biotherapeutic candidates.

One approach that shows promise for developing pharmaceutical protein products is the use of SIC for determining B_{22}, which shows significant utility in predicting protein precipitation and aggregation based on its correlation with relative protein solubility [12, 98]. Consequently, several screening methods have been developed based on measured B_{22} to predict the strength of protein self-interactions [12, 26]. The strong correlation between SIC and B_{22} values sets the stage for SIC to become an effective screening tool. The use of SIC offers several advantages, such as less time required and increased throughput over the SLS method. These features enable rapid screening of a large set of different candidate proteins or a large set of formulation conditions for a single biotherapeutic candidate.

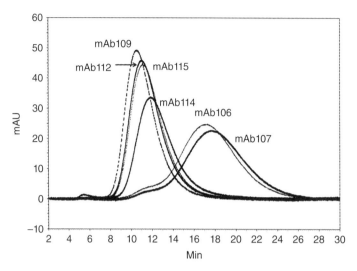

Figure 3.4. Retention of selected therapeutic antibody candidates on polyclonal human IgG column. Reproduced from Reference 12.

A related method to SIC, cross-interaction chromatography (CIC), has recently been developed and applied to screening antibody solubility [12]. This method is based on the principle of SIC. Instead of binding the candidate antibody to the support, polyclonal IgG is immobilized on the column matrix and candidate monoclonal antibodies are run through the column in the liquid phase. The high sequence homology of IgGs makes this an approximation for SIC. The retention time of individual antibodies on the column is reflective of its aggregation propensity. Most important for discovery application, this screen can be completed with as little as a few micrograms of recombinant protein. As described by Jacobs et al., this method was applied to a set of 19 therapeutic antibody candidates selected from phage display panning experiments targeting a specific antigen. All the antibodies showed potent activity in biological assays but their solubility was unknown. Figure 3.4 shows the chromatographic profiles of a subset of these antibodies. While most of the 19 antibodies eluted on the column rapidly (e.g., mAb109, mAb112, mAb114, and mAb115), two of them, mAb106 and mAb107, exhibited much longer retention times and broader elution profiles. These six antibodies were subsequently scaled up and their solubility was determined by the ultrafiltration method. The results confirmed that mAb106 and mAb107 indeed have drastically reduced solubility in comparison to the other antibodies tested (Table 3.3).

An alternative way to determine protein solubility that uses the PEG precipitation method was reported some time ago [18, 99]. PEG is thought to decrease the solubility of a protein in a quantifiable manner primarily via the excluded volume effect. A HTS procedure for relative protein solubility using PEG-induced precipitation was reported recently for facilitating formulation development [100]. Using this HTS approach, numerous combinations of buffer and excipients can be rapidly evaluated for their influence on solubility. The results are determined from PEG midpoint assessments

TABLE 3.3. Polyclonal IgG column screening of antibody candidates

Sample	T_r (min)	k'	Maximum observed concentration (mg/mL)
mAb106	17.15	0.99 ± 0.14	10
mAb107	17.74	1.02 ± 0.19	9
mAb109	10.54	0.21 ± 0.10	>108
mAb112	10.95	0.26 ± 0.12	>124
mAb114	11.86	0.33 ± 0.16	>109
mAb115	11.12	0.22 ± 0.10	>119

Source: Table adapted from Reference [12].
Error values for k' represent the standard deviations from three independent measurements.

using sigmoidal curve fitting to the entire data set. This method is amenable to screening multiple biotherapeutics and may be adapted for candidate selection in discovery stage using a small set of buffers and excipients.

Advances have also been made to increase the throughput of thermal stability measurements. ThermoFluor® is a miniaturized high-throughput (HT) protein stability assay that measures thermal stability using a greatly reduced amount of material (<5 µg) compared with DSC (~100 µg) [101]. ThermoFluor® measures the change in fluorescence of an environmentally sensitive dye on protein unfolding. The dye is quenched in aqueous environments but shows a large increase in quantum yield when bound to the hydrophobic interior of a protein as the protein denatures.

Another approach that uses fluorescence is differential scanning fluorescence (DSF), a method developed with HT capability using a polymerase chain reaction (PCR) instrument [102]. Using SYPRO Orange as the fluorescence probe, He et al. [103] demonstrated a good correlation between the fluorescence thermal transition data and DSC data for four antibodies in 84 buffers, thus establishing DSF as a reliable HT method for measuring thermal stability. A drawback of using extrinsic fluorescence probes is that they have the potential to interfere with protein folding and unfolding [104]. These methods may be adapted to assess multiple candidates in a small set of buffer and excipients in discovery stage.

Intrinsic fluorescence spectroscopy is another highly sensitive method for monitoring protein stability. Intrinsic tryptophan fluorescence has been assessed systematically for antibody conformational stability measurement [105]. Garidel and coworkers compared the data obtained from intrinsic tryptophan fluorescence spectroscopy with data derived from calorimetry and concluded that the average signal from antibodies with multiple tryptophan residues allows the derivation of global structural stability changes that correlate with calorimetry data. In the absence of sufficient number of tryptophans, information may be derived from tyrosine and phenylalanine fluorescence. A decisive advantage is that the intrinsic fluorescence assay uses a small amount of protein and is free of interference from external probes.

Assessing viscosity at high concentrations with a HT method is another area of active research. A recent report attempted to develop a qualitative screening tool to predict high-concentration viscosity using the interaction parameter obtained from dynamic light scattering and net molecular charge [106]. A qualitative trend was found between the interaction parameter and the viscosity at high concentrations, while an inconsistent relationship with net molecular charge was observed.

Experimental determination of posttranslational modifications can be a labor-intensive process. In order to assess the residue-specific chemical modification risk, detailed peptide-mapping data are required. Such analysis is so time consuming that it can become a bottleneck in the process. As highly automated technologies have become essential for the pharmaceutical industry, the huge volumes and complex dependencies of data produced by large-scale experiments have resulted in the development of novel and HT data analysis strategies (e.g., Reference 107). The advances in such data analysis tools hold promise that the peptide-mapping analysis and other complementary assays will be automated in the near future.

3.7 CONCLUDING REMARKS

In today's antibody discovery process lead candidates with the appropriate binding affinity and biological properties such as target neutralization are identified. This is followed by optimization, which often includes improving the binding affinity to increase the efficacy. These efforts, involving the screening and ranking of candidate therapeutic molecules, are evolving on two major fronts. The first is to transform the current methods into HT formats which will minimize the time, effort, and quantity of each protein required and to develop and adapt the screening methods so they can be applied to an increasing number of candidate molecules. The second front focuses on developing predictive methods and approaches that use structural knowledge and the biophysical and chemical properties of the candidates to rank and select potential therapeutics and increase the chances for successful development.

Today, after lead candidate molecules are identified, they undergo a much more rigorous screening process to establish not only the appropriate therapeutic benefit but also the optimal biophysical and chemical properties required for formulation, manufacturing, storage, and delivery. Such efforts usually produce candidates that go on to successful product development, but occasionally developability issues arise. Once the problem is identified the remedy involves protein engineering efforts to eliminate or minimize the unwanted chemical or biophysical properties. Candidate therapeutics with such optimized properties are key to accelerating the pharmaceutical development process while reducing the time and cost of development.

REFERENCES

1. Dani, B., Platz, R., and Tzannis, S.T. (2007) High concentration formulation feasibility of human immunoglobulin G for subcutaneous administration. *J. Pharm. Sci.*, **96**, 1504–1517.
2. Akers, M.J., Vasudevan, V., and Stickelmeyer, M. (2002) *Formulation Development of Protein Dosage Forms*. Kluwer Academic/Plenum Publishers, New York.

3. Arakawa, T. and Timasheff, S.N. (1985) Theory of protein solubility. *Methods Enzymol.*, **114**, 49–77.

4. Schein, C.H. (1990) Solubility as a function of protein structure and solvent components. *Biotechnology (NY)*, **8**, 308–317.

5. Trevino, S.R., Scholtz, J.M., and Pace, C.N. (2008) Measuring and increasing protein solubility. *J. Pharm. Sci.*, **97**, 4155–4166.

6. Bagby, S., Tong, K.I., and Ikura, M. (2001) Optimization of protein solubility and stability for protein nuclear magnetic resonance. *Methods Enzymol.*, **339**, 20–41.

7. Mosavi, L.K. and Peng, Z.Y. (2003) Structure-based substitutions for increased solubility of a designed protein. *Protein Eng. Des. Sel.*, **16**, 739–745.

8. Andrews, B.A., Schmidt, A.S., and Asenjo, J.A. (2005) Correlation for the partition behavior of proteins in aqueous two-phase systems: effect of surface hydrophobicity and charge. *Biotechnol. Bioeng.*, **90**, 380–390.

9. Trevino, S.R., Scholtz, J.M., and Pace, C.N. (2007) Amino acid contribution to protein solubility: Asp, Glu, and Ser contribute more favorably than the other hydrophilic amino acids in RNase Sa. *J. Mol. Biol.*, **366**, 449–460.

10. Strickler, S.S., et al. (2006) Protein stability and surface electrostatics: a charged relationship. *Biochemistry*, **45**, 2761–2766.

11. Shire, S.J., Shahrokh, Z., and Liu, J. (2004) Challenges in the development of high protein concentration formulations. *J. Pharm. Sci.*, **93**, 1390–1402.

12. Jacobs, S.A., et al. (2010) Cross-interaction chromatography: a rapid method to identify highly soluble monoclonal antibody candidates. *Pharm. Res.*, **27**, 65–71.

13. Wu, S.J., et al. (2010) Structure-based engineering of a monoclonal antibody for improved solubility. *Protein. Eng. Des. Sel.*, **23**, 643–651.

14. Moore, P.A. and Kery, V. (2009) High-throughput protein concentration and buffer exchange: comparison of ultrafiltration and ammonium sulfate precipitation. *Methods Mol. Biol.*, **498**, 309–314.

15. van Reis, R. and Zydney, A. (2001) Membrane separations in biotechnology. *Curr. Opin. Biotechnol.*, **12**, 208–211.

16. Haire, L.F. and Blow, D.M. (2001) A novel spin filter method for the measurement of solubility. *J. Cryst. Growth*, **232**, 17–20.

17. Qamar, S., Islam, M., and Tayyab, S. (1993) Probing the determinants of protein solubility with amino acid modification. *J. Biochem.*, **114**, 786–792.

18. Atha, D.H. and Ingham, K.C. (1981) Mechanism of precipitation of proteins by polyethylene glycols. Analysis in terms of excluded volume. *J. Biol. Chem.*, **256**, 12108–12117.

19. Middaugh, C.R., et al. (1979) Determination of the apparent thermodynamic activities of saturated protein solutions. *J. Biol. Chem.*, **254**, 367–370.

20. Demoruelle, K., et al. (2002) Correlation between the osmotic second virial coefficient and solubility for equine serum albumin and ovalbumin. *Acta Crystallogr. D Biol. Crystallogr.*, **58**, 1544–1548.

21. Neal, B.L., et al. (1999) Why is the osmotic second virial coefficient related to protein crystallization? *J. Cryst. Growth*, **196**, 377–387.

22. George, A. and Wilson, W.W. (1994) Predicting protein crystallization from a dilute solution property. *Acta Crystallogr. D Biol. Crystallogr.*, **50**, 361–365.

23. Ahamed, T., Ottens, M., van Dedem, G.W., and van der Wielen, L.A. (2005) Design of self-interaction chromatography as an analytical tool for predicting protein phase behavior. *J. Chromatogr. A*, **1089**, 111–124.

24. Gabrielsen, M., Nagy, L.A., DeLucas L.J., and Cogdell, R.J. (2010) Self-interaction chromatography as a tool for optimizing conditions for membrane protein crystallization. *Acta Crystallogr. D Biol. Crystallogr.*, **66**, 44–50.

25. Tessier, P.M., et al. (2002) Self-interaction chromatography: a novel screening method for rational protein crystallization. *Acta Crystallogr. D Biol. Crystallogr.*, **58**, 1531–1535.

26. Johnson, D.H., Parupudi, A., Wilson, W.W., and DeLucas, L.J. (2009) High-throughput self-interaction chromatography: applications in protein formulation prediction. *Pharm. Res.*, **26**, 296–305.

27. Tessier, P.M., Sandler, S.I., and Lenhoff, A.M. (2004) Direct measurement of protein osmotic second virial cross coefficients by cross-interaction chromatography. *Protein Sci.*, **13**, 1379–1390.

28. Le Brun, V., et al. (2010) A critical evaluation of self-interaction chromatography as a predictive tool for the assessment of protein-protein interactions in protein formulation development: a case study of a therapeutic monoclonal antibody. *Eur. J. Pharm. Biopharm.*, **75**, 16–25.

29. Kad, N.M. and Radford, S.E. (2001) Partial unfolding as a precursor to amyloidosis: a discussion of the occurrence, role, and implications. In: Lund, P. editor. *Molecular Chaperones in the Cell*. Frontiers in Molecular Biology Series. Oxford University Press, Oxford, pp. 257–278.

30. Kravchuk, Z.I., Chumanevich, A.A., Vlasov, A.P., and Martsev, S.P. (1998) Two high-affinity monoclonal IgG2a antibodies with differing thermodynamic stability demonstrate distinct antigen-induced changes in protein A-binding affinity. *J. Immunol. Methods*, **217**, 131.

31. Welflea, K., et al. (1999) Conformation, pH-induced conformational changes, and thermal unfolding of anti-p24 (HIV-1) monoclonal antibody CB4–1 and its Fab and Fc fragments. *Biochim. Biophys. Acta*, **1431**, 120–131.

32. Vermeer, A.W. and Norde, W. (2000) The thermal stability of immunoglobulin: unfolding and aggregation of a multi-domain protein. *Biophys. J.*, **78**, 394–404.

33. Garber, E. and Demarest, S.J. (2007) A broad range of Fab stabilities within a host of therapeutic IgGs. *Biochem. Biophys. Res. Commun.*, **355**, 751–757.

34. Ahrer, K., Buchacher, A., Iberer, G., and Jungbauer, A. (2006) Thermodynamic stability and formation of aggregates of human immunoglobulin G characterised by differential scanning calorimetry and dynamic light scattering. *J. Biochem. Biophys. Methods*, **66**, 73–86.

35. Pepinsky, R.B., et al. (2010) Improving the solubility of anti-LINGO-1 monoclonal antibody Li33 by isotype switching and targeted mutagenesis. *Protein. Sci.*, **19**, 954–966.

36. Cordoba-Rodriguez, R.V. (2008) Aggregates in MAbs and recombinant therapeutic proteins: a regulatory perspective. *BioPharm Int.*, **21**, 44–53.

37. Mahler, H.C., Friess, W., Grauschopf, U., and Kiese, S. (2009) Protein aggregation: pathways, induction factors and analysis. *J. Pharm. Sci.*, **98**, 2909–2934.

38. An, Z., et al. (2009) IgG2m4, an engineered antibody isotype with reduced Fc function. *MAbs*, **1**, 572–579.

39. Jiang, Y., et al. (2009) Tungsten-induced protein aggregation: solution behavior. *J. Pharm. Sci.*, **98**, 4695–4710.

40. den Engelsman, J., et al. (2011) Strategies for the assessment of protein aggregates in pharmaceutical biotech product development. *Pharm. Res.*, **28**, 920–933.

41. Shire, S.J., Shahrokh, Z., and Liu, J. (2004) Challenges in the development of high protein concentration formulations. *J. Pharm. Sci.*, **93**, 1390–1402.

42. Shire, S.J. (2009) Formulation and manufacturability of biologics. *Curr. Opin. Biotechnol.*, **20**, 708–714.

43. Harn, N., Allan, C., Oliver, C., and Middaugh, C.R. (2007) Highly concentrated monoclonal antibody solutions: direct analysis of physical structure and thermal stability. *J. Pharm. Sci.*, **96**, 532–546.

44. Demeule, B., et al. (2007) *Biochim. Biophys. Acta*, **1774**, 146–153.

45. Daugherty, A.L. and Mrsny, R.J. (2006) Formulation and delivery issues for monoclonal antibody therapeutics. *Adv. Drug. Deliv. Rev.*, **58**, 686–706.

46. Chumsae, C., Gaza-Bulseco G., Sun, J., and Liu, H. (2007) Comparison of methionine oxidation in thermal stability and chemically stressed samples of a fully human monoclonal antibody. *J. Chromatogr. B*, **850**, 285–294.

47. Jenkins, N., Murphy, L., and Tyther, R. (2008) Post-translational modifications of recombinant proteins: significance for biopharmaceuticals. *Mol. Biotechnol.*, **39**, 113–118.

48. Cacia, J., Keck, R., Presta, L.G., and Frenz, J. (1996) Isomerization of an aspartic acid residue in the complementarity-determining regions of a recombinant antibody to human IgE: identification and effect on binding affinity. *Biochemistry*, **35**, 1897–1903.

49. Harris, R.J., et al. (2001) Identification of multiple sources of charge heterogeneity in a recombinant antibody. *J. Chromatogr. B*, **752**, 233–245.

50. Wakankar, A.A., et al. (2007) Aspartate isomerization in the complementarity-determining regions of two closely related monoclonal antibodies. *Biochemistry*, **46**, 1534–1544.

51. Kerwin, B.A. and Remmele, R.L. Jr. (2007) Protect from light: photodegradation and protein biologics. *J. Pharm. Sci.*, **96**, 1468–1479.

52. Almagro, J.C., et al. (2011) Antibody modeling assessment. *Proteins*, **79**, 3050–3066.

53. Agrawal, N.J., et al. (2011) Aggregation in protein-based biotherapeutics: computational studies and tools to identify aggregation-prone regions. *J. Pharm. Sci.*, **100**, 5081–5095.

54. Fernandez-Escamilla, A.M., Rousseau, F., Schymkowitz, J., and Serrano, L. (2004) Prediction of sequence-dependent and mutational effects on the aggregation of peptides and proteins. *Nat. Biotechnol.*, **22**, 1302–1306.

55. Tartaglia, G.G., Cavalli, A., Pellarin, R., and Caflisch, A. (2005) Prediction of aggregation rate and aggregation-prone segments in polypeptide sequences. *Protein Sci.*, **14**, 2723–2734.

56. Tartaglia, G.G., et al. (2008) Prediction of aggregation-prone regions in structured proteins. *J. Mol. Biol.*, **380**, 425–436.

57. Chennamsetty, N., et al. (2009) Design of therapeutic proteins with enhanced stability. *Proc. Natl. Acad. Sci. U.S.A.*, **106**, 11937–11942.

58. Fowler, S.B., et al. (2005) Rational design of aggregation-resistant bioactive peptides: reengineering human calcitonin. *Proc. Natl. Acad. Sci. U.S.A.*, **102**, 10105–10110.

59. Wang, X., Das, T.K., Singh, S.K., and Kumar, S. (2009) Potential aggregation prone regions in biotherapeutics. *MAbs*, **1**, 1–14.

60. Wang, X., Singh, S.K., and Kumar, S. (2010) Potential aggregation-prone regions in complementarity-determining regions of antibodies and their contribution towards antigen recognition: a computational analysis. *Pharm. Res.*, **27**, 1512–1529.

61. Berman, H.M., et al. (2000) The protein data bank. *Nucleic Acids Res.*, **28**, 235–242.

62. Cunningham, B.C. and Wells, J.A. (1989) High-resolution epitope mapping of hGH-receptor interactions by alanine-scanning mutagenesis. *Science*, **244**, 1081–1085.

63. Bogan, A.A. and Thorn, K.S. (1998) Anatomy of hot spots in protein interfaces. *J. Mol. Biol.*, **280**, 1–9.

64. Moreira, I.S., Fernandes, P.A., and Ramos, M.J. (2007) Hot-spots – a review of the protein-protein interface determinant amino-acid residues. *Proteins*, **68**, 803–812.

65. Teplyakov, A., et al. (2009) Epitope mapping of anti-interleukin-13 neutralizing antibody CNTO607. *J. Mol. Biol.*, **389**, 115–123.

66. Pedelacq, J.D., et al. (2002) Engineering soluble proteins for structural genomics. *Nat. Biotechnol.*, **20**, 927–932.

67. Savchenko, A., et al. (2003) Strategies for structural proteomics of prokaryotes: quantifying the advantages of studying orthologous proteins and of using both NMR and X-ray crystallography approaches. *Proteins*, **50**, 392–399.

68. Zhou, P., Lugovskoy, A.A., and Wagner, G. (2001) A solubility-enhancement tag (SET) for NMR studies of poorly behaving proteins. *J. Biomol. NMR*, **20**, 11–14.

69. Lee, E.N., et al. (2005) Stabilizing peptide fusion for solving the stability and solubility problems of therapeutic proteins. *Pharm. Res.*, **22**, 1735–1746.

70. Dale, G.E., Oefner, C., and D'Arcy, A. (2003) The protein as a variable in protein crystallization. *J. Struct. Biol.*, **142**, 88–97.

71. Derewenda, Z.S. (2004) The use of recombinant methods and molecular engineering in protein crystallization. *Methods*, **34**, 354–363.

72. Shirai, T., et al. (1997) High-resolution crystal structure of M-protease: phylogeny aided analysis of the high-alkaline adaptation mechanism. *Protein Eng.*, **10**, 627–634.

73. Veronese, F.M. and Mero, A. (2008) The impact of PEGylation on biological therapies. *BioDrugs*, **22**, 315–329.

74. Pasut, G. and Veronese, F.M. (2009) PEGylation for improving the effectiveness of therapeutic biomolecules. *Drugs Today (Barc)*, **45**, 687–695.

75. Sinclair, A.M. and Elliott, S. (2005) Glycoengineering: the effect of glycosylation on the properties of therapeutic proteins. *J. Pharm. Sci.*, **94**, 1626–1635.

76. Elliott, S., et al. (2003) Enhancement of therapeutic protein in vivo activities through glycoengineering. *Nat. Biotechnol.*, **21**, 414–421.

77. Basu, A., et al. (2006) Structure-function engineering of interferon-beta-1b for improving stability, solubility, potency, immunogenicity, and pharmacokinetic properties by site-selective mono-PEGylation. *Bioconjug. Chem.*, **17**, 618–630.

78. Rajan, R.S., et al. (2006) Modulation of protein aggregation by polyethylene glycol conjugation: GCSF as a case study. *Protein Sci.*, **15**, 1063–1075.

79. Goochee, C.F., et al. (1991) The oligosaccharides of glycoproteins: bioprocess factors affecting oligosaccharide structure and their effect on glycoprotein properties. *Biotechnology (NY)*, **9**, 1347–1355.

80. Bause, E. (1983) Structural requirements of N-glycosylation of proteins. Studies with proline peptides as conformational probes. *Biochem. J.*, **209**, 331–336.

81. Narhi, L.O., et al. (1991) The effect of carbohydrate on the structure and stability of erythropoietin. *J. Biol. Chem.*, **266**, 23022–23026.

82. Endo, Y., et al. (1992) Heat-induced aggregation of recombinant erythropoietin in the intact and deglycosylated states as monitored by gel permeation chromatography combined with a low-angle laser light scattering technique. *J. Biochem.*, **112**, 700–706.

83. Elliott, S., et al. (2004) Control of rHuEPO biological activity: the role of carbohydrate. *Exp. Hematol.*, **32**, 1146–1155.

84. Braun, A., Kwee, L., Labow, M.A., and Alsenz, J. (1997) Protein aggregates seem to play a key role among the parameters influencing the antigenicity of interferon alpha (IFN-alpha) in normal and transgenic mice. *Pharm. Res.*, **14**, 1472–1478.

85. Ryan, M.E., Webster, M.L., and Statler, J.D. (1996) Adverse effects of intravenous immunoglobulin therapy. *Clin. Pediatr. (Phila)*, **35**, 23–31.

86. Wang, W., et al. (2007) Antibody structure, instability, and formulation. *J. Pharm. Sci.*, **96**, 1–26.

87. Golovanov, A.P., Hautbergue, G.M., Wilson, S.A., and Lian, L.Y. (2004) A simple method for improving protein solubility and long-term stability. *J. Am. Chem. Soc.*, **126**, 8933–8939.

88. Serno, T., Geidobler, R., and Winter, G. (2011) Protein stabilization by cyclodextrins in the liquid and dried state. *Adv. Drug Deliv. Rev.*, **63**, 1086–1106.

89. Brewster, M.E., Hora, M.S., Simpkins, J.W., and Bodor, N. (1991) Use of 2-hydroxypropyl-beta-cyclodextrin as a solubilizing and stabilizing excipient for protein drugs. *Pharm. Res.*, **8**, 792–795.

90. Serno, T., Carpenter, J.F., Randolph, T.W., and Winter, G. (2010) Inhibition of agitation-induced aggregation of an IgG-antibody by hydroxypropyl-beta-cyclodextrin. *J. Pharm. Sci.*, **99**, 1193–1206.

91. Duddu, S.P. and Dal Monte, P.R. (1997) Effect of glass transition temperature on the stability of lyophilized formulations containing a chimeric therapeutic monoclonal antibody. *Pharm. Res.*, **14**, 591–595.

92. Chang, L.L., et al. (2005) Effect of sorbitol and residual moisture on the stability of lyophilized antibodies: implications for the mechanism of protein stabilization in the solid state. *J. Pharm. Sci.*, **94**, 1445–1455.

93. Andya, J.D., Hsu, C.C., and Shire, S.J. (2003) Mechanisms of aggregate formation and carbohydrate excipient stabilization of lyophilized humanized monoclonal antibody formulations. *AAPS Pharm. Sci.*, **5**, E10.

94. Liu, J., Nguyen, M.D., Andya, J.D., and Shire, S.J. (2005) Reversible self-association increases the viscosity of a concentrated monoclonal antibody in aqueous solution. *J. Pharm. Sci.*, **94**, 1928–1940.

95. Wang, N., et al. (2009) Opalescence of an IgG1 monoclonal antibody formulation is mediated by ionic strength and excipients. *BioPharm Int.*, **22**, 1–9.

96. Shenoy, B., Wang, Y., Shan, W., and Margolin, A.L. (2001) Stability of crystalline proteins. *Biotechnol. Bioeng.*, **73**, 358–369.

97. Harris, L.J., Skaletsky, E., and McPherson, A. (1995) Crystallization of intact monoclonal antibodies. *Proteins*, **23**, 285–289.

98. Saluja, A., et al. (2010) Diffusion and sedimentation interaction parameters for measuring the second virial coefficient and their utility as predictors of protein aggregation. *Biophys. J.*, **99**, 2657–2665.

99. Ingham, K.C. (1990) Precipitation of proteins with polyethylene glycol. *Methods Enzymol.*, **182**, 301–306.

100. Gibson, T.J., et al. (2011) Application of a high-throughput screening procedure with PEG-induced precipitation to compare relative protein solubility during formulation development with IgG1 monoclonal antibodies. *J. Pharm. Sci.*, **100**, 1009–1021.

101. Matulis, D., Kranz, J.K., Salemme, F.R., and Todd, M.J. (2005) Thermodynamic stability of carbonic anhydrase: measurements of binding affinity and stoichiometry using ThermoFluor. *Biochemistry*, **44**, 5258–5266.

102. Beck, A., Wurch, T. Bailly, C., and Corvaia, N. (2010) Strategies and challenges for the next generation of therapeutic antibodies. *Nat. Rev. Immunol.*, **10**, 345–352.

103. He, F., et al. (2010) High throughput thermostability screening of monoclonal antibody formulations. *J. Pharm. Sci.*, **99**, 1707–1720.

104. Smoot, A.L., et al. (2001) The binding of bis-ANS to the isolated GroEL apical domain fragment induces the formation of a folding intermediate with increased hydrophobic surface not observed in tetradecameric GroEL. *Biochemistry*, **40**, 4484–4492.

105. Garidel, P., Hegyi, M., Bassarab, S., and Weichel, M. (2008) A rapid, sensitive and economical assessment of monoclonal antibody conformational stability by intrinsic tryptophan fluorescence spectroscopy. *Biotechnol. J.*, **3**, 1201–1211.

106. Yadav, S., Shire, S.J., and Kalonia, D.S. (2012) Viscosity behavior of high-concentration monoclonal antibody solutions: correlation with interaction parameter and electroviscous effects. *J. Pharm. Sci.*, **101**, 998–1011.

107. Fischer, H.P. and Heyse, S. (2005) From targets to leads: the importance of advanced data analysis for decision support in drug discovery. *Curr. Opin. Drug Discov. Dev.*, **8**, 334–346.

SECTION 2

FIRST-IN-HUMAN AND UP TO PROOF-OF-CONCEPT CLINICAL TRIALS

4

BIOPHYSICAL AND STRUCTURAL CHARACTERIZATION NEEDED PRIOR TO PROOF OF CONCEPT

Angela W. Blake-Haskins, Yen-Huei Lin, Zhuchun Wu, Melissa D. Perkins, and Thomas M. Spitznagel

BioPharmaceutical Development, Human Genome Sciences, Inc., Rockville, MD, USA

4.1	Introduction	96
4.2	Biophysical Methods for Elucidation of Protein Structure and Physiochemical Properties	96
	4.2.1 Protein Primary Structure	97
	4.2.2 Protein Secondary and Tertiary Structures	99
	4.2.3 Quaternary Structure	102
	4.2.4 Posttranslational Modifications	104
4.3	Biophysical and Structural Characterization Data	106
4.4	Case Study—Characterization of Higher Order Structure of a Fusion Protein with Biophysical Methods	108
4.5	Biophysical and Structural Characterization Data in Analytical Comparability Assessments	109
	4.5.1 Case Study—Product Formulation Change	115
	4.5.2 Case Study—Cell Line and Process Change	116
4.6	Summary and Future Perspectives	118
	Acknowledgments	120
	References	120

Biophysical Methods for Biotherapeutics: Discovery and Development Applications, First Edition. Edited by Tapan K. Das.
© 2014 John Wiley & Sons, Inc. Published 2014 by John Wiley & Sons, Inc.

4.1 INTRODUCTION

Characterization of primary, secondary, and higher order protein structure and analysis of potential posttranslational modifications and degradation pathways are important components of biopharmaceutical development. Drug development of a biopharmaceutical comprises a series of phases including molecule selection, determination of mechanism of action, preclinical toxicology studies, first-in-human (FIH) and pre-proof-of-concept (POC) clinical trials, and late stage clinical safety and efficacy studies. Biophysical methodologies are utilized throughout these development phases to elucidate structure (primary and higher order) and physiochemical properties including posttranslational modifications (e.g., carbohydrate structure) and product-related substances and impurities. However, the degree and extent of biophysical and structural characterization of the protein drug product will generally vary with phase of development.

Regulatory guidelines from the United States Food and Drug Administration (FDA) and European Medicines Agency (EMEA) recognize that the time period during FIH clinical trials is unique and structural information about the molecule may be limited [1–3]. The guidance documents provide limited detail regarding the amount or depth of biophysical and structural characterization data to be submitted and recognize that data will be supplemented throughout the phases of clinical development. In general, the level of information submitted depends on the phase of clinical investigation, state of development, patient population, and type/duration of the clinical studies. Safety is of utmost importance and level of risk should be considered when designing structural and biophysical characterization studies that support human clinical trials.

This chapter will discuss approaches for the characterization of protein structure and physiochemical properties using biophysical methodology performed prior to initiating phase 3 clinical trials (i.e., in pre-POC stage). A description of commonly employed biophysical methods utilized during preclinical development to understand primary and higher order protein structure, posttranslational modifications, and physiochemical properties will be presented. Important elements of the biophysical and structural characterization studies to be included in the clinical trial applications (CTAs) prior to FIH clinical studies will also be discussed. The uses of biophysical and structural characterization studies to support ongoing clinical studies, for example, through analytical comparability testing, will be reviewed. Case studies are also included throughout the chapter to illustrate the use of biophysical and structural characterization studies to support clinical development pre-POC.

4.2 BIOPHYSICAL METHODS FOR ELUCIDATION OF PROTEIN STRUCTURE AND PHYSIOCHEMICAL PROPERTIES

A wide variety of high-resolution biophysical techniques are currently available for characterization of protein molecules. Prior to FIH clinical trials (i.e., during preclinical stage), the focus should be on understanding how the structure of the molecule and its basic properties relate to its overall function. In support of clinical studies in humans, a more thorough evaluation of structure, biophysical properties, and degradation products should be performed. Biophysical methods are also utilized throughout the clinical development phase to assess whether process changes affect these protein properties through analytical comparability studies. This section describes the various techniques available to the biophysical scientist for characterizing protein drug products with focus on understanding the molecule and its mechanism of action. While any one specific methodology may not be absolutely required to enter FIH trials, these tests provide the framework for establishing what attributes of the molecule to monitor as development proceeds. Furthermore, the information from these methods can provide the basis for doing an initial risk assessment as part of a quality by design (QbD) approach.

4.2.1 Protein Primary Structure

Proteins rely on their specific spatial conformations to function. The biological properties of a protein are directly linked to its structure. A protein's primary amino acid sequence defines its three-dimensional structure, which is driven by a number of noncovalent interactions and is influenced by solution and environmental conditions. Thus, the examination of therapeutic protein function and properties starts with the examination of the primary structure.

Prior to production of the protein, the DNA (deoxyribonucleic acid) construct sequence of the master cell bank (MCB) for a specific recombinant protein is verified. Transcription and translation are processes with high fidelity and seldom generate incorrect sequences [4, 5]. However, it is necessary to directly confirm the amino acid sequence of the recombinant protein, thus, completely ruling out any deletion, insertion, repeat, or excursion. Although complete amino acid sequence verification is generally not required in early stage development, companies have put more effort on sequence confirmation earlier in development to catch either unintended amino acid sequence changes or low-level sequence variants [6, 7]. If detected, changes can be made with a different codon, sequence, or a more stable cell line. In some cases, it may be well worth the effort of confirming the amino acid sequence early in development, considering the benefits in later stage development of avoiding potential product-related impurities and an inconsistent manufacturing process due to an unstable cell line. In addition, sequence confirmation provides valuable information when characterizing the reference standard. This section describes two commonly used methods for determining the primary amino acid sequence of a protein molecule: peptide mapping with mass spectrometry detection and terminal amino acid sequencing.

4.2.1.1 Peptide Mapping/Mass Spectrometry. Currently, peptide mapping and mass spectrometry are used to confirm and verify the amino acid sequence by most companies. The peptide map is a "fingerprint" of the protein primary structure generated through protein fragmentation and subsequent separation of resultant peptides. A well-developed peptide map covers most of the sequence and is very sensitive to changes in the sequence. There are several types of mass spectrometers classified according to their principle of measuring the mass of molecules (e.g., quadrupole, sector instruments,

time-of-flight (Tof), Fourier transform ion cyclotron resonance (FT-ICR), and ion trap (three-dimensional quadrupole and linear quadrupole)) [8]. The most frequently used mass spectrometers for protein sequencing and characterization are Tof and ion trap mass spectrometers. The basic principle of using mass spectrometry to sequence the protein is based on the different masses of the 20 amino acids. The two amino acids glutamine (Gln) and lysine (Lys) differ only by 0.036 Da, which can pose challenges for low-resolution and less accurate mass spectrometers. Nevertheless, Gln and Lys can be distinguished with additional information in peptide mapping (e.g., when using trypsin or Lys-C as the enzyme for digestion, the Lys residue is always located at the C-terminus of the peptide fragments). As leucine (Leu) and isoleucine (Ile) have exactly the same mass, MS/MS with high-energy collision-induced dissociation (CID) or N-terminal amino acid sequencing by Edman degradation can be utilized to differentiate these amino acids.

Peptide mapping combined with mass spectrometry is a very useful tool for confirming the amino acid sequence or detecting one or more incorrect residues, even at a very low level. Peptide mapping with liquid chromatography-mass spectrometry (LC-MS) detection often covers more than 95% of the sequence. The subtle sequence differences in the fragment peptides often result in a hydrophobicity change, thus, giving a different elution time and peak position in the profile. As mentioned above, the change in amino acid sequence also results in a mass change in the corresponding peptides, except the change between Leu and Ile. The mass accuracy of modern mass spectrometers has reached below 10 ppm. With such a mass accuracy, peptides are identified with a very high confidence. The likelihood of mis-assignment of a peak to a specific peptide is very low. In addition, MS/MS can be used to further verify the sequence of the peptides [9, 10]. Alternatively, a second digestion with another enzyme can be used to confirm the remaining sequences and to cross-confirm the overlapping sequences.

4.2.1.2 Terminal Amino Acid Sequencing. N-terminal sequencing by Edman degradation has been used to sequence proteins for more than a half century [11, 12]. Edman degradation is based on the reaction of phenylisothiocyanate (PITC) with the free amino group of the N-terminal residue, and subsequently acid hydrolysis of acid-labile phenylthiocarbamyl derivative, resulting in the release of the amino acid residue from a protein at the N-terminus. Automation of Edman reactions led to N-terminal sequencer systems that can repeat each degradation cycle, sequentially releasing the N-terminal amino acid from proteins. The released amino acid is further modified to a more stable phenylthiohydantoin (PTH) derivative and identified by its unique retention time in reversed phase high-performance liquid chromatography (RP-HPLC). Because chemical reactions in each degradation cycle do not always go to completion, sequencing beyond 50–60 residues is difficult. In practice, most reliable sequencing data are obtained within 30 amino acid residues. Since its invention over a half century ago, N-terminal sequencing by Edman degradation has been one of the most effective tools to sequence peptides and proteins. It provides direct amino acid sequence information for all commonly occurring amino acids, including Ile/Leu and Lys/Gln.

4.2.2   Protein Secondary and Tertiary Structures

The function of a protein hinges on its native higher order structure. The primary amino acid sequence must fold into appropriate secondary and tertiary structures to assume biological activity. While atomic-level, three-dimensional structure information using X-ray crystallography may illustrate the relationship between structure and function, it is rarely feasible to perform this analysis in early phase biopharmaceutical development due to technical and resource limitations. Biophysical methods, however, can be easily established in biopharmaceutical development labs and generate structure information in the product-relevant formulation matrix. Many biophysical methods have been well established to characterize changes in higher order structure. Some are based on molecular spectroscopy such as ultraviolet (UV) absorbance, fluorescence, and circular dichroism (CD), while others are based on thermodynamic principles (e.g., differential scanning calorimetry (DSC)). These methods can be used to probe the secondary and tertiary structural profiles of the protein in its product-relevant matrix. Once these baseline profiles are established, it becomes feasible to detect changes in higher order structure in support of real-time/accelerated stability studies, forced degradation studies, comparability assessments, and so on. An extensive technical discussion is not within the scope of this chapter but can be found elsewhere, including other chapters in this book and in Jiskoot et al. [13]. Some of the frequently used biophysical methods for characterizing secondary and tertiary structures in proteins are briefly described in the following paragraphs.

4.2.2.1   Circular Dichroism Spectroscopy. CD spectroscopy measures the difference between left- and right-handed polarized light. The difference may originate from inherently asymmetric groups in a biological macromolecule (e.g., α-carbon of the peptide bond), cysteine, tryptophan, tyrosine, and phenylalanine. The CD signal may also reflect the asymmetric chemical environment that exists in a protein.

Far-UV CD spectroscopy (wavelength $\leq \sim 250$ nm) has been applied to estimate the secondary structure content in proteins, based on the assumption that the spectral feature of a protein is the linear sum of the components of its secondary structure. Many studies with samples of pure secondary structure elucidated the characteristic far-UV CD profile (e.g., absorption minima at 222/208 nm for α-helix and 216 nm for β-sheet). Computing algorithms were developed to deconvolute the CD profile and estimate the content of each secondary structure. One application of far-UV CD spectroscopy for biopharmaceutical development is to monitor the thermal stability of proteins. For example, changes in the absorption minima specific for α-helix or β-sheet can be used to monitor changes in secondary structure due to heat and/or pH [14, 15]. One limitation of the application of far-UV CD spectroscopy is interference from some frequently used excipients (e.g., histidine and chloride salts).

Near-UV CD spectroscopy measures the signal originated from the Coulomb interaction of nearby chromophores, which is a valuable tool to monitor changes in global tertiary structure. Buffer interference is not a problem with near-UV CD spectroscopy in general. However, the relatively weaker signal and lower signal/noise ratio usually requires higher protein concentration (1–5 mg/mL) for near-UV CD experiments compared to far-UV CD experiments ($\leq$ 1 mg/mL).

4.2.2.2 Fourier Transform Infrared Spectroscopy.

Fourier transform infrared (FTIR) spectroscopy analysis for protein secondary structure determination assumes that the vibration of amide carbon=oxygen (C=O) and carbon–hydrogen (C–H) bonds is influenced by the characteristic hydrogen bonding, which determines the secondary structure of a protein. Characteristic absorption bands in amide I (1600–1700 cm^{-1}) and amide II (1500–1600 cm^{-1}) have been associated with specific secondary structures (e.g., 1649–1658 cm^{-1} for α-helix and 1620–1635 cm^{-1} for β-sheet). The FTIR spectral data are expressed as a function of wavenumber (cm^{-1}) for conventional and practical reasons. One advantage of FTIR spectroscopy is that it can be used to assess structural changes for either liquid or solid dosage forms. Therefore, FTIR spectroscopy is a common method to evaluate the impact of lyophilization on protein structure. FTIR spectroscopy is a good alternative and orthogonal method for far-UV CD spectroscopy to assess secondary structure as long as interference from water ($\sim$1650 cm^{-1}) in the amide I band is well controlled and subtracted during analysis.

4.2.2.3 UV Absorbance Spectroscopy.

UV absorption spectroscopy is conceivably the most prevalent biophysical method for protein analysis. The amide group and aromatic amino acid chromophores form the basis of the characteristic UV absorption pattern for proteins. The zero-order spectral data can be used to measure protein concentration with the appropriate absorption coefficient or to assess protein aggregation by light scattering signal present at >320 nm. When the absorption spectrum is measured at high resolution (e.g., by averaging data over an extended timeframe; >30 seconds vs. <1 second, at 1 nm increments), high-resolution derivative UV absorption data at sub-nanometer precision can be generated by taking derivatives of zero-order data [16]. The resulting derivative UV absorption data can be used to assess changes in the polarity of the environment of aromatic amino acid residues [17]. The peak positions of aromatic residues shift to a shorter wavelength (blue shift) signaling a more polar local environment, which may implicate exposure of hydrophobic residues due to unfolding. The maximum theoretically possible shift produced by a hydrophobic microenvironment of a protein interior to completely solvent-exposed state is approximately 5 nm [17]. Such shifts in peak positions of single or multiple chromophores can be used to assess the impact of pH, heat, or solvent on the protein structure.

4.2.2.4 Fluorescence Spectroscopy.

Proteins are natural fluorophores owing to the presence of aromatic amino acids. Tryptophan, tyrosine, and phenylalanine have max absorption/emission at 280/350, 275/304, and 258/282 nm, respectively, in an aqueous environment at neutral pH. Therefore, intrinsic fluorescence spectroscopy has been used to study changes in the tertiary structure of proteins. The sensitivity of fluorescence to the environment is due to the relatively long time (10^{-9} to 10^{-8} seconds compared to 10^{-15} seconds for absorption) that a molecule stays in the excited state before deexcitation [18]. Tryptophan fluorescence is the most frequently used for structural characterization of proteins due to its highest absorption coefficient and quantum yield among the three amino acid fluorophores. Typically, a red shift (moving toward longer wavelength) in the emission maximum and/or fluorescence intensity increase is interpreted as the aromatic amino acid moving to a more polar environment [19]. These changes are sometimes interpreted as unfolding of the aromatic amino acid-containing domain(s). As one can see, such interpretation is highly dependent on the distribution of the aromatic amino acids. Furthermore, intrinsic fluorescence data should be examined carefully as many factors such as interaction with oxygen and charge quenching can affect the fluorescence signal. In addition, a concomitant decrease of tryptophan emission signal with an increase of emission at $\sim$450 nm may signal the oxidation of tryptophan residues [20].

Protein fluorescence can also be measured using fluorescent dyes. In this case, the fluorescence signal is independent of aromatic amino acids and proportional to the binding (generally non-covalent) of the extrinsic fluorophore to the protein. An unfolded protein typically has more hydrophobic surface and binds more hydrophobic dye (e.g., 1-anilinonaphthalene-8-sulfonic acid (ANS), bis-ANS, Nile red, and SYPRO® Orange) than a folded protein. Therefore, the extrinsic fluorescence signal tends to increase when a protein unfolds and exposes internal hydrophobic amino acids [21]. Extrinsic fluorescence assays should be optimized to maximize the signal to noise ratio and reduce background noise.

4.2.2.5 *Differential Scanning Calorimetry.*

The secondary and tertiary structures of a protein are mostly maintained by the intramolecular, non-covalent bonding among amino acid residues, as well as intramolecular disulfide bonds. The energy required to unfold or denature a protein can be a good indicator of the structural stability of a protein. DSC directly measures the energy change associated with the unfolding of a protein independent of the distribution of amino acid residues, in contrast to most spectroscopic methods where local environment of the amino acids influences signal. The heat capacity of the protein sample is measured as a function of the temperature during the thermal unfolding of the protein in the DSC instrument. The complete energetic profile (Gibbs free energy, ΔG; enthalpy, ΔH; and entropy, ΔS) can be calculated from the experimental data. Changes in this energetic profile represent changes in the thermal stability and/or conformational stability. It has been demonstrated that long-term data evaluating the protein's physical stability (e.g., aggregation) correlate with the thermal stability, although exceptions to this correlation have been noted [22]. However, it is important to bear in mind that all the parameters generated from the DSC experiment, including melting temperature (T_m), ΔG, ΔH, and ΔS should all be considered in order to assess protein stability at various temperatures. In addition, the stability curve, ΔG as a function of temperature, can be used to compare the stability of different proteins or the stability of the same protein under different formulation conditions [23].

4.2.2.6 *Hydrogen Deuterium Exchange Mass Spectrometry.*

A protein exists as a dynamic breathing molecule as opposed to a rigid frozen model. Various parts of a protein are exposed to solvent differently due to their location in the molecule and folding dynamics/energy state around the site. Hydrogen deuterium exchange mass spectrometry (H/DX-MS) technique utilizes this dynamic nature of protein structure to exchange the backbone amide hydrogen with deuterium by replacing water (H_2O) with deuterium oxide (D_2O) in the solution. Such exchange increases the mass of the protein over time. If a protein after H/DX is fragmented by enzyme and subsequently subjected to a chromatographic separation with mass spectrometer detection, the mass increase in peptide fragments of 3–10 amino acids due to H/DX can be mapped back to the protein sequence. The timing and extent of the deuteration indicate the extent of solvent exposure and folding dynamic of the sequence represented by the peptide fragments. Although H/DX-MS was developed in the early 1990s, recent advances in chromatographic methods, peptide identification algorithms, and data processing make H/DX-MS a viable method in biopharmaceutical development labs. The main advantages of H/DX-MS include low sample requirement (sub-nanomole level), compatibility with common product formulation matrices, and reasonable operation time.

H/DX-MS does not directly measure secondary or tertiary structure. Instead, it monitors the conformational dynamics and structural changes of a protein at local and global levels. It has been used to detect the structural changes/conformational dynamic due to glycosylation on IgG1 and due to pegylation of interferon (IFN)-β-1b [24, 25]. H/DX-MS has also demonstrated to be a useful tool for qualitative and quantitative assessment of comparability for IFN-β-1b made from different processes [25]. A quantitative assessment of H/DX-MS comparison data was performed by Houde et al. [25] by calculating the difference between reference and experimental data sets. Replicates (at least three measurements) were collected over different days allowing calculation of uncertainty in the difference to be about ±0.14 Da. For comparability evaluations, Houde et al. used this estimate of uncertainty with a 98% confidence limit for the difference data to determine a threshold limit (±0.5 Da) to determine a statistically significant difference in deuterium levels.

The main drawbacks of H/DX-MS include that it is a manual procedure (sample handling, measurements, etc.) and that data analysis is a highly labor-intensive and potentially error-prone process. Some ideas for improved data presentation have been proposed [25].

4.2.3 Quaternary Structure

As described in the previous section, a protein's function hinges on its native higher order structure and for some proteins the formation of subassemblies is necessary for optimal biological activity. For example, insulin in its native, active form exists as a hexamer. In other cases, proteins may be prone to self-association under certain solution conditions. This self-association behavior can result in stable, long-lived protein aggregates or short-term assemblies that readily disassociate. Long-lived protein–protein interactions have the potential to form large assembled protein aggregates. These ordered protein arrays may lead to protein subvisible and visible particles and there are concerns that these particles have the potential to affect biological activity, immunogenicity, and pharmacokinetics of the drug. Therefore, characterization of protein self-association is an important component of biopharmaceutical development. A description of the commonly used methods for measuring protein self-association and subvisible particulate matter are provided below.

4.2.3.1 Analytical Ultracentrifugation. In solution under a centrifugal field, macromolecules with different sizes and molecular weights move with the centrifugal force at different speeds until equilibrium is reached to balance the various forces that act upon the molecules. Analytical ultracentrifugation (AUC) utilizes ultrahigh centrifugal force generated by spinning a rotor at very high speed and precise measurement of the sample concentration at a time and position as it sediments by either UV light absorption or interference optical refractive index. Sedimentation velocity experiments study the movement of macromolecules in the entire time course as they move toward the cell bottom, while the equilibrium sedimentation experiment determines the time-independent final concentration distribution after equilibrium is reached. Recent advances in instrumentation and data analysis have made AUC a powerful technique to characterize solution properties of proteins, including the molecular mass, shape, self-association, and interactions [26, 27]. In particular, the size variants and amounts of protein aggregation in solution can be studied as a complementary and orthogonal method to size-exclusion chromatography (SEC), addressing the accuracy questions in SEC caused by potential dilution, precipitation, or nonspecific interactions of the protein and its aggregates with the chromatographic column material and HPLC hardware [27, 28].

4.2.3.2 Subvisible Particles. Subvisible particulate matter in biopharmaceuticals has been a topic with heightened attention, due to a potential risk of immunogenicity caused by protein aggregates [29] and the acknowledged "gap" of characterization in the 0.1–10 μm size range [30, 31]. Typically, at a minimum for drug product release, SEC is used to monitor heterogeneity in the <0.1 μm size range and light obscuration or microscopy, following United States Pharmacopoeia (USP) <788>, is used to monitor particle counts ≥10 and ≥25 μm. Several methods are now being used and are in development for characterizing and sometimes monitoring 0.1–10 μm particles. The performance and limitations of available measurement techniques have been reviewed [31, 32]. Microscopic methods, although allowed in some instances per USP <788>, are not widely used because of the subjective and labor-intensive nature and concerns about proteinaceous particulate matter loss during filtration [33]. Two of the most commonly used methods for the 1–10 μm size range are flow imaging microscopy and light obscuration. A description of these methods is provided below.

Flow Imaging Microscopy. Flow microscopy technology captures particle concentration and images as a sample stream passes through a flow cell. The images allow one to determine the nature of the particle population in the sample. Because protein aggregates have some unique characteristics (nonuniform, flexible, and semitransparent), proteinaceous particles can be differentiated from extrinsic particles such as dust, air bubbles, and silicone, even when combined within one solution. Images are compiled to obtain count, size, concentration, transparency, and shape parameters. Flow imaging microscopy offers advantages over light obscuration methods when attempting to determine particle type/origin. Particles of approximately 1 μm and above can be quantified [34].

Light Obscuration.   Light obscuration operates on the principle of light blockage as a sample stream passes through a flow cell, which allows an automatic determination of the size of particles and the number of particles according to size. The instrument is commonly used and in many ways simpler to use than flow imaging microscopy. Particles of 2 μm size and above can be quantified [33, 35]. Challenges have been identified when measuring particles with similar refractive index to the solution (similar issues have also been observed with flow imaging microscopy measurements). In addition, testing of individual low-volume products is currently not covered in USP <788>, although lower volume light obscuration methods have been developed [35].

4.2.4   Posttranslational Modifications

Posttranslational modification is a common phenomenon for proteins occurring during the manufacturing process and storage, resulting in a heterogeneous product profile. Posttranslational modification not only changes the protein structure and physicochemical properties, many posttranslational modifications in nature are designed to mediate biological functions. In addition to enzyme-catalyzed posttranslational modifications, proteins are subjected to various attacks by chemicals in solution. Nonenzymatic modifications can occur as a result of stress and/or storage. Understanding, characterizing, monitoring, and controlling the protein modifications to maintain a consistent and desirable analytical profile are primary focuses of bioprocess development for protein therapeutics. Although changes in the product analytical profile are expected during development, a complete understanding of the product profile is generally not finalized until late stage development. Obtaining a general understanding of the analytical profile early in development is beneficial for bioprocess development and clinical trials. Frequently used techniques for measuring posttranslational chemical modifications including carbohydrate structure are described below in the following subsections.

4.2.4.1   Peptide Mapping/Mass Spectrometry.   Most posttranslational modifications result in a change in mass and can be identified by mass spectrometry. When using peptide mapping to confirm the amino acid sequence, posttranslational modifications can be studied and identified using the same set of data. Common modifications in recombinant proteins, such as deamidation, oxidation, glycation, alkylation, acetylation, carbamylation, disulfide bonding, and modifications resulting from reactions with unpaired cysteines, may also be investigated. It is frequently observed that recombinant proteins exhibit carboxyl(C)-and amino(N)-terminal heterogeneity, such as C-terminal amidation, N-terminal pyroglutamate formation, N-terminal extensions due to incomplete removal of the leader sequence, and terminal truncation due to the presence of aminopeptidase or carboxypeptidase. A search of these modifications in peptide-mapping data (with MS detection) can be carried out. Once the modification is detected, MS/MS can be used to locate the precise residue that has been modified and to identify the nature of modification [9, 36]. Finally, the intact mass profile of the recombinant protein should be obtained. The mass of the intact protein should match the sum of all amino acids and posttranslational modifications. The exact mass and mass profile reveal the overall integrity of the primary structure and mass heterogeneity of the

protein. With the identification of protein modifications, appropriate analytical methods can be developed to measure the protein purity and heterogeneity.

4.2.4.2 Glycosylation and Carbohydrate Structure.

Among posttranslational modifications, glycosylation generates most of the heterogeneities. O-linked glycosylation and N-linked glycosylation are two common types of glycosylation. Each type of glycosylation can generate many different glycan structures. For example, the N-linked glycan may have either high mannose or complex, hybrid type of structures. Glycan structures within each type may be different by number of antennary, length and sequence of each antennary, linkage, and anomericity. Sialic acids can attach to the nonreduced ends of glycans, generating heterogeneity. Glycosylation can affect bioactivity, influence clearance and half-life, and raise immunogenicity concerns [37]. Although the dominant glycan structure types are very much determined by the specific cell line used, glycan structural variations and distribution may be altered, depending on the specific clone selected and cell culture conditions. Thus, monitoring and controlling glycosylation are necessary during early stage process and product development. A complete glycosylation characterization includes identifying glycosylation sites, measuring the glycosylation site occupancy, and determining the glycan structures and distribution.

4.2.4.3 Glycosylation Site Identification and Occupancy.

Glycosylation sites and site occupancy can be determined by peptide mapping and mass spectrometry. Due to the hydrophilic property of glycans, the glycopeptides usually elute earlier than the non-glycosylated peptide. The unique mass pattern of glycopeptides caused by glycan heterogeneity (i.e., glycans with mass difference of 162 Da) lends the glycopeptides themselves to be easily identified. The site occupancy can be estimated by comparing the intensities of glycopeptides and the non-glycosylated peptide. However, this estimation may be far from accurate due to glycan heterogeneity that produces many glycopeptides in one glycosylation site. It is more accurate to perform deglycosylation and compare the intensities of deglycosylated peptide and non-glycosylated peptide. Deglycosylation removes the glycan heterogeneity, resulting in a homogeneous molecule that can be better detected and quantified. In addition to removing the sugars, deglycosylation by PNGase F converts the asparagine residue to aspartic acid with a 1 Da mass increase. This 1 Da mass difference between deglycosylated and non-glycosylated peptides can be detected by most mass spectrometers. The structural glycans are packed looser than proteins, thus are much larger in size. A size-based method can be used to determine the amount of non-glycosylation. For example, sodium dodecyl sulfate-capillary gel electrophoresis (SDS-CGE) has been widely used to measure the amount of non-glycosylation in recombinant IgG antibodies [38].

4.2.4.4 Glycan Structure and Distribution Determination.

For glycan structure and distribution determination, glycans are usually released from the glycoproteins by either enzymes or chemicals and analyzed directly by matrix-assisted laser desorption/ionization (MALDI)-Tof mass spectrometry and or by LC-MS after derivatization. Although there are many possible glycan structures, the types of glycan structures from a specific cell line are usually known. With the knowledge of potential glycan structure in the specific cell line, the glycan structures can be assigned using the mass spectrometry data. Glycan structures may be further elucidated by exoglycosidase reactions, monosaccharide analysis, lectin microarrays, MS/MS after derivatization, and nuclear magnetic resonance (NMR) [39]. A complete verification of glycan structures may only be necessary at later stages of product development. However, the structures and amounts of glycans that are known to affect the efficacy and safety should be characterized and quantified early in development. In the case of immunoglobulin G, isotype 1 (IgG_1) monoclonal antibodies, absence of core fucosylation at the Fc-conserved N-linked glycosylation site results in a substantial increase in antibody-dependent cell-mediated cytotoxicity (ADCC) activity [37].

To determine the distribution and amount of each glycan, normal phase HPLC and capillary electrophoresis laser-induced fluorescence (CE-LIF) analysis of derivatized glycans are usually employed, which have better precision, accuracy, and robustness than MALDI-Tof and high-performance anion exchange chromatography with pulsed amperometric detection (HPAEC-PAD), although direct analysis without derivatization can be done for the two latter methods. Fluorescent labeling before the HPLC or CE analysis is necessary, since glycans lack functional groups that have absorbance in the UV region.

Sialic acid content can be determined by acid hydrolysis and HPAEC-PAD analysis of released sialic acids [40] or RP-HPLC analysis after 1, 2-diamino-4, 5-methyleneoxybenzene (DMB) labeling [41]. Sialic acids in many glycoproteins were found to affect the half-life and clearance [42]. Certain types of sialic acids, for example, N-glycolylneuraminic acid (NGNA), may raise immunogenicity concerns [43].

4.3 BIOPHYSICAL AND STRUCTURAL CHARACTERIZATION DATA

While both FDA and EMEA regulatory guidance documents express understanding that the amount of structure information may be limited in the early stages of drug development [2, 44, 45] data to support the physical, chemical, and biological characteristics of the drug product should be provided in the submission supporting the POC trials. The guidance documents indicate that sponsors should describe any chemistry and manufacturing differences between the clinical drug product and that used in animal toxicology trials which provide the basis for safety conclusions [2]. In each phase of the investigation, sufficient information should be submitted to assure proper identification, quality, purity, and strength of the investigational drug [1]. In terms of characterization, some evidence to support the proposed chemical structure is needed although it is understood that the amount of structural information will be limited in the early stages of drug development [2]. As the clinical program progresses, new data for the primary and higher order structure of the drug product may be included in regulatory filing updates. Throughout clinical development, it is expected that extensive characterization is performed, a reference standard is available and appropriately characterized, new analytical technologies/methodologies are introduced as appropriate, and specification acceptance criteria are appropriately established. An awareness of the expanding requirements from phase 1 through 3 and ultimately license applications is valuable [46, 47].

Like characterization data for a protein molecule, it is also recognized that analytical methodology and acceptance criteria for those methods will progress throughout the clinical trial phases, although safety should be ensured. Proper references should be made to appropriate Pharmacopoeia, such as the European Pharmacopoeia (Ph. Eur.), the Pharmacopoeia of a European Union Member State, the USP, and the Japanese Pharmacopoeia (JP) [2, 3]. Following FDA's guidance [2], acceptable limits should be provided although method validation data and established specifications ordinarily need not be submitted at the initial stage of drug development, unless needed to ensure safety. Following EMEA's guideline [3], for phase 1 clinical trials, the suitability of the analytical methods should be confirmed. The acceptance limits and parameters for performing validation should be presented. Where applicable, reference standard characterization data should be submitted [3].

The CTA typically includes a wide variety of analytical data collected both through routine quality control testing of the batch(s) to be used in the clinical study as well as data collected as part of characterization of the structure and biophysical properties. Protein size purity is typically measured by SEC. The covalent size purity under nonreducing and reducing conditions is measured by sodium dodecyl sulfate-polyacrylamide gel electrophoresis (SDS-PAGE) with coomassie blue stain and/or SDS-CGE. SDS-CGE is gaining popularity for its superior resolution, quantification, sensitivity, and automation. RP-HPLC has been used to determine extent of deamidation, oxidation, and any other variants that change the hydrophobicity. IE-HPLC, capillary zone electrophoresis (CZE), and capillary isoelectric focusing (cIEF) are commonly employed methods for measuring charge heterogeneity.

Prior to FIH studies and CTA, a product batch is typically manufactured for use in preclinical animal/toxicology studies. This material is generally produced using a process similar to what will be used to manufacture clinical supplies for FIH studies. In some cases, material from the toxicology lot is also used to produce the initial reference standard for use during current good manufacturing practices (cGMP), manufacturing, and analytical testing. Therefore, it is often the most ideal material to perform a thorough initial characterization of protein structure and biophysical properties for inclusion in the CTA. The methods described in the previous section for measuring primary (e.g., peptide mapping with MS detection), secondary (e.g., CD spectroscopy), tertiary (e.g., DSC), and higher order (e.g., AUC) structure are not typically part of batch release testing and, therefore, are generally only utilized to examine these attributes on the reference standard.

Due to the underlying nature of the methods and limitations in technology, no single biophysical method provides a complete profile of the protein's structure. Therefore, it is prudent to establish orthogonal biophysical methods for the therapeutic protein in development. Orthogonal methods can be used to complement each other's deficiencies and to provide a comprehensive profile of the structure, as well as provide valuable information in developing the initial formulation. For example, far-UV CD spectroscopy has high sensitivity to probe the secondary structural profile with relatively high sensitivity and relatively low sample concentration requirement. However, this method does not work well for product in buffers containing histidine or chloride salts. The noise appears to be problematic starting at 205 nm and at lower wavelengths in the presence of these compounds. In this case, the secondary structural profile can be probed with FTIR spectroscopy. Second-derivative UV and intrinsic fluorescence spectroscopic techniques can also be used to probe perturbations in tertiary structure. However, these methods rely on the abundance and spatial arrangement of the aromatic amino acids for signal. The distribution of these amino acids will dictate the interpretation of profile changes. DSC, which is independent of abundance and distribution of aromatic amino acid residues, can provide a sequence-independent analysis, complementing the methods based on UV absorbance or intrinsic fluorescence.

The development and use of biochemical and biophysical methods for therapeutic proteins should be justified on a case-by-case basis for their intended use. Due to the nature and purpose of their use, the assays for structural characterization typically are categorized as characterization methods. However, the peptide-mapping method can be used as a release method for identity. Justification of the use of these methods should be based on the actual assay capability.

Structural characterization methods are typically proceduralized to add consistency in data collection, equipment parameters, and system suitability evaluations. However, qualification of biophysical methods for structural characterization can present special challenges and is increasingly being discussed [1] but not yet "typical" industry practice in early phase development. The importance of making non-qualitative conclusions from structural characterization, however, is increased if the methods are used to make definitive claims on whether higher order structural modifications have been made. The methods, for instance, can be very supportive of the evaluation of changes and any related comparability assessments. Although not desired during early phase clinical studies, major process and/or formulation changes can be a necessity. In the special instances where biophysical-related comparability claims are warranted to support such changes, qualification is typically performed.

4.4 CASE STUDY—CHARACTERIZATION OF HIGHER ORDER STRUCTURE OF A FUSION PROTEIN WITH BIOPHYSICAL METHODS

Biophysical methods can be valuable tools for determination of higher order structure in fusion proteins and possible alterations that may occur as a result of the fusion itself. This case study describes the utility of spectroscopic and calorimetric methods in characterizing the albinterferon alfa-2b fusion protein during early development. Interferons (IFN) are a class of cytokines that play an important role in the innate immune response to viral infections. Various subtype/subfamilies of IFN (Alferon® N, Infergen®, Pegasys®, Peg-Intron™, Rebif®, and Roferon®-A) have been used to treat chronic hepatitis C infections. First-generation IFN products suffered from short half-life. Pegylation of the IFN proved to be one way to extend the serum half-life for IFN molecules (Pegasys® and Peg-Intron™). Stabilization by fusion with albumin also proved to be an innovative way to extend the serum half-life of therapeutic proteins. When the C-terminus of human serum albumin (HSA) is genetically fused to the N-terminus of interferon alfa-2b (IFNα-2b), the resulting fusion protein, albinterferon alfa-2b, has a dramatically extended serum half-life (approximately 200 hours) compared to IFNα-2b alone (approximately 2–3 hours) while still maintaining the biological activity of IFN [48–50].

When creating a fusion protein, the potential for structural perturbation leading to alternations in biological activity and/or increased immunogenicity is always a concern. Biological activity may be inadvertently affected due to disruption of the native conformation. Alternative folding could also occur due to the proximity of the primary sequences created by the fusion. Alternative folding might cause a change in potency or induce immunogenicity to the parent sequences. Two proteins fused by genetic engineering should ideally maintain the same conformation and structure as the parent molecules with no new conformation/structural components. Although *in vitro* activity of albinterferon alfa-2b had been demonstrated, it was also important to provide additional assurance that the albumin and IFN moieties remain independent in their native states.

In this case study, second-derivative UV spectroscopy, intrinsic fluorescence spectroscopy, far- and near-UV CD spectroscopy, and DSC were used to demonstrate that the secondary and tertiary structures of IFN and HSA remain native in the recombinant fusion form. If the HSA and IFN remain in their native structure and do not form additional or alternative folding, the biophysical characterization should provide supporting data for the following two hypotheses: (1) the spectral and thermographic profiles of albinterferon alfa-2b should resemble the summation of spectral or thermographic profiles for IFNα-2b and HSA measured independently and (2) the spectral and thermographic profiles of albinterferon alfa-2b should resemble IFNα-2b and HSA measured in equimolar mixture.

Indeed, the spectral profiles measured with second-derivative UV spectroscopy, intrinsic fluorescence spectroscopy, far- and near-UV CD spectroscopy (Figures 4.1, 4.2, and 4.3), and DSC (Figure 4.4) for albinterferon alfa-2b closely resemble the sum of the two independent spectra/thermograms as well as the profile for the equimolar mixture of the two proteins. Therefore, biophysical characterization provided supporting data that the fusion of IFNα-2b and HSA did not cause structural perturbation of the individual proteins.

4.5 BIOPHYSICAL AND STRUCTURAL CHARACTERIZATION DATA IN ANALYTICAL COMPARABILITY ASSESSMENTS

Comparability assessments are required to support changes implemented during biopharmaceutical product development per regulatory guidance to demonstrate that the changes will not have an adverse impact on the quality, safety, and efficacy of the drug product [51]. Early in development, changes in the manufacturing process itself including cell line, raw materials, and equipment may occur; however, facility and/or scale changes are often more common to accommodate flexibility for manufacturing clinical material. Alternate formulations including changes in excipient types and/or

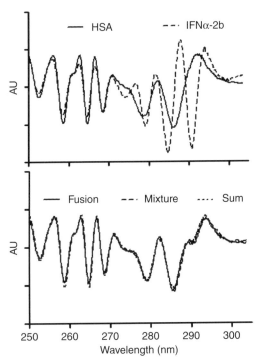

Figure 4.1. Second-derivative UV spectra of human serum albumin and interferon alfa-2b (top panel); second-derivative UV spectra of albinterferon alfa-2b, equimolar mixture of human serum albumin and interferon alfa-2b, and mathematical summation of human serum albumin and interferon alfa-2b (bottom panel).

amounts may be introduced to provide enhanced stability during manufacturing, product handling, and long-term storage. Container closure changes and changes in product concentration are sometimes desired to accommodate changes in clinical dosing and/or indication.

Approaches for assessing the comparability of pre- and post-change materials at any stage of development are still evolving [52]. Various factors should be considered in designing the comparability assessment including number of changes being implemented and extent of the change(s). At minimum, analytical comparability assessments at all stages of development typically consist of an evaluation of the quality attributes monitored as part of routine quality control testing on at least one batch each of pre- and post-change materials. Methods for measuring product appearance, components of the formulation (e.g., pH and osmolality), size purity (e.g., SEC, SDS-PAGE with coomassie blue stain), charge heterogeneity (e.g., ion exchange (IE)-HPLC), product concentration (e.g., absorbance at 280 nm), identity (e.g., enzyme-linked immunosorbent assay (ELISA)), biological activity (e.g., cell proliferation, inhibition of binding), process residuals (e.g., host cell proteins), and microbiological assays (bacterial endotoxin, sterility) that are elements of batch release testing are performed as part of the

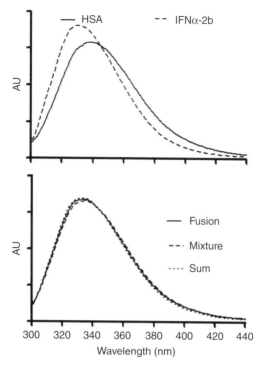

Figure 4.2. Intrinsic fluorescence spectra of human serum albumin and interferon alfa-2b (top panel); intrinsic fluorescence spectra of albinterferon alfa-2b, equimolar mixture of human serum albumin and interferon alfa-2b, and mathematical summation of human serum albumin and interferon alfa-2b (bottom panel).

assessment. Multiple batches (e.g., $n = 3$ each from pre- and post-change product) should be utilized in comparability assessments whenever possible. It is also recommended that purity and potency methods be performed in a side-by-side fashion to reduce inter-assay variation that could impact interpretation of the comparability assessment.

Nonroutine, characterization methods measuring primary and higher order protein structure and relevant posttranslational modifications (e.g., glycosylation) are also generally included in a comparability evaluation. Similar to routine batch release tests, the characterization assays are usually performed side by side on all pre- and post-change lots. Early in development when product knowledge is still being gathered, peptide mapping, mass spectrometry, and N-terminal sequencing are useful to minimally verify that the amino acid sequence has not been altered as a result of the change. N-linked oligosaccharide analysis using CE-LIF and sialic acid content would also be included in the comparability evaluation for glycoproteins. A measurement of the protein's higher order structure is also typically included in comparability assessments, particularly for the final drug product. Secondary structure can be evaluated using far-UV CD spectroscopy or FTIR spectroscopy. Near-UV CD spectroscopy, fluorescence, dual-beam UV spectroscopy, and DSC are all potential options for evaluating tertiary structure in

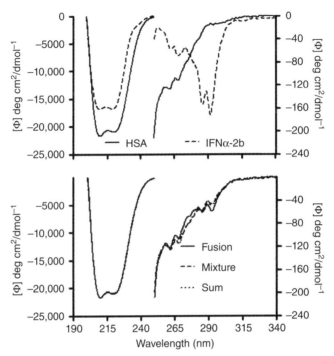

Figure 4.3. CD spectra of human serum albumin and interferon alfa-2b (top panel); CD spectra of albinterferon alfa-2b, equimolar mixture of human serum albumin and interferon alfa-2b, and mathematical summation of human serum albumin and interferon alfa-2b (bottom panel).

pre- and post-change product. Protein self-association and/or higher order aggregates can be assessed using AUC. Techniques for measuring subvisible particles, such as light obscuration and flow imaging microscopy (Micro-Flow Imaging™, MFI™), in the drug product should also be considered when designing comparability studies.

Characterization methods for evaluating the protein structure (primary and higher order) are often not qualified early in development, but one may consider performing a limited qualification study prior to using these methods for comparability assessments in order to appropriately (and quantitatively) interpret any differences in pre- and post-change product. Precision (repeatability, intermediate), sensitivity, and specificity for detecting minor structural differences in the presence of native protein are examples of method attributes to consider for the qualification experiments.

Depending on the degree of product and method knowledge gathered during early development, peptide mapping and mass spectrometry can also be used to compare types and amounts (generally semi-quantitatively) of various posttranslational modifications including oxidation, deamidation, and glycation as well as identification of affected residues. Primary degradation pathway knowledge may also be limited early in development; however, if specific methods for measuring degradation products are available,

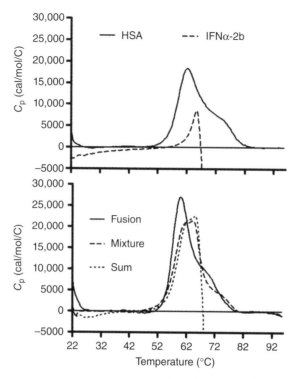

Figure 4.4. DSC thermograms of human serum albumin and interferon alfa-2b (top panel); DSC thermograms of albinterferon alfa-2b, equimolar mixture of human serum albumin and interferon alfa-2b, and mathematical summation of human serum albumin and interferon alfa-2b (bottom panel).

they should be considered in designing the comparability study. These may include hydrophobicity assays such as RP-HPLC and hydrophobic interaction chromatography (HIC) for measuring deamidation and oxidation, Ellman's reagent for measuring free thiol content, and boronate affinity methods for measuring glycation. Characterization methods included in comparability assessments may also represent orthogonal techniques to those methods used as part of routine release testing. Examples include alternate methods for measuring charge heterogeneity including CZE and cIEF. AUC could also be used as an orthogonal method for measuring size heterogeneity (e.g., dimer and higher molecular weight aggregates).

Stability and degradation profile assessments are also important considerations for analytical comparability studies. At minimum, data from stability-indicating methods collected as part of the stability monitoring program for clinical material (drug substance and drug product) should be evaluated for comparison between pre- and post-change product. Data at the intended storage condition (e.g., 2–8°C) and accelerated storage conditions (e.g., 25°C and 40°C) should be utilized for the assessments. In general, the focus of these assessments is on the commonly known degradation pathways (e.g., aggregation detected by SEC; deamidation detected by IE-HPLC); however, data from all methods on the stability protocol including potency, protein concentration, and appearance should also be assessed. Comparison of rates of degradation (using a minimum of three data points) using a statistical approach (e.g., t-test, one-way analysis of variance (ANOVA)) can be useful in assessing any potential differences observed in stability profiles. If limited data are available at the time of the comparability assessment (i.e., less than three data points), comparison to historical stability ranges over the same storage period of the pre-change product can be made. In the case of limited available stability data, greater emphasis is placed on demonstrating similar stability profiles at accelerated conditions.

Stress and forced degradation studies may also be useful when performing comparability assessments. Early in development, the effect of stress conditions (e.g., pH extremes, oxidants, shaking, light) on the protein may not be fully understood, therefore, the choice of which stress conditions to include as well as the degree of the stress should be well thought through and based on available product knowledge. Temperature stress (beyond what is already included in stability monitoring of clinical material) can be particularly valuable when performing comparability assessments due to changes in the manufacturing process of the drug substance, whose intended storage condition is typically frozen ($-20°C$ to $-80°C$). Stress stability assessments of drug substance in the liquid state at $25°C$ for several weeks, for example, may be capable of discerning various degradation pathways detectable by stability methods. A similar approach may also be useful for lyophilized drug products (i.e., storage of reconstituted product at ambient conditions) relevant to patient in-use conditions.

Light and shaking stress studies are also commonly considered when designing comparability assessments. Facility changes to enhance flexibility for manufacturing can be common early in development and lighting conditions may be different site to site (or even suite to suite) justifying the use of a photostability assessment in the comparability study. The sensitivity of the protein to light should be considered when determining the amount of light exposure to use in the study. International Conference on Harmonisation (ICH) guidance document Q1B [53] (written primarily for small molecules) recommended levels for fluorescent (1.2 million lux hours) and UV (200 watt hr/m^2) light can be used as a basis for the study design. Alternatively, the amount of light exposure chosen for the comparability stress study could be justified based on light exposure during processing in the two different facilities. Evaluating the effect of shaking stress on pre- and post-change product during a comparability assessment may also be justified in particular for formulation changes (e.g., change in surfactant type and/or concentration) and/or aggregation-prone proteins.

Early in development, a comparability study plan describing the methods (release and characterization) and whether they will be performed in a side-by-side fashion is typically assembled. It is generally expected that the data from side-by-side testing using release test methods as well as long-term stability data of clinical material will conform to the acceptance criteria listed in the product specification in determining whether comparability of pre- and post-change product has been established. Pre-defined acceptance criteria for characterization methods are not typically assigned for comparability studies performed early in development. It is expected that historical product knowledge as well as method qualification data would be used to interpret if any differences observed impact demonstration of comparability.

4.5.1 Case Study—Product Formulation Change

As mentioned above, alternate formulations with different buffering agents, excipient types and levels, and pH may be introduced during development to provide desired product stability to ensure a suitable shelf life and decrease supply chain risk. A formulation change was implemented for the production of "Protein X" drug substance and drug product that included a change in the protein concentration (greater than twofold increase), buffering agent, pH, and stabilizers. The comparability approach included assessment of one lot each of drug substance and drug product in the prior formulation compared to one lot each of drug substance and drug product in the new formulation. The lots were tested side by side for purity (SE-HPLC), charge heterogeneity (IE-HPLC), and potency. Additional characterization methods were performed side by side including orthogonal size (AUC) and charge (cIEF) heterogeneity methods. Characterization of the primary structure was performed on the drug substance using N-terminal sequencing and peptide mapping. Secondary and tertiary structures were compared to the drug product using FTIR spectroscopy and DSC, respectively. In addition, a temperature stress study (storage at 25°C for 6 weeks) comparing the drug substance in the two formulations was also conducted.

Results for purity, charge heterogeneity, and potency were essentially identical between pre- and post-change product. Primary structure and stress stability data were also deemed comparable. Differences were observed in both FTIR spectroscopy and DSC due to the change in the formulation. Figure 4.5 shows the second-derivative FTIR

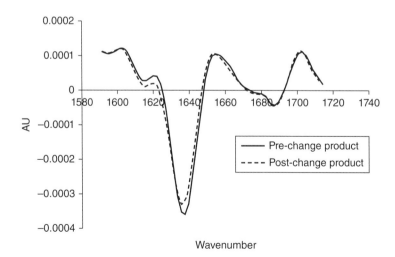

Figure 4.5. Second-derivative FTIR spectra of product in previous and new formulation buffers at intended product concentrations.

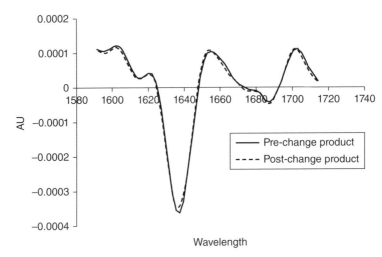

Figure 4.6. Second-derivative FTIR spectra of product in previous and new formulation buffers at pre-change product concentration.

spectrum of drug product in the prior formulation compared to that in the new formulation. Although the band position of the two batches is identical at 1637 cm^{-1}, a significant difference in the intensity of this band is observed between the two batches. In this case, the FTIR method had not been qualified prior to its use in the comparability study, therefore, making the interpretation of the difference in signal intensities challenging. The formulation change included a greater than twofold increase in product concentration and the lots were initially analyzed at their intended concentrations since dilution is not necessary for FTIR measurements. When post-change product was diluted to match pre-change product concentration and reanalyzed by the FTIR method, the signal intensities at 1637 cm^{-1} were found to be similar indicating that differences in concentration and not formulation buffer caused the difference in intensities (Figure 4.6).

A significant difference in the thermal transition temperatures by DSC was also observed for the product in the old and new formulations, with a 3°C shift (first thermal transition temperature) for the product in the new formulation (Figure 4.7). DSC measurements performed during formulation development for each buffer indicated that the main formulation parameter influencing the observed shift in thermal transition temperature is the buffering agent (i.e., differences in pH and stabilizers had negligible effects). This case study example highlights the need to understand how formulation variables may influence the interpretation of biophysical data for higher order structure characterization in comparability assessments.

4.5.2 Case Study—Cell Line and Process Change

Process changes are common in early phase development as manufacturers gain production experience. Process changes could be made to boost the yield to supply for higher

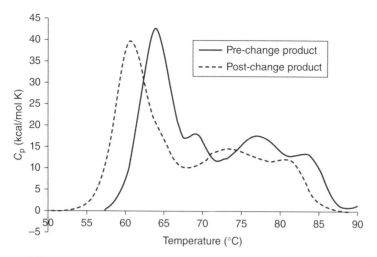

<u>Figure 4.7.</u> DSC thermogram of product in previous and new formulation buffers.

dose clinical trials, to alter the product profile to eliminate certain impurities, to increase process robustness, and/or to reduce manufacturing expenses. It is rare to change the parental cell line during clinical development, since such a change affects all subsequent processes and is considered a totally new production process. However, altering the cell line (through sub-cloning, amplification, etc.) is often the primary approach to achieving a substantial increase in product yield and/or a desired change in product attribute. Comparability requirements for a cell line change are very high. A thorough study including every aspect of structural, physicochemical, and biological properties should be performed.

For "protein C," multiple changes were made in manufacturing the bulk drug substance during early clinical development. A new MCB was made from a fully adapted clone of the initial MCB, which was used for manufacturing the FIH material. Cell culture process modifications including changes to process parameters, feed media, and feed strategy were also implemented. The affinity chromatography column resin, resin load, filter material and size, and elution gradient in the downstream process were also modified. The comparability approach consisted of evaluation of three bulk drug substance (BDS) lots from each of the pre- and post-change processes. All lots were analyzed by a full set of release assays. Of these, the purity and potency methods were performed side by side, with the remainder being obtained at the time of release testing. The six lots were also evaluated by a variety of characterization methods to assess the structural profile and physicochemical and biological properties. The primary structure was further compared by a co-mixture study using a peptide-mapping method with 1:1 mixture of early and later lots. In addition, a temperature stress study was also performed on early and later lots.

The comparability study showed that the primary structures were identical. The degradation profiles were highly similar. The content of various process-related impurities was either similar or reduced in the new process. All lots met the specification

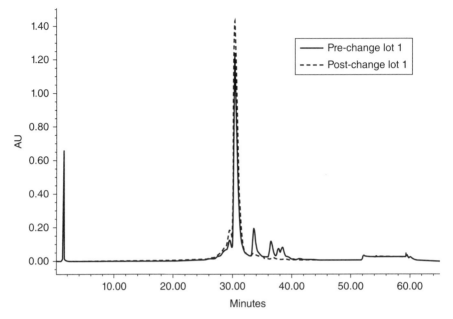

Figure 4.8. IE-HPLC profiles of "Protein C."

for the release assays and the acceptance criteria for the characterization methods that were outlined in a comparability protocol. Differences were observed in charge variants. Material from the new process had significantly less basic variants, including N-terminal extensions and lysine variants, resulting in a more homogeneous product (Figure 4.8). In addition, a change in N-linked glycan distribution, a decrease in G0 ($\sim$16%), and an increase of sialic acid (NGNA) ($\sim$5%) content (Figure 4.9) were also observed, although the types of glycans observed in the two processes were identical. Based on mechanism of action, such changes were not expected to affect the safety or efficacy, which was verified by the subsequent clinical trials.

4.6 SUMMARY AND FUTURE PERSPECTIVES

Biochemical and biophysical data should be included in the CTAs for therapeutic proteins. The data are critical to support the description of primary structure, characterization of higher order structure, and determination of posttranslational modifications and physiochemical properties. In practice, these data should be included in the general information and characterization of drug substance sections and/or the pharmaceutical development section [54, 55]. As the development of the new drug moves into later stages of development, the quantity and quality of biophysical data supporting structural characterization are expected to increase and be updated in the regulatory filings.

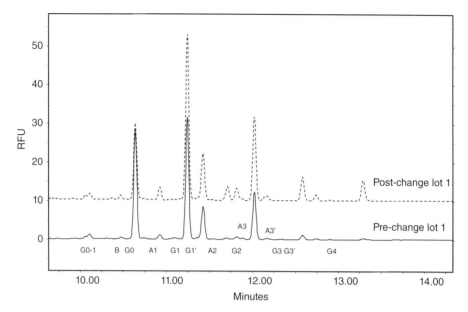

Figure 4.9. N-linked glycan profiles of "Protein C" by CE-LIF.

While most of the information in this chapter has focused on the requirements for FIH studies, there is tremendous opportunity to utilize biophysical and structural characterization as the industry more fully embraces the concepts of QbD. While many of the larger companies have started down this path, it is likely that the use of this information will continue to evolve. Much of the attention with QbD tends to focus on late stage development, since traditionally characterizing design space of a process is a late stage activity. However, one of the cornerstones of QbD is the determination of the critical quality attributes (CQA) of a molecule. While this understanding evolves over the life cycle of the development process, the early stage characterization can be a powerful tool to ensuring the right molecule gets developed.

One of the first uses of the primary structural characterization can start even before the initial cell line is developed. Mass spectroscopy and CE-LIF can be utilized to help screen potential cell lines for their carbohydrate profile and other important chemical attributes. Spectroscopic methods can be used to evaluate whether different sequences within a fusion protein impact the secondary or tertiary structure. By using these tools it is possible to engineer desired attributes, such as a more native folded structure, or elimination or enhancement of a chemical attribute. Examples include eliminating a deamidation hot spot or a secondary glycosylation site in the complementary determining region of an antibody. One can also use these tools to screen for cell lines that generate more desirable carbohydrate profiles that can either enhance safety (lower levels of nonhuman variants such as α-gal or NGNA) or modulate clearance (fucosylation, etc.). While these activities may require more resources than typically devoted this early in the life cycle of a program, the elimination of these problematic residues can streamline development later on.

Once the cell line is selected, the early characterization to support a FIH filing can also be used to form the basis for the first risk assessment associated with a QbD approach to development. While at this stage of development, there is little to no knowledge of whether a product-specific variant impacts safety or efficacy, one can often leverage prior knowledge and literature to assign initial assessments, especially for monoclonal antibodies. Spectroscopic methods, in addition to any binding and potency assays, can also be used to determine if any isolated variants retain their native structure. This information can be a useful surrogate to estimating the impact of these variants on efficacy. Using these initial characterizations, a risk assessment can be performed for all the identified variants, and an initial control strategy and specifications can be rationally developed.

ACKNOWLEDGMENTS

The authors thank Richard Affleck, Michael Byrne, Angie Deng, and Mark Hesselberg for their assistance in assembling this manuscript.

REFERENCES

1. 21 Code of Federal Regulations 312.23, IND content and format.
2. Food and Drug Administration (FDA), Center for Drug Evaluation and Research (CDER) and Center for Biologics Evaluation and Research (CBER). (1995) *Guidance for Industry, Content and Format of Investigational New Drug Applications (INDs) for Phase 1 Studies of Drugs, Including Well-Characterized, Therapeutic, Biotechnology-derived Products.*
3. European Medicines Agency Inspections (EMEA), Committee for Medicinal Products for Human Use (CHMP). (2006) *Guideline on the Requirements to the Chemical and Pharmaceutical Quality Documentation Concerning Investigational Medicinal Products in Clinical Trials.*
4. Ogle, J.M. and Ramakrishnan, V. (2005) Structural insights into translational fidelity. *Annu. Rev. Biochem.*, **74**, 129–177.
5. Zaher, H.S. and Green, R. (2009) Fidelity at the molecular level: lessons from protein synthesis. *Cell*, **136**(4), 746–762.
6. Guo, D., et al. (2010) Mechanisms of unintended amino acid sequence changes in recombinant monoclonal antibodies expressed in Chinese Hamster Ovary (CHO) cells. *Biotechnol. Bioeng.*, **107**(1), 163–171.
7. Yang, Y., et al. (2010) Detecting low level sequence variants in recombinant monoclonal antibodies. *MAbs*, **2**(3), 285–298.
8. Aebersold, R. and Mann, M. (2003) Mass spectrometry-based proteomics. *Nature*, **422**(6928), 198–207.
9. Medzihradszky, K.F. (2005) Peptide sequence analysis. *Methods Enzymol.*, **402**, 209–244.

10. Seidler, J., Zinn, N., Boehm, M.E., and Lehmann, W.D. (2010) De novo sequencing of peptides by MS/MS. *Proteomics*, **10**(4), 634–649.

11. Thoma, R.S., et al. (2009) The ABRF Edman Sequencing Research Group 2008 Study: investigation into homopolymeric amino acid N-terminal sequence tags and their effects on automated Edman degradation. *J. Biomol. Tech.*, **20**(4), 216–225.

12. Findlay, J.B.C. and Geisow, M.J., editors. (1989) *Protein Sequencing: A Practical Approach (Practical Approach Series)*. IRL Press, Oxford.

13. Jiskoot, W. and Crommelin, D., editors. (2005) *Methods for Structural Analysis of Protein Pharmaceuticals*. American Association of Pharmaceutical Scientists, Arlington, VA.

14. Fan, H., et al. (2007) Effects of pH and polyanions on the thermal stability of fibroblast growth factor 20. *Mol. Pharmacol.*, **4**(2), 232–240.

15. Harn, N., Allan, C., Oliver, C., and Middaugh, C.R. (2007) Highly concentrated monoclonal antibody solutions: direct analysis of physical structure and thermal stability. *J. Pharm. Sci.*, **96**(3), 532–546.

16. Mach, H., Thomson, J.A., Middaugh, C.R., and Lewis, R.V. (1991) Examination of phenylalanine microenvironments in proteins by second-derivative absorption spectroscopy. *Arch. Biochem. Biophys.*, **287**(1), 33–40.

17. Mach, H. and Middaugh, C.R. (1994) Simultaneous monitoring of the environment of tryptophan, tyrosine, and phenylalanine residues in proteins by near-ultraviolet second-derivative spectroscopy. *Anal. Biochem.*, **222**(2), 323–331.

18. Cantor, C.R. and Schimmel, P.R. (1980) *Biophysical Chemistry Part II: Techniques for the Study of Biological Structure and Function*. W.H. Freeman and Company, New York.

19. Jiskoot, W., Hlady, V., Naleway, J.J., and Herron, J.N. (1995) Application of fluorescence spectroscopy for determining the structure and function of proteins. *Pharm. Biotechnol.*, **7**, 1–63.

20. Stroop, S.D., et al. (2011) Photosensitizers form in histidine buffer and mediate the photodegradation of a monoclonal antibody. *J. Pharm. Sci.*, **100**(12), 5142–5155.

21. He, F., et al. (2010) Detection of IgG aggregation by a high throughput method based on extrinsic fluorescence. *J. Pharm. Sci.*, **99**(6), 2598–2608.

22. Remmele, R.L., Jr., Nightlinger, N.S., Srinivasan, S. and Gombotz, W.R. (1998) Interleukin-1 receptor (IL-1R) liquid formulation development using differential scanning calorimetry. *Pharm. Res.*, **15**(2), 200–208.

23. Becktel, W.J. and Schellman, J.A. (1987) Protein stability curves. *Biopolymers*, **26**(11), 1859–1877.

24. Houde, D., et al. (2009) Characterization of IgG1 conformation and conformational dynamics by hydrogen/deuterium exchange mass spectrometry. *Anal. Chem.*, **81**(7), 2644–2651.

25. Houde, D., Berkowitz, S.A., and Engen, J.R. (2011) The utility of hydrogen/deuterium exchange mass spectrometry in biopharmaceutical comparability studies. *J. Pharm. Sci.*, **100**(6), 2071–2086.

26. Cole, J.L., Lary, J.W., P Moody, T., and Laue, T.M. (2008) Analytical ultracentrifugation: sedimentation velocity and sedimentation equilibrium. *Methods Cell. Biol*, **84**, 143–179.

27. Philo, J.S. (2009) A critical review of methods for size characterization of non-particulate protein aggregates. *Curr. Pharm. Biotechnol.*, **10**(4), 359–372.

28. Gabrielson, J.P. (2011) Measuring low levels of protein aggregation by sedimentation velocity. *Methods*, **54**(1), 83–91.

29. Rosenberg, A.S. (2006) Effects of protein aggregates: an immunologic perspective. *AAPS J.*, **8**, E501–E507.

30. Carpenter, J.F., et al. (2009) Overlooking subvisible particles in therapeutic protein products: gaps that may compromise product quality, *J. Pharm. Sci.*, **98**(4), 1201–1205.

31. Singh, S.K., et al. (2010) An industry perspective on the monitoring of subvisible particles as a quality attribute for protein therapeutics, *J. Pharm. Sci.*, **99**(8), 3302–3321.

32. Narhi, L.O., et al. (2009) A critical review of analytical methods for subvisible and visible particles. *Curr. Pharm. Biotechnol.*, **10**(4), 373–381.

33. Cao, S., Jiang, Y., and Narhi, L.O. (2010) A light-obscuration method specific for quantifying subvisible particles in protein therapeutics. *Pharmacopeia Forum.*, **36**(3), 824–834.

34. Sharma, D., et al. (2007) Flow microscopy for particulate analysis in parenteral and pharmaceutical fluids. *Eur. J. Parenter. Pharma. Sci.*, **12**(4), 97–101.

35. Kiese, S., Papppenberger, A., Friess, W., and Mahler, H.C. (2008) Shaken, not stirred—mechanical stress testing of an IgG1 antibody. *J. Pharm. Sci.*, **97**(10), 4347–4366.

36. Wiesner, J., Premsler, T., and Sickmann, A. (2008) Application of electron transfer dissociation (ETD) for the analysis of posttranslational modifications. *Proteomics*, **8**(21), 4466–4483.

37. Jefferis, R. (2009) Recombinant antibody therapeutics: the impact of glycosylation on mechanisms of action. *Trends Pharmacol. Sci.*, **30**(7), 356–362.

38. Salas-Salano, O., Gennaro, L., and Felten, C. (2008) Optimization approaches in the routine analysis of monoclonal antibodies by capillary electrophoresis. *LCGC Europe*, **21**(12), 615–622.

39. Bielik, A.M. and Zaia, J. (2010) Historical overview of glycoanalysis. *Methods Mol. Biol.*, **600**, 9–30.

40. Rohrer, J.S., Thayer, J., Weitzhandler, M., and Avdalovic, N. (1998) Analysis of the N-acetylneuraminic acid and N-glycolylneuraminic acid contents of glycoproteins by high-pH anion-exchange chromatography with pulsed amperometric detection. *Glycobiology*, **8**(1), 35–43.

41. Hara, S., et al. (1987) Fluorometric high-performance liquid chromatography of N-acetyl and N-glycolylneuraminic acids and its application to their micro determination in human and animal sera, glycoproteins, and glycolipids. *Anal. Biochem.*, **164**, 138–145.

42. Walsh, G. and Jefferis, R. (2006) Post-translational modifications in the context of therapeutic proteins. *Nat. Biotechnol.*, **24**(10), 1241–1252.

43. Ghaderi, D., et al. (2010) Implications of the presence of N-glycolylneuraminic acid in recombinant therapeutic glycoproteins. *Nat. Biotechnol.*, **28**(8), 863–867.

44. European Medicines Agency. (2004) *Guideline on the Requirements to the Chemical and Pharmaceutical Quality Documentation Concerning Investigational Medicinal Products in Clinical Trials.*

45. United States Food and Drug Administration. (2003) *Guidance for Industry: INDs for Phase 2 and Phase 3 Studies, Chemistry, Manufacturing, and Controls Information.*

46. Food and Drug Administration (FDA), Center for Drug Evaluation and Research (CDER) and Center for Biologics Evaluation and Research (CBER). (1999) *Guidance for Industry, Q6B Specifications: Test Procedures and Acceptance Criteria for Biotechnological/Biological Products.*

47. Food and Drug Administration (FDA), Center for Drug Evaluation and Research (CDER) and Center for Biologics Evaluation and Research (CBER). (1996) *Guidance for Industry for the Submission of Chemistry, Manufacturing, and Controls Information for a*

Therapeutic Recombinant DNA-Derived Product or a Monoclonal Antibody Product for In Vivo Use.

48. Rustgi, V.K. (2009) Albinterferon alfa-2b, a novel fusion protein of human albumin and human interferon alfa-2b, for chronic hepatitis C. *Curr. Med. Res. Opin.*, **25**(4), 991–1002.

49. Riggs M.M., et al. (2012) Population pharmacokinetics and exposure-response of albinterferon alfa-2b. *J. Clin. Pharmacol.*, **52**(4), 475–486.

50. Available at http://www.accessdata.fda.gov/drugsatfda_docs/label/2011/103132s5161lbl.pdf. Accessed July 9, 2012.

51. International Conference on Harmonization of Technical Requirements for Registration of Pharmaceuticals for Human Use. (2004) *ICH Harmonized Tripartite Guideline Q5E Comparability of Biotechnological/Biological Products Subject to Changes in their Manufacturing Process (CPMP/ICH/5721/03).*

52. Putnam, W.S., et al. (2010) Pharmacokinetic, pharmacodynamics, and immunogenicity comparability assessment strategies for monoclonal antibodies. *Trends Biotechnol.*, **28**(10), 509–516.

53. International Conference on Harmonization of Technical Requirements for Registration of Pharmaceuticals for Human Use. (1998) *ICH Harmonized Tripartite Guideline Q1B Photostability Testing of New Active Substances and Medicinal Products (CPMP/ICH/279/95).*

54. Food and Drug Administration (FDA). (2001) *Guidance for Industry: M4: Organization of the CTD.*

55. Food and Drug Administration. (2001) *Guidance for Industry: M4Q: The CTD—Quality.*

5

NUCLEATION, AGGREGATION, AND CONFORMATIONAL DISTORTION

Christopher J. Roberts

Department of Chemical Engineering, University of Delaware, Newark, DE, USA

Biophysical Methods for Biotherapeutics: Discovery and Development Applications, First Edition. Edited by Tapan K. Das.
© 2014 John Wiley & Sons, Inc. Published 2014 by John Wiley & Sons, Inc.

5.1 INTRODUCTION

Unwanted aggregation of biotherapeutics is often mediated via some degree of conformational change(s) in protein structure, either prior to or as part of forming inter-protein contacts. In many cases, these changes must be inferred or determined indirectly rather than monitored directly. This is due to instrumental limitations and the inherently transient nature and low populations of the relevant species in pharmaceutical products. This chapter provides an overview of the multistage nature of nonnative aggregation, arguments regarding the relative importance of thermodynamics versus dynamics for different stages, and gaps in our current knowledge regarding which step or steps are most important in controlling aggregation of therapeutic proteins. The subsequent discussion and examples highlight the importance of conformational distortions in facilitating aggregation via a number of pathways, as well as aspects in which conformational changes may be of little importance. Experimental approaches to measure the extent of conformational changes are briefly reviewed, as is the importance of considering competing avenues for aggregate nucleation and/or growth and the issues and considerations for aggregation mediated by adsorption to bulk interfaces.

Aggregation of therapeutic proteins can be broadly categorized based on whether the process is easily reversible upon dilution or a small change in solvent conditions or temperature (or pressure). Examples of such reversible processes include protein crystallization, salting out, and self-assembly of natively multimeric proteins; a common feature in such systems is that the monomers that compose the resulting aggregates are fully or predominantly folded and involve inter-protein contacts between residues on the surface of the native or fully folded monomer state(s), with little or no distortion of the monomer secondary or tertiary structure [1–3].

Net-irreversible aggregation often involves some degree of conformational change in the constituent proteins within aggregates, relative to folded monomer state(s), with or without covalent changes in the proteins [4, 5]. Those conformational changes are at least putatively needed in order to allow two or more proteins to form inter-protein contacts that are of similar or greater stability than intra-protein contacts in the folded state, thereby requiring denaturing solvent conditions in order to dissociate them and thus making them effectively irreversible under the conditions that such nonnative aggregates form [5, 6].

Chapters elsewhere in this book provide principles and applications of an array of biophysical techniques for characterizing protein structure and protein–protein interactions. The present chapter focuses on underlying principles of nonnative aggregation (for brevity, hereafter typically referred to as aggregation in this chapter) and the current understanding of different aggregation mechanisms, to then motivate selected examples of the use of different biophysical assays to aid in the control, characterization, and monitoring of aggregation. As aggregation in this context is net irreversible, the kinetics or rates of formation for aggregates of different sizes are necessarily important, as are the paths by which different types of aggregates form. These considerations are also briefly reviewed, as needed to help illustrate the application of, and challenges for, current and future methods for experimental study of nonnative aggregation within the context of product development of biotherapeutics.

5.2 NONNATIVE AGGREGATION INVOLVES MULTIPLE COMPETING PROCESSES

Figure 5.1 provides an overview of a range of potentially important steps or stages in the process of creation or nucleation of new aggregates from proteins that would otherwise exist or easily revert to monomer states, as well as propagation or growth of aggregates once they form. In Figure 5.1, steps that have both forward and reverse arrows are ones that are expected to be reasonably reversible, although they may occur far from equilibrium. The steps with only a single, forward arrow are those that represent molecular "nucleation" or "initiation" steps for forming nuclei—defined as the smallest effectively irreversible species. In this context, aggregates even as small as dimers can constitute a nucleus, in that they are stable and easily separable, for example, in chromatography or other fractionation methods [7–10].

Within the context of Figure 5.1, a pre-nucleus is a reversible aggregate (e.g., dimer or small oligomer) that is not stable and would likely dissociate upon dilution, changes in temperature or pressure, or as part of common fractionation methods. It may share the same number of monomer units (i.e., stoichiometry) with the nuclei, but has not yet "committed" to the irreversible aggregate pathway(s). Such species are necessarily problematic to isolate or characterize directly, as they are transient and unstable, and therefore also tend to exist at low levels compared to other species. As such, there is little known conclusively about the nature of the transitions that ultimately commit such pre-nuclei to become net irreversible. Therefore the transitions from pre-nuclei to nuclei that are shown in Figure 5.1 should be considered schematic at best.

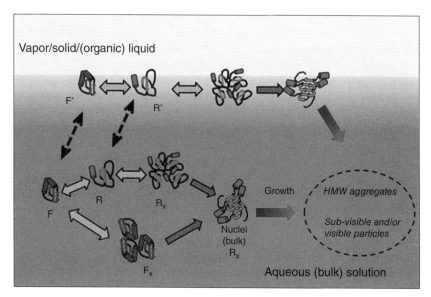

Figure 5.1. Schematic overview of the multistage nature of aggregate nucleation, including multiple steps that potentially involve conformational changes. (See color insert)

As discussed in Section 5.2.2, the term nucleation can also be used to describe the process of creating macroscopically phase-separated particles, prior to those particles growing in a manner akin to droplet or crystallite growth in protein or small-molecule phase separation. This latter process is better described as a continuum, macroscopic transport-limited phenomenon and occurs at length scales well beyond those captured in Figure 5.1.

That notwithstanding, Figure 5.1 is intended to highlight that molecular-scale growth of aggregates—via addition of monomer and/or via coalescence of aggregates with one another—can be important mechanisms by which the so-called high molecular weight (HMW) soluble aggregates form. Figure 5.1 also explicitly shows that aggregation need not occur solely in bulk solution. Adsorption of proteins to bulk interfaces can and does occur for vapor–liquid, liquid–liquid, and solid–liquid interfaces (see also Section 5.5).

The more speculative steps in Figure 5.1 are those in which proteins unfold and self-assemble at the surface, as well as how they desorb and possibly refold or further aggregate once back in bulk solution. Experimentally, little is currently known conclusively about those putative stages of the aggregation process; they are included in Figure 5.1 for the sake of completeness and to highlight the potential complexities that adsorption phenomena add to the overall process.

Once one considers bulk interfaces, all of the steps that must at least putatively occur in bulk solution now have a counterpart within the interface between bulk solutions. For liquid products, this minimally includes the interface between the liquid and the container (e.g., glass, plastic, and metal), as well as any headspace and any silicone oil that may be present. For freeze-dried products and frozen products or bulk drug substance, this often includes liquid–ice interfaces for at least some time during production and/or storage. In addition, each surface species can potentially exchange reversibly with the corresponding bulk species, necessarily involving convective and/or diffusive mass transport between the bulk and surface regimes. Finally, the depiction in Figure 5.1 is for a system with static interfaces. The implications of that simplifying assumption are briefly discussed later in this chapter (cf. Section 5.5).

5.2.1 Aggregation Rates, Pre-Equilibration, and Rate-Determining Step(s)

For any process in which there are multiple serial and/or parallel steps, such as that depicted in Figure 5.1, the observed rate or rates that one measures will be determined by at least two key factors: (1) what species is/are being measured; (2) where is (are) the rate-determining step(s) in the process that creates the species that are being measured. As reviewed elsewhere [11], most commercially available measurement techniques fall into one of a few categories based on what aspect of aggregation they measure: (i) changes in protein structure at the level of secondary structure or local solvent exposure; (ii) changes in the amount of protein that exists as a monomer; (iii) changes in time-average magnitude of scattered light or X-rays, per unit volume of solution, and possibly as a function of scattering angle; (iv) changes in the optical properties and spatial distribution for bulk solution and/or microscopic sub-volumes within the bulk; (v) changes in the average or distribution of molecular mobility based on time correlations in spectroscopic or light-scattering signals.

Techniques that fall into categories (i) and (ii) focus essentially on the amount and/or properties of the parent monomers as they shift from being free in solution to being incorporated into (nonnative) aggregates. Examples include [9–14] binding of certain fluorescent dyes such as thioflavin T (ThT) and Sypro Orange (SyprO), near- and far-UV circular dichroism (CD), intrinsic fluorescence (FL) of Trp residues that are sensitive to changes in polarity of their local environment, size-exclusion chromatography, and other fractionation techniques. Thus, the rate-determining steps (RDS) that are potentially relevant for such assays are typically those involved in the conversion from folded or native monomer (F in Figure 5.1) to nuclei (A_x, the irreversible counterpart to R_x in Figure 5.1), and the subsequent recruitment of additional monomers to form larger nonnative aggregates (HMW aggregates in Figure 5.1) by monomer addition.

The step or steps that occur ahead of the RDS in any given process are effectively pre-equilibrated. That is, although the overall process is under kinetic control, any "upstream" steps that occur much faster than the RDS will equilibrate on the time scale of the RDS. For example, in the context of Figure 5.1, if formation of a nucleus from a pre-nucleus is rate determining, then steps such as folding/unfolding, self-association, and/or transport to/from the bulk or interface to create the pre-nuclei (R_x and F_x) will be effectively governed by the thermodynamics of those steps, rather than the dynamics. A similar argument can be made, for example, if self-association and nucleation occur at the air–water interface, but they occur more rapidly than protein can diffuse to that interface from the bulk. In this case, the slow step in the process becomes transport to that interface.

Table 5.1 lists each of the main steps or stages depicted in Figure 5.1, along with typical experimental variables that are expected to strongly influence overall aggregation rates if a given step is rate determining. For example, adsorption to liquid–vapor, liquid–liquid, or liquid–solid interfaces may be at equilibrium or under mass-transfer control. If the former, then mixing the sample (e.g., gentle stirring without entraining vapor) is not expected to affect the concentration of adsorbed species. If the latter, then mixing is expected to enhance adsorption and/or desorption rates via convective mass transfer.

Similarly, for most liquids over the practical working temperature range, changes in temperature *per se* do not strongly influence convective or diffusive mass transfer. This can be seen by realizing that, in this context, the effects of temperature (T) on transfer rates are primarily due to changes in the viscosity (η) of the system; for water, d ln η/d($1/T$) $\sim$ 4 kcal/mol, and this is one to two orders of magnitude less than the effective activation energies or enthalpy changes associated with other steps in Figure 5.1 (e.g., unfolding/refolding steps), with smaller activation energies and equilibrium enthalpy changes perhaps expected for reversible self-association steps (e.g., F to F_2). In addition, when the observed temperature dependence of aggregation rates (e.g., monomer loss) via nonnative aggregation displays much greater temperature sensitivity (effective activation energies on the order of hundreds of kilocalories per mole), it becomes very

TABLE 5.1. Summary of potentially important steps for initiating nonnative aggregation and experimental variables that are expected to cause large (e.g., order of magnitude) changes in associated rates and/or equilibria for those steps over reasonable experimental operating ranges

	Temperature	Pressure	Bulk protein concentration	Bulk mixing[a]	Solvent composition	Bulk surfaces
Bulk solution						
Unfolding	X	X	–	–	X	–
Diffusion	–	–	X	–	X	–
Pre-nuclei formation	X	X	X	–	X	–
Rearrangement/ nucleation[b]	X	X	–	–	X	–
Interfaces						
Transport to/from bulk solution	–	–	X	X	X	X
Adsorption/ desorption[c]	X	–	X	–	X	X
Surface diffusion	–	–	[d]	–	X	X
Unfolding	X	X	–	–	X	X
Pre-nuclei formation	X	X	X	–	X	X
Rearrangement/ nucleation[b]	X	X	–	–	X	X

[a]Gentle mixing without causing increases in the amount of air–water interfacial area.
[b]Referring to structural rearrangement of otherwise reversible oligomers.
[c]Referring to changes in surface capacity, for example, as reflected in binding isotherms.
[d]This depends on whether the surface concentration is saturated relative to its maximum capacity.

difficult to argue that liquid-state transport steps (diffusion, convection) are rate limiting [15–18]. However, if one is instead considering supercooled liquids and glasses (e.g., in the context of freeze- and/or spray-drying), then changes in η can be dramatic over just a few degrees as the glass transition is approached—indeed, bulk vitrification to arrest mass transfer and molecular diffusion is one of the key aspects to slowing aggregation for such systems [19–21].

Overall, including experimental plans that test each of the potentially "accelerating" variables in Table 5.1 can help to identify the RDS for a given system or at least help to eliminate potential candidates from a scheme such as Figure 5.1. Doing so can help to elucidate whether one should be primarily concerned with the dynamics or the thermodynamics of a given step, and therefore whether one should expect experimental measures of the dynamics versus the thermodynamics of that step to be useful in assessing product stability [11, 19–22].

Finally, Table 5.1 and the discussion above suggest it is possible that the identity of the RDS may depend on the protein of interest and the sample conditions (solvent, temperature, pressure, protein concentration, etc.). For a more detailed discussion of this topic, the interested reader is referred to specific examples in the literature that address this question experimentally and/or in more general detail [4, 15, 17, 23]. For the purposes of this chapter, it is sufficient to simply highlight that protein aggregation is expected to proceed via the "path of least resistance"—that is, the RDS is expected to change as a function of temperature, protein concentration, mixing rates, extent of exposure to different surfaces, and so on, because changing these quantities can change the quantities of reactants, catalysts, and/or free energy barriers for a given step. At present, it is not possible to practically predict where the RDS lie in a given aggregation process. Until that limitation is overcome, they must be determined on a case-by-case basis [22].

5.2.2 Nucleation versus Growth in the Context of Stability of Biotherapeutics

Once "nuclei" have formed, they have the potential to grow to (much) larger sizes either via the addition of monomers or via coalescence of the aggregates that have formed up to that point in the process. The former step will deplete monomers directly, while the latter does not. The former step also does not alter the number of aggregates per unit volume, but does increase their average molecular weight (M_w). The latter step necessarily decreases the number of aggregates per unit volume while increasing M_w (e.g., taking two dimers to form a tetramer decreases the number of aggregates by a factor of 2, but increases the M_w) [4]. Of course, if "nuclei" do not grow significantly then the term "nucleus" is somewhat misleading if one is accustomed to nucleation necessarily requiring subsequent growth (such as in bulk phase separation), and in that case "nucleus" should be used simply to denote that this is the smallest stable species. Notably, if one is interested in HMW soluble aggregates and larger particles such as what are currently delineated as subvisible and visible particles, then some form of growth from the (molecular-scale) nuclei must occur in order for aggregates to reach such larger length scales.

Experimental techniques within categories (iii) to (v), as listed above, include ones such as static (multi-angle) laser scattering [24], dynamic light scattering [25] and some other forms of correlation spectroscopy [26], light obscuration [27], Coulter counters [28], subvisible and visible particle counting and tracking [29], and direct visual observation [27, 29, 30]. These techniques are primarily ineffective in detecting nucleation in the context of Figure 5.1. This is because the length scales they probe are too large compared to the size of the species in Figure 5.1 (for most proteins of interest) and/or because they are ensemble-averaged measurements, and therefore are not sensitive to species that are only present at small weight or volume fractions compared to the remainder of the protein in the system. Section 5.3.2.1 includes further discussion on this point.

Alternative techniques such as neutron or X-ray scattering can probe the length scales of interest for nucleation in the context of Figure 5.1, but often have practical limitations of lack of commercial instrumentation and limited time at government facilities, as well as necessarily being ensemble-averaged techniques. Section 5.5 includes a brief discussion of the pros and cons of different techniques within the context of interface-mediated aggregate formation, including those that specifically focus on the interfacial region between bulk phases.

The above concerns notwithstanding, the techniques that focus on longer length scales can be well suited for monitoring aggregate growth. In the case of growth via monomer addition, SLS can be particularly powerful, as it allows for real-time *in situ* or ex situ monitoring and quantification of the weight-average molecular weight (M_w) of the aggregate size distribution [8, 31]. This can be used to track aggregation starting essentially when the population of oligomers (dimers, trimers, etc.) reaches concentrations within at least a few percent of the starting monomer concentration, and up to sizes of very large HMW aggregates ($\sim 10^2$ nm); DLS can be used in a similar fashion, although one must often assume a relationship between the average mobility and the size/shape of the growing aggregates [8, 31, 32].

Many of the other techniques do not detect below length scales of relatively large aggregates ($\sim 10^2$ nm or larger). As such, when such techniques detect a "nucleus," this is more correctly termed the smallest detectable aggregate within the size range of the instrument [28]. It therefore becomes clear that there is nothing necessarily fundamental, *per se*, about the "nuclei" in such a context. Rather, the use of multiple instruments that overlap the different size ranges and detection limits is helpful in quantifying and tracking the growth of aggregates across a wide range of sizes that is typically untenable with a single instrument [27, 29].

The fundamental initiation process for such subvisible and visible particles may or may not be tied to nucleation or small aggregates formed via mechanisms such as in Figure 5.1 [8, 33–35]. In at least some cases, it is clear that aggregates that nucleate as in Figure 5.1 are then able to grow until they ultimately coalesce into macroscopic particles [8, 36, 37]. Thus for at least some systems it is desirable to suppress the formation of small soluble aggregates (nuclei in Figure 5.1), as well as to suppress their subsequent growth.

5.3 IMPORTANCE OF CONFORMATIONAL CHANGES IN FORMING/NUCLEATING AGGREGATES

From the previous sections, it is apparent that conformational changes can occur in a number of stages in the aggregation process. It remains an open question as to which conformational changes, if any, are absolutely required in order to facilitate and/or stabilize the resulting aggregates [22]. Within the context of Figure 5.1, there are at least three qualitatively distinct stages where (partial) unfolding is typically argued to be important. The first is unfolding of monomer protein in bulk solution (i.e., species R in Figure 5.1). Such unfolding may expose sufficient hydrophobic surface area to help drive self-association, and ultimately stable aggregate formation. A second candidate is unfolding of protein adsorbed to a bulk interface (i.e., species R' in Figure 5.1). This is analogous to the role of R in the bulk, but may be more important if the interface is effective at lowering the free energy of unfolding—for example, if the interface is reasonably hydrophobic [38, 39].

A third stage where conformational changes are potentially important is that for reversible dimers and oligomers of R and/or R′ undergoing additional conformational changes as a group or cluster (i.e., the pre-nucleus → nucleus stage in Figure 5.1). As discussed in Section 5.1, these additional changes putatively allow strong, effectively irreversible, inter-protein contacts to form such as extended intermolecular β-sheets, stabilized by both hydrophobic and hydrogen bonding attractions [22, 40]. In cases where it is possible to fully populate the (partially) unfolded monomer prior to aggregate formation, clear differences between the secondary and tertiary structures of R and the resulting aggregates have been found (see also Section 5.4) [41].

Most models for nonnative aggregation of folded or foldable proteins are based on mechanisms in which unfolding occurs prior to self-association or the final step in nucleation (see, e.g., Reference 4 and references therein). However, it is also possible that under highly native-favoring conditions the probability of two or more R or R′ molecules self-associating is impractically low on experimental time scales. In such a situation it is not unreasonable that an alternative pathway is, for folded monomers (F) or adsorbed folded monomers (F′), to first self-associate or cluster to form an analogue to the pre-nucleus R_x (or $R_x′$, by analogy to R and R′ in Figure 5.1.). The nucleation step would still then entail a significant conformational change to form nuclei, and the free energy barriers for the $F_x → A_x$ or for the $F′_x → A_x$ step would presumably be much larger than that for pre-nuclei composed of R or R′ species. It would therefore become the net fastest route for nucleation when the concentration of R and/or R′ species became sufficiently small. Quantitatively, it is difficult to predict the precise regime of conditions (e.g., temperature and protein concentration) when this will occur (C.J. Roberts, unpublished). That notwithstanding, this thought experiment illustrates again that some degree of conformational change or distortion in the underlying monomer secondary/tertiary structure would be required in order to initiate or nucleate new, net-irreversible aggregates.

Experimentally, the primary evidence that indicates that some degree of unfolding is required to form stable aggregates is mostly phenomenological or indirect and can be summarized in terms of at least two main observations [5, 22, 42, 43]: (a) increasing the concentration of partially or fully unfolded species via changes in temperature, pressure, and/or pH has a very strong correlation with increased aggregation rates; (b) a posteriori characterization of soluble, and sometimes insoluble, aggregates often shows some detectable conformational change.

Examples of the former and latter have been extensively documented [5, 14, 22, 36, 41–49]. Table 5.2 summarizes some examples, spanning from disease-related proteins, to enzymes, to therapeutic proteins. It also illustrates that not all experimental techniques are effective at detecting conformational distortions (e.g., relative to native monomers) of the secondary and/or tertiary structures of the constituent protein chains within aggregates. At the very least, the physical state of the aggregates is important; if the aggregates are "soluble" or molecularly dispersed then they are amenable to techniques such as CD, FL, and UV-VIS spectroscopy; if they are "insoluble," sediment, or large enough or smaller but concentrated enough to scatter incident beams for spectroscopy, then they are more amenable to techniques such as FTIR, that are less prone to

TABLE 5.2. Illustrative examples of therapeutic proteins and sample conditions where aggregation rates correlate with unfolding and/or comparison of aggregate or adsorbed protein and native (bulk) structures indicates detectable conformational changes upon aggregation and/or surface adsorption

Protein	Technique(s) for monitoring conformational changes[a]	Reference
IgG1 and IgG2 antibodies	Dye binding	[9]
IgG1 and IgG2 antibodies	Dye binding; DSC	[12]
IgG1 antibodies	UV abs; FTIR; CD; DSC	[13]
Factor VIII	Dye binding; FTIR; DSC	[14]
bG-CSF	CD; intrinsic FL	[15, 46]
IgG1 and IgG4 antibodies	UV abs; intrinsic FL; dye binding	[16]
IgG1 fusion proteins; cytokines	FTIR (solid state)	[20]
Interferon-γ	IR; CD; DSC	[47]
IgG1	CD; dye binding; DSC	[36]
α-Chymotrypsinogen A	CD; intrinsic FL; dye binding; DSC; HDX-MS	[41, 46, 50, 51]
α-Lactalbumin	Dye binding; CD; FTIR	[48]
Light-chain antibodies	CD; FTIR; NMR; SAXS; intrinsic FL	[44]
Bevacizumab	HDX-MS	[45]
γD-Crystallin	Dye binding; DSC	[49]
Interferon-γ	HDX-MS	[52]
Serum albumin	FTIR-HDX	[39]
β-Lactoglobulin; α-lactalbumin; serum albumin	HDX-MS	[38]

[a]See Table 5.3 for explanations of abbreviations.

artifacts from scattering. Section 5.4 addresses the question of whether conformational distortions are an important part of aggregate growth. In what follows in this section, conformational changes (in bulk solution) at the level of individual monomers, and at the level of pre-nucleus → nucleus, are first addressed. Analogous conformational changes for proteins adsorbed to interfaces are discussed in Section 5.5.

5.3.1 Measuring Global and Local Unfolding/Conformational Changes

Most available methods to monitor conformational changes in proteins can be categorized according to whether they are sensitive to small, local conformational changes or sensitive to relatively large changes involving many residues. Table 5.3 summarizes many of the most commonly used experimental techniques for monitoring unfolding, including whether a given technique is most sensitive to local versus more global structural changes and the type(s) of changes it is most sensitive to.

TABLE 5.3. Summary of common commercially available experimental techniques (with abbreviations in Table 5.2) for monitoring changes in global and local conformations of proteins in solution and/or solid states

Technique	Abbreviation	Level of structure probed
Circular dichroism	CD	2° (far-UV), 3° (near-UV)
Intrinsic fluorescence	Intrinsic FL	Solvent exposure of fluorophore side chains
Binding by ThT[a], Sypro Orange, ANS, bis-ANS, or related dyes	Dye binding	Solvent/dye exposure of certain structural motifs
Differential scanning calorimetry	DSC	Any conformational changes that involve heat consumption or release
Infrared spectroscopy (IR) and/or Fourier transform IR	IR; FTIR	Networks of hydrogen bonding, and corresponding 2°, 3°, or higher-order structure
Ultraviolet absorbance	UV abs	Solvent exposure of Trp, Tyr, and Phe side chains
Hydrogen–deuterium exchange monitored by mass spectrometry	HDX-MS	Percentage of solvent exposure of amino acids across entire protein or local regions[a]
Nuclear magnetic resonance	NMR	Changes in local and global unfolding
Small-angle X-ray scattering[b]	SAXS	Changes in radius of gyration; local changes in folding[c]
Infrared spectroscopy with H–D exchange	FTIR-HDX (or IR-HDX)	Changes in solvent exposure of bulk or adsorbed proteins

[a]If using proteolysis to isolate peptides or specific amino acids.
[b]Analogous information available from SANS; both SANS and high-resolution SAXS require access to a nuclear facility, while low-resolution SAXS is available commercially.
[c]If aided by molecular modeling.

Inspection of Table 5.3 shows that there are a wide range of techniques available, with some overlap in terms of the type(s) of information they give regarding structure. An increasing body of evidence indicates that aggregates can be stabilized by contacts between relatively small stretches of amino acids [42,50,53]. As such, local measures of conformational change may appear to be more appealing than global measures. However, the location or identity of the "hot spots" for aggregation is not typically known for most therapeutic proteins. While computational approaches exist to attempt to predict such hot spots, the large majority of experimental evidence that seeks to validate those predictions is at best indirect, as the actual contacts in the resulting aggregates have been characterized in only a few cases thus far [40,50,52,54].

Furthermore, almost all of the techniques in Table 5.3 that are sensitive to local unfolding are biased toward detecting only certain types of residues or regions of a protein—for example, intrinsic FL is sensitive to only those changes that alter the environment around aromatic rings in a protein (primarily Trp and Tyr, rather than Phe), while changes in ligand affinity are primarily a measure of changes in the active site of a protein. As such, it is often not possible to know whether a given experimental technique will be sensitive to the conformational changes or distortions that are actually involved in aggregate formation. At present, two of the most useful techniques for characterizing local conformational changes, and for doing so across the full protein sequence, are H–D exchange (HDX) coupled with NMR and/or mass spectrometry (MS) with or without partial proteolysis [38, 44, 45, 50, 52]. NMR has the limitation that it often requires high protein concentrations, but this is perhaps less of a concern with the current industry trend of seeking high-concentration dosage forms for self-administered products. HDX-MS requires relatively little protein material, but the technique as it applies to conformational changes and hot spots in aggregation-prone proteins has historically been used primarily by academic laboratories, although this is likely changing as commercial packages for more routine measurement and data interpretation are now becoming more widely available.

Some of the techniques in Table 5.3 can be used in either a (pseudo-)equilibrium mode or a dynamic one. Figure 5.2 and the discussion below illustrate this for the case of intrinsic Trp fluorescence. More generally, it is an open question whether equilibrium or dynamic measurements are more useful in assessing changes relevant to aggregation. On one hand, proteins are dynamic molecules, and the process of aggregation necessarily involves kinetic step(s). On the other hand, the overall time scales of interest for aggregate formation in most pharmaceutical settings are very long compared to the time scales probed by many of the techniques in Table 5.3. It remains an open question as to what stages in Figure 5.1 are pre-equilibrated or not, as the discussion above regarding RDS highlighted. As such, it also remains a matter of some debate as to how best to select and utilize information about equilibrium versus dynamic structural changes in order to best select conditions for processing, storing, and/or shipping protein products. All of the techniques in Table 5.3 suffer from the limitation discussed below, in that they are ensemble-averaged methods.

Industrially, the most common approaches to test the effects of different sample conditions on the conformational state(s) of proteins is to either monitor isothermally—usually at temperatures corresponding to refrigerated, ambient, and slightly accelerated conditions—or monitor during continuous temperature ramping. The advantage of the former is that it sometimes can be done without convolution between conformational changes of monomers before aggregation with those changes occurring as aggregates are formed or as they grow. The disadvantage is that the temperatures may be too low for discernable differences to occur between many of the sample conditions of interest. Temperature ramping is more likely to highlight such differences, but is also more likely to report a convolution with aggregation [13].

Figure 5.2 provides two contrasting examples where both equilibrium and dynamics of fluorophore exposure were determined experimentally for aggregation-prone proteins. The first example is α-chymotrypsinogen A (aCgn). Figure 5.2a shows the equilibrium FL curves as a function of urea for a series of relatively low temperatures. Figure 5.2b shows the corresponding results from stopped-flow FL measurements, in which the concentration of urea was rapidly "jumped" from low to high levels (or vice versa), so as to monitor the dynamics of unfolding or refolding of aCgn, respectively. The

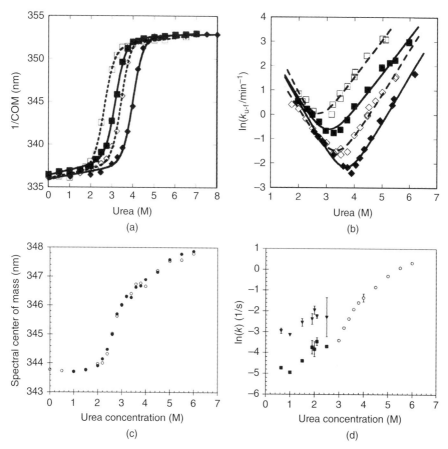

Figure 5.2. Equilibrium unfolding monitored by spectral center of mass for intrinsic fluorescence (panels a and c) and dynamics of unfolding/refolding monitored by the intensity of intrinsic fluorescence (panels b and d) for aCgn at pH 3.5, 10 mM sodium citrate (a and b) and the bhx domain of P22 tailspike at pH 7, 100 mM sodium phosphate (c and d). Panels (a) and (b) are from Reference 41, while panels (c) and (d) are from Reference 55). The temperature values for (a) and (b) are 26°C (closed diamonds), 30°C (open diamonds), 34°C (closed squares), 38°C (filled squares). Spectral center of mass (COM) is defined as COM $= \left(\sum_{\lambda} I_{\lambda}/\lambda\right)\Big/\sum_{\lambda} I_{\lambda}$ and is a measure of the emission peak position that is weighted proportional to higher energy (lower wavelength) emission values.

decay or rise in FL signal showed simple first-order kinetics, as expected for this two-state protein [41]. The *y*-axis in Figure 5.2b is the effective rate coefficient (k or $k_{u\text{-}f}$) for the kinetics of the relatively global conformational changes in going from folded to unfolded states, or vice versa. In the case of aCgn, this conformational change corresponds to a loss in tertiary structure, with little or no concomitant loss in secondary structure [41].

The behavior in Figure 5.2b follows classic expectations for a so-called Chevron plot. The low-urea branch of k versus [urea] is dominated by the rate coefficient for folding (k_f), while the high-urea branch is dominated by that for unfolding (k_u). As for any reversible monomolecular process, the overall rate at which the equilibrium between folded and unfolded states is established is set by the faster of k_u and k_f. This follows because $k = k_u + k_f$. As a result, if the time scales for aggregation are much longer than $1/k$, then it is the equilibrium population of folded/unfolded states that is most relevant to aggregation. Further experiments for aCgn showed that to be the case across a range of elevated temperatures [41,56]. In such cases, the dynamics of unfolding are not expected *per se* to be relevant to the aggregation process, although it may be possible that they serve as a surrogate measure of other steps in the process, and so might still correlate loosely with aggregation behavior [11].

Figures 5.2c and 5.2d show a contrasting example, using the β-helix (bhx) domain of P22 tailspike, with the same *y*-axis labels as in panels (a) and (b), but now for a single temperature (15°C). In this case, the equilibrium unfolding profile shows a hysteresis loop (Figure 5.2c), as it is kinetically difficult to fully refold the bhx domain once it is allowed to be fully unfolded for extended periods of time (approximately hours to days). In terms of stopped-flow unfolding/refolding (Figure 5.2d), a simple Chevron-like behavior is evidenced for only the unfolding branch. Separate experiments using double-jump stopped-flow kinetics, as well as proline isomerase, indicated that the kinetic bottleneck(s) for refolding involved isomerization of one or more Pro in bhx. The slow isomerization of at least one of those Pro residues resulted in partially refolded intermediates being kinetically trapped as one jumped from high to low urea. With the urea removed, there was sufficient driving force for aggregation of the partially folded intermediates, and aggregation occurred more readily [55]. This example illustrates in a different way that the kinetics of folding play a key role in determining whether equilibrium versus dynamics of (un)folding is likely to be most important in controlling aggregate formation. For bhx, it clearly would be helpful to identify which Pro residues were most important in controlling the conformational changes promoting aggregation—unfortunately that was a nontrivial task, since bhx has 16 Pro residues and it was not clear that one of those alone was responsible for the behavior.

5.3.2 Measuring Nonnative Structures in Aggregates and Detecting Nuclei

Each of the techniques in Table 5.3 are, in principle, applicable to measuring the conformational state of the protein chains within aggregates. This is potentially useful if one can use that information to better inform as to which changes in monomer conformational state(s) might be important in populating aggregation-prone intermediates. The major issues with determining nonnative structure(s) in aggregate often relate to whether the size and/or optical properties of the aggregates preclude accurate measurement or interpretation of the results from techniques that were designed to measure primarily un-aggregated systems. Table 5.3 also indicates some of these issues for those techniques most prone to such artifacts. Assuming one is using only techniques that are appropriate for use with aggregates of the size and optical properties of interest, as well as solution versus solid-state samples in general, the discussion below holds equally well for the question of detecting aggregate nuclei as it does for the question of detecting poorly populated or rare monomer conformational states.

5.3.2.1 *Ensemble-Averaged Detection Methods.*

All of the techniques in Table 5.3 measure an ensemble average of the conformations that proteins adopt in solution or the solid state. That is, the signal that is detected is an average over all the different conformers in the volume of the sample that is probed by the instrument in question. Thus, the signal is dominated by those conformers or species that are most highly populated in the sample for the duration of the measurement, unless certain species provide a much stronger signal than others for the technique in question. Pre-nuclei are, by definition, expected to be relatively high in free energy compared to the monomers that self-associate to form them. In addition, once a nucleus forms it may grow rapidly to much larger species. Therefore, it is not surprising that pre-nuclei will typically be very poorly populated compared to the un-aggregated and irreversibly aggregated species in a sample. As such, it is highly unlikely that ensemble-averaged techniques will be useful in a general sense for detecting or characterizing transient, relatively unstable key intermediates involved in nucleating new aggregates, unless one is fortunate enough that the nuclei provide a much stronger (i.e., orders of magnitude) signal than the background protein population. This reasoning holds similarly for detecting rare conformers in the monomer state.

5.3.2.2 *Alternative Approaches.*

A potentially lucrative approach to overcoming the limitations of ensemble-averaged methods is to employ techniques in which a ligand is introduced to the system that is relatively specific to the monomer and/or self-associated structural states that are "on pathway" to aggregates. Doing so unambiguously can be difficult, in that it is not always readily apparent which structures (monomer or otherwise) are "on pathway." That limitation not withstanding, there are some recent examples that at least indicate some promise for this type of strategy.

One example is the case in which amyloid-specific antibodies have been developed to bind specifically to conformational motifs that are common to amyloid structures that derive from different polypeptide sequences [57,58]. Such antibodies might in principle also be tagged with labels that make them detectable at very low concentrations (e.g., via radioactive labeling). Therefore rare conformers and/or transient states could, in principle, be detectable against the otherwise dominant background signal of other monomer and aggregate states. To the best of the author's knowledge, this approach has not yet been successfully used with pharmaceutical proteins.

An alternative example, and one with an arguably much less specific "ligand," is that of Sypro Orange (SyprO), one of a family of dye molecules that have been screened in recent years to produce dyes (ligands) that tend to bind and concomitantly alter their fluorescence in the presence of nonnative protein aggregates. While ThT has been known for decades to serve in this capacity for amyloid aggregates, SyprO was recently shown to serve as a means to track aggregate formation—both qualitatively and quantitatively—for therapeutic proteins such as monoclonal antibodies [9,12]. In current practice, the intensity of the dye signal may not be sufficient to accurately detect low-population species (or order 1 in 1000 or lower), but this could potentially be overcome by using laser-induced fluorescence or alternative means to increase the intensity of the signal.

Other dyes have also been used in a qualitative sense in this regard for therapeutic proteins [59], but in many cases the dyes are less discriminating and bind generally to any exposed hydrophobic region on a protein—independent of whether that region is ultimately involved in aggregate formation. Those dyes may still be of potential interest, as changes in the exposure of hydrophobic regions can also occur during native aggregation, and thus differences in the amount of hydrophobic solvent-exposed surface for native proteins may be an important factor, for example, in mitigating the solubility of a given protein in its folded state.

5.4 CONFORMATIONAL CHANGES DURING AGGREGATE GROWTH

As noted earlier, this section provides some illustrative examples of aggregate growth within the context of detectable conformational changes or distortions as growth proceeds. The first example is that of aCgn, under acidic conditions and low ionic strength. Figure 5.3 shows an overlay of the normalized signals from a number of the techniques in Table 5.3 as a function of the time that samples were incubated at elevated temperature

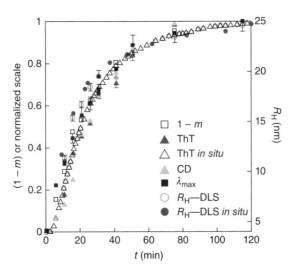

Figure 5.3. Comparison of time course of the fractional extent of monomer loss on a mass basis $(1-m)$ to the change in a variety of spectroscopic signals and laser light scattering (see figure legend and y-axis labels) during aggregate growth versus time (t), under solution conditions where aggregate growth occurs primarily via monomer addition. From Reference 51.

to accelerate aggregation. Also plotted on the same scale is the fraction (mass basis) of monomer that has been consumed by aggregation. Inspection of Figure 5.3 shows that all of the spectroscopic signals that indicate changes in secondary and/or tertiary structure follow a quantitative correlation with the consumption of monomer. It was shown elsewhere in that work [51] that the large majority of monomer loss in this case was due to aggregate growth, rather than nucleation of new aggregates.

A similar observation was made for an IgG1 antibody when tracking aggregation quantitatively using size-exclusion chromatography and ThT binding under conditions where the aggregates remained soluble, and grew primarily via monomer loss [8, 36]. By contrast, when aggregation for aCgn or the IgG proceeded primarily via aggregate–aggregate coalescence, spectroscopic signals such as CD, intrinsic FL, or dye binding did not correlate well with the progression of aggregation. Rather, light scattering was able to clearly show that aggregate growth was occurring rapidly, with little or no detectable structural changes at the level of secondary/tertiary structure of the constituent protein chains within the aggregates. In these cases it was concluded that aggregate–aggregate coalescence was relatively nonspecific in terms of structural aspects and was mitigated mostly by electrostatic effects—with greater coalescence when charge–charge repulsions were screened or reduced by changing pH and/or adding salt [8, 37, 51].

Interestingly, there is also evidence to suggest that the conformational state/structure of monomers within aggregates can evolve even after the aggregate has formed. Kayser et al. [16] reported a case for a different IgG than that described above, in which fluorescence showed that the structure within aggregates was detectably changing over time, after the aggregates had been formed. There was no evidence for aggregates breaking and reforming, so this appeared to be akin to an annealing process, whereby the internal structure of the aggregates "relaxed" slowly over time. Together, this example and those above highlight that tracking conformational changes may or may not be useful if one is seeking to track the progress of aggregation. It is not recommended, in general, that one rely heavily on such techniques to monitor aggregation, unless they are at least first validated with other methods that are more directly applicable [11]. As noted earlier, one must also be careful when inferring changes in conformational state(s) when large/insoluble aggregates are present, at least because of issues such as scattering of the incident and/or emitted light that underlie many of the techniques in Table 5.3.

5.5 SURFACE-MEDIATED UNFOLDING AND ASSEMBLY

This section briefly reviews surface-mediated protein aggregation, from the perspective of experimental signatures and current methods and challenges for monitoring the relevant species and intermediates that are involved. It is beyond the scope of this chapter to exhaustively review protein adsorption to bulk interfaces. Experimentally, it is clear that proteins can and do adsorb to vapor–liquid [60], liquid–liquid [35, 61], and solid–liquid [33, 34, 38, 54, 62–64] interfaces.

However, it is less clear how adsorption *per se* helps to promote aggregation. Figure 5.1 depicts scenarios in which proteins adsorb to an interface, then unfold at that interface, and those unfolded species can then assemble as at least the first stages of aggregate formation. However, much of those or analogous models for surface-mediated aggregation are based on indirect evidence. Some of the relevant experimental observations that motivate hypothesized schemes such as those in Figure 5.1 include the following:

1. Most proteins are amphiphilic and will adsorb and form stable films (either monolayers or multilayers) at air–water interfaces, solid–liquid interfaces, and interfaces between water and organic liquids.
2. If left quiescent, such protein films are relatively stable and can develop a significant modulus to resist flow.
3. Agitation appears to be necessary in many cases to promote surface-mediated aggregation—in this context, agitation may result in convective mass transfer as well as generation of new vapor–liquid interfacial area.
4. Creation of vapor–liquid interfaces is not required for agitation to promote aggregation.
5. Aggregates, including visible and subvisible particles, that are promoted by exposure to surfaces are often not sequestered at the bulk interface(s)—they are readily observable in bulk solution; as such, monomers and/or aggregates must desorb from the interface(s) at some point in the aggregation process.
6. Except in extreme cases, the magnitude of shear forces and the small length scales relevant for most therapeutic proteins make it difficult to justify that shear *per se* is what induces aggregation via monomer unfolding.

5.5.1   Additional Challenges Presented by Interface-Mediated Aggregation

Based on the above observations that are common to many protein systems, the schematic in Figure 5.1 highlights that there are at least three aspects that distinguish this process from aggregation in bulk solution. The first is that mass transfer between bulk and interfacial "phases" or regions may become important in determining the rates of aggregation. The second is that the physical and chemical nature of the interface may play a significant role in the process. For example, vapor–liquid interfaces are more deformable, and new interfacial area can easily be formed during agitation. Alternatively, the roughness and chemical nature of solid surfaces affect aggregation rates under accelerated (high mixing rate) conditions [64].

The third aspect is that even in cases where the amount of interfacial area is deliberately increased, the total volume or mass of material that the interface constitutes is a small fraction of the overall sample. This is problematic when using ensemble-averaged techniques such as those noted earlier, as those assays are dominated by signals from the bulk of the sample, rather than the interfacial regions. The number of techniques available or in common use for monitoring the conformational and/or assembly state(s) of proteins at interfaces is notably smaller than those designed for characterization of proteins in bulk solution.

In terms of monitoring conformational changes and/or the assembly of proteins adsorbed to interfaces, most methods can be categorized as direct or indirect. Examples of direct methods include X-ray and neutron reflectivity [65–69], HDX (possibly coupled to FTIR) [38, 39], as well as total internal reflectance methods [70]—although the latter may measure a mix of contributions from interface and bulk regions, depending on the penetration depth for the experimental configuration of interest. Similar to the use of HDX for bulk solution, it also provides information regarding changes in solvent exposure for proteins adsorbed to surfaces [38, 39]. X-ray and neutron reflectivity methods are not in common use in industrial pharmaceutical laboratories, but do potentially provide a means to probe relatively large changes in protein conformation at interfaces, as well as the evolution of assembled structures.

Examples of indirect methods include tensiometry [60, 61], Langmuir troughs and surface pressure measurements [71], and seeding experiments in which controlled amounts of interfacial area are introduced (e.g., stainless steel or glass particles) and the response of the system (e.g., aggregate mass in bulk solution) is monitored as a function of the amount of interfacial area [33, 34, 64]. Each of the techniques provides a practical means to assess whether a change in solvent conditions or the protein of interest is affecting some aspect of the vapor–liquid, liquid–liquid, or solid–liquid interface. However, on their own, each cannot discriminate between a number of possible changes that may be occurring. For example, a change in surface tension or pressure may be due to a change in the amount of protein, a change in how the proteins are packed, and/or a change in the conformational state(s) of proteins at the interface. Similar ambiguities arise for interpreting aggregation kinetics of "surface seeding" experiments, as surfaces with different properties could affect multiple steps in the surface-mediated aggregation process (cf., Figure 5.1).

5.5.2 Potential Roles of Surfactants

By far, the most common approach to attempt to circumvent or at least mitigate surface- and/or agitation-mediated aggregation for liquid products is to add surfactants [72–74]. For solid-state products, agitation *per se* is not likely to be a factor in promoting aggregation, but surfactants may still be helpful in mitigating adsorption of proteins to ice–water interfaces [63].

Surfactants may potentially affect a number of steps in the overall aggregation process. Like any added cosolute or solvent, they can potentially adsorb preferentially (relative to water) to the protein surface [75, 76]. If they do so to a greater extent for the partially or fully unfolded state(s) of a protein (relative to folded states), then addition of surfactant can significantly lower the conformational stability of a given protein, and therefore lead to measurable conformational distortions [77]; indeed, this is a desired effect in the case of strongly denaturing surfactants such as sodium dodecyl sulfate.

Pharmaceutical products more typically employ nonionic, polysorbate-based surfactants such as Tween 20 and Tween 80, as well as other nonionic surfactants such as the Triton family [27, 72, 74]. Empirically, these surfactants do not typically denature proteins even when the surfactants are well above their respective critical micelle concentrations (cmc). This can be advantageous, in that if one can operate above the cmc for a given surfactant then this will maximize the amount of "free" surfactant in solution. This then maximizes the driving force for surfactant to adsorb to bulk interfaces, particularly for aqueous phases in contact with hydrophobic interfaces such as air or organic liquids.

In the simplest picture of how surfactants affect surface-mediated aggregation, surfactants "coat" the interface, with their hydrophilic segments pointing inward toward the aqueous phase and minimizing the driving force for hydrophobic portions of proteins to adsorb to those interfaces [23, 72]. Notably, adsorption to glass or metal interfaces from aqueous solutions has a significant contribution from electrostatic interactions. As such, the role of nonionic surfactants in mitigating aggregation may be anticipated to be more complex in such situations. In addition, for any of the interfaces considered above, rather than thinking of surfactants as making a static "coating" that completely excludes protein from the surface, it is arguably more accurate to consider this as a flexible and dynamic layer, where proteins may be able to compete with the surfactant for coverage. Overall, the precise structures and interactions involving therapeutic proteins and surfactants in bulk solution and interfaces remain relatively open questions. Particular examples include how conformational changes are mitigated by interfaces and how or if this impacts the key intermediates in aggregate nucleation. These remain outstanding challenges in the field, particularly as such surfaces and interfaces are inherently part of dosage forms and manufacturing environments for protein therapeutics.

5.6 SUMMARY

Conformational changes are key to the initiation and propagation of nonnative aggregates for a large range of protein systems. Many of the current challenges regarding the role of conformational changes in aggregation stem from difficulties with detecting small changes, or small populations with significant changes, using conventional experimental techniques. This holds both in bulk solution and when considering aggregation mediated by vapor–liquid, liquid–liquid, and/or solid–liquid interfaces that are inherent throughout the production, storage, shipping, and administration of therapeutic proteins. Some potentially lucrative approaches to complement current techniques include structurally specific ligands and advanced scattering techniques. Although beyond the scope of this chapter, improved molecular modeling approaches to discern the early stages of aggregation (i.e., nucleation) will also presumably improve our understanding of the key intermediates in the process—particularly those that are too poorly populated and/or short lived to be directly discernable in experiment.

REFERENCES

1. Zubay, G. (1993) *Biochemistry,* 3rd edn. McGraw-Hill Higher Ed.
2. Creighton, T.E. (1992) *Proteins: Structures and Molecular Properties*, 2nd edn. W.H. Freeman.

3. Curtis, R.A. and Lue, L. (2005) A molecular approach to bioseparations: protein-protein and protein-salt interactions. *Chem. Eng. Sci.*, **61**, 907–923.

4. Roberts, C.J. (2007) Non-native protein aggregation kinetics. *Biotechnol. Bioeng.*, **98**, 927–938.

5. Fink, A.L. (1998) Protein aggregation: folding aggregates, inclusion bodies and amyloid. *Fold. Des.*, **3**, R9–R23.

6. Laurence, J.S. and Middaugh, C.R. (2010) Fundamental structures and behaviors of proteins. In: Wang, W. and Roberts, C.J. editors. *Aggregation of Therapeutic Proteins*. John Wiley & Sons, Inc., pp. 1–61.

7. Bartkowski, R., Kitchel, R., Peckham, N., and Margulis, L. (2002) Aggregation of recombinant bovine granulocyte colony stimulating factor in solution. *J. Protein Chem.*, **21**, 137–143.

8. Brummitt, R.K., et al. (2011) Nonnative aggregation of an IgG1 antibody in acidic conditions, part 2: nucleation and growth kinetics with competing growth mechanisms. *J. Pharm. Sci.*, **100**, 2104–2119.

9. He, F., et al. (2010) Detection of IgG aggregation by a high throughput method based on extrinsic fluorescence. *J. Pharm. Sci.*, **99**, 2598–2608.

10. Liu, J., Andya, J.D., and Shire, S.J. (2006) A critical review of analytical ultracentrifugation and field flow fractionation methods for measuring protein aggregation. *AAPS J.*, **8**, E580–E589.

11. Weiss, W.F., Young, T.M., and Roberts, C.J. (2009) Principles, approaches, and challenges for predicting protein aggregation rates and shelf life. *J. Pharm. Sci.*, **98**, 1246–1277.

12. Goldberg, D.S., Bishop, S.M., Shah, A.U., and Sathish, H.A. (2011) Formulation development of therapeutic monoclonal antibodies using high-throughput fluorescence and static light scattering techniques: role of conformational and colloidal stability. *J. Pharm. Sci.*, **100**, 1306–1315.

13. Harn, N., Allan, C., Oliver, C., and Middaugh, C.R. (2007) Highly concentrated monoclonal antibody solutions: direct analysis of physical structure and thermal stability. *J. Pharm. Sci.*, **96**, 532–546.

14. Grillo, A.O., et al. (2001) Conformational origin of the aggregation of recombinant human factor VIII. *Biochemistry*, **40**, 586–595.

15. Roberts, C.J., Darrington, R.T., and Whitley, M.B. (2003) Irreversible aggregation of recombinant bovine granulocyte-colony stimulating factor (bG-CSF) and implications for predicting protein shelf life. *J. Pharm. Sci.*, **92**, 1095–1111.

16. Kayser, V., et al. (2011) Evaluation of a non-Arrhenius model for therapeutic monoclonal antibody aggregation. *J. Pharm. Sci.*, **100**, 2526–2542.

17. Perico, N., Purtell, J., Dillon, T.M., and Ricci, M.S. (2009) Conformational implications of an inversed pH-dependent antibody aggregation. *J. Pharm. Sci.*, **98**, 3031–3042.

18. Brummitt, R.K., Nesta, D.P., and Roberts, C.J. (2011) Predicting accelerated aggregation rates for monoclonal antibody formulations, and challenges for low-temperature predictions. *J. Pharm. Sci.*, **100**, 4234–4243.

19. Franks, F. (2003) Scientific and technological aspects of aqueous glasses. *Biophys. Chem.*, **105**, 251–261.

20. Wang, B., et al. (2009) Impact of sucrose level on storage stability of proteins in freeze-dried solids: II. Correlation of aggregation rate with protein structure and molecular mobility. *J. Pharm. Sci.*, **98**, 3145–3166.

21. Wang, B., Cicerone, M.T., Aso, Y., and Pikal, M.J. (2010) The impact of thermal treatment on the stability of freeze-dried amorphous pharmaceuticals: II. aggregation in an IgG1 fusion protein. *J. Pharm. Sci.*, **99**, 683–700.

22. Roberts, C.J., Das, T.K., and Sahin, E. (2011) Predicting solution aggregation rates for therapeutic proteins: approaches and challenges. *Int. J. Pharm.*, **418**, 318–333.

23. Cleland, J.L., Powell, M.F., and Shire, S.J. (1993) The development of stable protein formulations: a close look at protein aggregation, deamidation, and oxidation. *Crit. Rev. Ther. Drug Carrier Syst.*, **10**, 307–377.

24. Zimm, B.H. (1948) Apparatus and methods for measurement and interpretation of the angular variation of light scattering; preliminary results on polystyrene solutions. *J. Chem. Phys.*, **16**, 1099–1116.

25. Pecora, R. (2000) Dynamic light scattering measurement of nanometer particles in liquids. *J. Nanopart. Res.*, **2**, 123–131.

26. Haustein, E., and Schwille, P. (2007) Fluorescence correlation spectroscopy: novel variations of an established technique. *Annu. Rev. Biophys. Biomol. Struct.*, **36**, 151–169.

27. Mahler, H., Fischer, S., Randolph, T.W., and Carpenter, J.F. (2010) Protein aggregation and particle formation: effects of formulation, interfaces, and drug product manufacturing operations. In: Wang, W. and Roberts, C.J., editors. *Aggregation of Therapeutic Proteins*. John Wiley & Sons, Inc., pp. 301–331.

28. Chi, E.Y., Kendrick, B.S., Carpenter, J.F., and Randolph, T.W. (2005) Population balance modeling of aggregation kinetics of recombinant human interleukin-1 receptor antagonist. *J. Pharm. Sci.*, **94**, 2735–2748.

29. Sharma, V.K. and Kalonia, D.S. (2010) Experimental detection and characterization of protein aggregates. In: Wang, W. and. Roberts, C.J., editors. *Aggregation of Therapeutic Proteins*. John Wiley & Sons, Inc., pp. 205–256.

30. Weinberg, W.C., Ha, L., Kirshner, S.L., and Verthelyi, D.I. (2010) Regulatory perspective on aggregates as a product quality attribute. In: Wang, W. and Roberts, C.J., editors. *Aggregation Therapeutic Proteins*. John Wiley & Sons, Inc., pp. 435–451.

31. Andersen, C.B., et al. (2010) Aggregation of a multidomain protein: a coagulation mechanism governs aggregation of a model IgG1 antibody under weak thermal stress. *Protein Sci.*, **19**, 279–290.

32. Pallitto, M.M. and Murphy, R.M. (2001) A mathematical model of the kinetics of β-amyloid fibril growth from the denatured state. *Biophys. J.*, **81**, 1805–1822.

33. Hoehne, M., et al. (2011) Adsorption of monoclonal antibodies to glass microparticles. *J. Pharm. Sci.*, **100**, 123–132.

34. Bee, J.S., et al. (2009) Aggregation of a monoclonal antibody induced by adsorption to stainless steel. *Biotechnol. Bioeng.*, **105**, 121–129.

35. Ludwig, D.B., et al. (2011) Flow cytometry: a promising technique for the study of silicone oil-induced particulate formation in protein formulations. *Anal. Biochem.*, **410**, 191–199.

36. Brummitt, R.K., et al. (2011) Nonnative aggregation of an IgG1 antibody in acidic conditions: part 1. Unfolding, colloidal interactions, and formation of high-molecular-weight aggregates. *J. Pharm. Sci.*, **100**, 2087–2103.

37. Li, Y., Ogunnaike, B.A., and Roberts, C.J. (2010) Multi-variate approach to global protein aggregation behavior and kinetics: effects of pH, NaCl, and temperature for α-chymotrypsinogen A. *J. Pharm. Sci.*, **99**, 645–662.

38. Gospodarek, A.M., Smatlak, M.E., O'Connell, J.P., and Fernandez, E.J. (2011) Protein stability and structure in HIC: hydrogen exchange experiments and COREX calculations. *Langmuir*, **27**, 286–295.

39. Smith, J.R., Cicerone, M.T., and Meuse, C.W. (2009) Tertiary structure changes in albumin upon surface adsorption observed via Fourier transform infrared spectroscopy. *Langmuir*, **25**, 4571–4578.

40. Eisenberg, D., et al. (2006) The structural biology of protein aggregation diseases: fundamental questions and some answers. *Acc. Chem. Res.*, **39**, 568–575.

41. Andrews, J.M., and Roberts, C.J. (2007) Non-native aggregation of α-chymotrypsinogen occurs through nucleation and growth with competing nucleus sizes and negative activation energies. *Biochemistry*, **46**, 7558–7571.

42. Uversky, V.N., and Fink, A.L. (2004) Conformational constraints for amyloid fibrillation: the importance of being unfolded. *Biochim. Biophys. Acta, Proteins Proteomics*, **1698**, 131–153.

43. Chi, E.Y., Krishnan, S., Randolph, T.W., and Carpenter, J.F. (2003) Physical stability of proteins in aqueous solution: mechanism and driving forces in nonnative protein aggregation. *Pharm. Res.*, **20**, 1325–1336.

44. Khurana, R., et al. (2001) Partially folded intermediates as critical precursors of light chain amyloid fibrils and amorphous aggregates. *Biochemistry*, **40**, 3525–3535.

45. Zhang, A., et al. (2011) Distinct aggregation mechanisms of monoclonal antibody under thermal and freeze-thaw stresses revealed by hydrogen exchange. *Pharm. Res.*, **28**, 1179–93.

46. Weiss, W.F., et al. (2007) Nonnative protein polymers: structure, morphology, and relation to nucleation and growth. *Biophys. J.*, **93**, 4392–4403.

47. Kendrick, B.S., et al. (1998) Aggregation of recombinant human interferon gamma: kinetics and structural transitions. *J. Pharm. Sci.*, **87**, 1069–1076.

48. Goers, J., et al. (2002) Conformational prerequisites for α-lactalbumin fibrillation. *Biochemistry*, **41**, 12546–12551.

49. Sahin, E., et al. (2011) Computational design and biophysical characterization of aggregation-resistant point mutations for γD crystallin illustrate a balance of conformational stability and intrinsic aggregation propensity. *Biochemistry*, **50**, 628–639.

50. Zhang, A., et al. (2010) Molecular level insights into thermally induced α-chymotrypsinogen a amyloid aggregation mechanism and semiflexible protofibril morphology. *Biochemistry*, **49**, 10553–10564.

51. Li, Y., Weiss, W.F., and Roberts, C.J. (2009) Characterization of high-molecular-weight nonnative aggregates and aggregation kinetics by size exclusion chromatography with inline multi-angle laser light scattering. *J. Pharm. Sci.*, **98**, 3997–4016.

52. Tobler, S.A., Holmes, B.W., Cromwell, M.E.M., and Fernandez, E.J. (2004) Benzyl alcohol-induced destabilization of interferon-γ: a study by hydrogen-deuterium isotope exchange. *J. Pharm. Sci.*, **93**, 1605–1617.

53. Caflisch, A. (2006) Computational models for the prediction of polypeptide aggregation propensity. *Curr. Opin. Chem. Biol.*, **10**, 437–444.

54. Zhang, A., Qi, W., Singh, S.K., and Fernandez, E.J. (2011) A new approach to explore the impact of freeze-thaw cycling on protein structure: hydrogen/deuterium exchange mass spectrometry (HX-MS). *Pharm. Res.*, **28**, 1179–1193.

55. Spatara, M.L., Roberts, C.J., and Robinson, A.S. (2009) Kinetic folding studies of the P22 tailspike beta-helix domain reveal multiple unfolded states. *Biophys. Chem.*, **141**, 214–221.

56. Andrews, J.M., Weiss, W.F., and Roberts, C.J. (2008) Nucleation, growth, and activation energies for seeded and unseeded aggregation of α-chymotrypsinogen A. *Biochemistry*, **47**, 2397–2403.

57. Kayed, R., et al. (2003) Common structure of soluble amyloid oligomers implies common mechanism of pathogenesis. *Science (Washington, DC, U.S.)*, **300**, 486–489.

58. Tessier, P.M. and Lindquist, S. (2009) Unraveling infectious structures, strain variants and species barriers for the yeast prion [PSI+]. *Nat. Struct. Mol. Biol.*, **16**, 598–605.

59. Maddux, N.R., et al. (2011) Multidimensional methods for the formulation of biopharmaceuticals and vaccines. *J. Pharm. Sci.*, **100**, 4171–4197.

60. Svitova, T.F., Wetherbee, M.J., and Radke, C.J. (2003) Dynamics of surfactant sorption at the air/water interface: continuous-flow tensiometry. *J. Colloid Interface Sci.*, **261**, 170–179.

61. Beverung, C.J., Radke, C.J., and Blanch, H.W. (1999) Protein adsorption at the oil/water interface: characterization of adsorption kinetics by dynamic interfacial tension measurements. *Biophys. Chem.*, **81**, 59–80.

62. Fogle, J.L., and Fernandez, E.J. (2006) Amide hydrogen-deuterium exchange: a fast tool for screening protein stabilities in chromatography. *LCGC North Am.*, 96–101.

63. Webb, S.D., et al. (2002) Surface adsorption of recombinant human interferon-γ in lyophilized and spray-lyophilized formulations. *J. Pharm. Sci.*, **91**, 1474–1487.

64. Biddlecombe, J.G., et al. (2009) Factors influencing antibody stability at solid-liquid interfaces in a high shear environment. *Biotechnol. Prog.*, **25**, 1499–1507.

65. Yano, Y.F., et al. (2009) Driving force behind adsorption-induced protein unfolding: a time-resolved X-ray reflectivity study on lysozyme adsorbed at an air/water interface. *Langmuir*, **25**, 32–35.

66. Evers, F., Steitz, R., Tolan, M., and Czeslik, C. (2009) Analysis of Hofmeister effects on the density profile of protein adsorbates: a neutron reflectivity study. *J. Phys. Chem. B*, **113**, 8462–8465.

67. Horne, D.S., et al. (1998) Neutron reflectivity study of competitive adsorption of β-lactoglobulin and nonionic surfactant at the air-water interface. *Int. Dairy J.*, **8**, 73–77.

68. Yano, Y.F., et al. (2011) Protein salting out observed at an air-water interface. *J. Phys. Chem. Lett.*, **2**, 995–999.

69. Holt, S.A., Henderson, M.J., and White, J.W. (2002) Thermal denaturation of interfacial protein layers. *Aust. J. Chem.*, **55**, 449–459.

70. Shibata, C.T. and Lenhoff, A.M. (1992) TIRF [total internal reflectance fluorescence] of salt and surface effects on protein adsorption. I. Equilibrium. *J. Colloid Interface Sci.*, **148**, 469–484.

71. Singh, A., Konovalov, O., Novak, J., and Vorobiev, A. (2010) The sequential growth mechanism of a protein monolayer at the air-water interface. *Soft Matter*, **6**, 3826–3831.

72. Wang, W. (2005) Protein aggregation and its inhibition in biopharmaceutics. *Int. J. Pharm.*, **289**, 1–30.

73. Wang, W. and Warne, N.W. (2010) Approaches to managing protein aggregation in product development. In: Wang, W. and Roberts, C.J., editors. *Aggregation of Therapeutic Proteins*. John Wiley & Sons, Inc., pp. 333–365.

74. Kiese, S., Papppenberger, A., Friess, W., and Mahler, H. (2008) Shaken, not stirred: mechanical stress testing of an IgG1 antibody., *J. Pharm. Sci.*, **97**, 4347–4366.

75. Timasheff, S.N. (1995) Solvent stabilization of protein structure. *Methods Mol. Biol. (Totowa, N.J.)*, **40**, 253–269.

76. Schellman, J.A. (2005) Destabilization and stabilization of proteins. *Q. Rev. Biophys.*, **38**, 351–361.

77. Hamill, A.C., Wang, S., and Lee, C.T. (2005) Probing lysozyme conformation with light reveals a new folding intermediate., *Biochemistry*, **44**, 15139–15149.

6

UTILIZATION OF CHEMICAL LABELING AND MASS SPECTROMETRY FOR THE BIOPHYSICAL CHARACTERIZATION OF BIOPHARMACEUTICALS

Justin B. Sperry[1] and Lisa M. Jones[2,3]

[1]*Analytical Research and Development, Biotherapeutics Pharmaceutical Sciences, Pfizer Inc., Chesterfield, MO, USA*
[2]*Department of Chemistry, Washington University in St. Louis, St. Louis, MO, USA*
[3]*Department of Chemistry and Chemical Biology, Indiana University-Purdue University Indianapolis, Indianapolis, IN, USA*

6.1 MASS SPECTROMETRY OF BIOPHARMACEUTICALS

The characterization of protein biopharmaceuticals is important for understanding their complex biochemical and biophysical properties, degradation products, and process-dependent heterogeneities including posttranslational modifications. Mass spectrometry is a useful tool in studying protein systems because it is sensitive, is accurate, consumes low amounts of sample, and is generally applicable to biopharmaceuticals [1–3]. The molecular mass of proteins and their linear sequences of amino acids can be determined experimentally by using electrospray ionization (ESI), accurate mass measurement, and product-ion sequencing. Although characterization of biopharmaceuticals by mass spectrometry has mostly concentrated on primary structure analysis, recent developments over the last few decades have led to the expansion to higher order structure analyses. The combination of these traditional mass spectrometry-based approaches with chemical labeling approaches such as amide hydrogen/deuterium (H/D) exchange and/or covalent labeling creates a powerful means to monitor conformational changes in proteins, to monitor the effects of degradation, to locate antigen-binding sites, to follow binding dynamics, and, more recently, to determine the binding affinity of a ligand to a protein. The partnership of mass spectrometry with H/D exchange and covalent labeling to study biopharmaceuticals is the subject of increasing interest within the biotechnology industry.

The only way to measure the mass of a neutral analyte by mass spectrometry is to form an ion, whether it is an even-electron species or odd-electron species. The introduction of ESI mass spectrometry in the 1980s by John Fenn led to one of the most significant breakthroughs in analytical sciences in recent decades [4, 5]. This development makes polypeptides accessible to ionization for mass analysis. ESI enabled a direct interface between the solution phase, where polypeptides are found, and the gas phase, where ions are detected [6]. Ionization by electrospray is accomplished by applying high voltage (~3 kV for positive-ion mode) on a metal needle through which the solution containing the protein flows at a desired rate. At the tip of the needle, highly charged droplets are formed as a result of an electric potential removing negative ions from solution (i.e., Cl^- and OH^-). The inclusion of desolvation gas and a high capillary temperature aids in removing additional solvent molecules, thus decreasing the size of the charged droplets. As the droplets decrease in size, Coulombic interactions increase owing to the high density of charge on the droplet, leading to the ejection of ions into the gas phase.

The benefit of ESI is that it produces ions of an analyte with multiple charges: $[M + nH^+]^{n+}$ in positive-ion mode and $[M - nH^+]^{n-}$ in the negative-ion mode, where M is the molecular mass of the analyte of interest. This is useful for protein analysis because ESI produces ions in a mass-to-charge (m/z) range compatible with most mass spectrometers (typically 100–3000 m/z). Upon formation of the multiply charged species, the ions are guided through the mass spectrometer from high potential to low potential for detection. Electrospray is considered a "soft" ionization technique, whereby chemical bonds are not broken upon ion formation (often via proton transfer) and weak noncovalent interactions present in protein–ligand complexes are retained [7,8]. ESI can be coupled to all types of mass spectrometers, including ion traps [9, 10], time-of-flight (ToF) [11], Fourier transform ion cyclotron resonance (FTICR) [12], quadrupole/triple quadrupole [13], and, more recently, any hybrid made up by combining members of the list, including quadrupole time-of-flight (Q-ToF) [14–16] and ion trap-FTICR instruments [17].

6.2 INTRODUCTION TO HYDROGEN/DEUTERIUM EXCHANGE

Shortly after the discovery of the α-helix and β-sheet secondary structures in proteins [18–22], the first H/D exchange studies were conducted by Linderstrøm-Lang [23–28] using relatively primitive devices to record the increase in mass of a protein. The use of H/D exchange to study protein dynamics, conformations, and folding became more tractable when nuclear magnetic resonance (NMR) was used as the measurement strategy. Now, H/D exchange, as followed by NMR and MS, has become a common biophysical strategy, continuing for nearly 60 years; the subject has been widely reviewed [27–37].

Proteins contain three classes of hydrogens throughout their linear sequence of amino acids. The alkyl hydrogens (C-H) present on side chains and C_α positions exchange with solvent slowly, and thus they are not monitored in H/D exchange experiments. The hydrogens considered acidic or basic and present on side chains, such as on glutamic acid, aspartic acid, lysine, threonine, tyrosine, and arginine, exchange too rapidly to be measured in H/D exchange experiments. Hydrogens located at amide sites constitute the final class and can exchange rapidly (<1 second) or very slowly (for several months) depending on the local environment and degree of hydrogen bonding to which they are submitted. When they become involved in hydrogen bonding, their amide exchange rates can decrease by a factor of 10^8 compared to the amide exchange rate in an unfolded state, underscoring the importance of hydrogen bonding in protein structures [34].

Most hydrogen exchange experiments follow pseudo first-order kinetics because the concentration of D_2O present at equilibrium is much greater than that of the protein. Hydrogen exchange in solution can be acid, base, or water catalyzed depending on the pH. The effect of water catalysis is minimal compared to acid and base catalysis in unstructured peptides [35, 38]. The rate of water catalysis is 6 orders of magnitude smaller than that of acid catalysis and 13 orders of magnitude smaller than that of base catalysis, thus this term is typically not regarded in H/D exchange experiments. The hydrogen exchange rate reaches a minimum between pH 2 and 3, where the acid- and base-catalyzed rates become equal. For every increase of one pH unit above 3, the base-catalyzed rate increases by 1 order of magnitude. For every decrease of one pH unit below 2, the acid-catalyzed rate increases by 1 order of magnitude. This makes mass spectrometry-based experiments with protein samples ideal because near pH 7 the base-catalyzed reaction abstracts amide hydrogens and exchanges them with deuterium on timescales of seconds to days. To measure the extent of deuterium uptake, the reaction can then be quenched by decreasing the pH to 2–3. An additional decrease in the intrinsic hydrogen exchange rate can be implemented by a drop in temperature. A decrease in the reaction temperature from 25°C to 0°C decreases the exchange rate by another order of magnitude. The total drop in rate from pH 7 and 25°C to pH 2–3 and 0°C is nearly 5 orders of magnitude.

The "gold standard" means of site-specific amide exchange rate constant measurement is NMR [39, 40]. The rate of exchange is affected by two different parameters: the inductive effects of the neighboring side chains and the degree of amide hydrogen bonding [41]. The inductive effects from neighboring side chains can decrease or increase the acidity of the amide hydrogen by donating or removing, respectively, electron density from the site [42–45]. Hydrogen bonding can increase or decrease the amide exchange rate depending on the strength of the bond. When a transient separation of the H-bond donor and acceptor occurs, the exchange event can occur with the solvent D_2O. Literature refers to this as the structural opening and structural closing rate constants, k_{op} and k_{cl}, respectively [32].

There are two types of hydrogen exchange regimes for proteins: EX1 and EX2 [27, 33, 46–48], also described using the terms "local motions" and "unfolding" [49]. Proteins are dynamic in solution and are constantly undergoing structural fluctuations. The rate of partial unfolding during structural fluctuations can compete with the rate of hydrogen exchange according to Equation 6.1:

$$N_H \underset{k_{-1}}{\overset{k_1}{\rightleftharpoons}} U_H \overset{k_{ex}}{\longrightarrow} U_D \underset{k_1}{\overset{k_{-1}}{\rightleftharpoons}} N_D \tag{6.1}$$

where N_H represents the native, undeuterated state of the protein, k_1 represents the rate of unfolding in solution, U_H represents the unfolded, undeuterated state of the protein, k_{ex} is the rate of exchange with deuterated solvent, U_D represents the unfolded, deuterated state of the protein, k_{-1} is the rate of refolding in solution, and N_D is the folded, deuterated state of the protein.

In EX1 exchange the rate of deuteration is much faster than the rate of refolding ($k_{ex} \gg k_{-1}$). This results in mass spectra having multiple populations of deuterated species, typically a distinct low and a high mass signal. EX1 kinetics can be induced by a change in pH or chemical denaturants [48, 50, 51], but can also occur under physiological conditions [52, 53]. In the EX2 exchange regime, the rate of deuteration is much slower than the rate of refolding ($k_{-1} \gg k_{ex}$). This results in mass spectra having a single population and showing a gradual increase in mass with time [42]. Almost all proteins undergo exchange in the EX2 regime at physiological conditions; however, both of the exchange regimes can occur at the same time in different regions of one protein [54].

6.3 APPLICATIONS OF HYDROGEN/DEUTERIUM EXCHANGE AND MASS SPECTROMETRY TO PROTEINS

A mass spectrometric technique that reveals changes in solvent accessibility of protein surfaces, monitors shifts in the hydrogen bonding network owing to ligand binding, and/or affords information regarding protein folding and dynamics would be particularly valuable and advantageous. Such an approach could take advantage of the speed and sensitivity of MS. One evolving approach combines H/D exchange and mass spectrometry. The use of H/D exchange and mass spectrometry was first demonstrated by Katta and Chait [44] and has been reviewed extensively over the past decade [42, 45, 55–58]. There is a twofold benefit in combining H/D exchange and mass spectrometry: (1) amide sites with potentially exchangeable hydrogens are located at every residue in a protein, except the N-terminus and at proline residues, and (2) for each amide hydrogen that exchanges with deuterium, the protein mass increases by 1 Da. If a protein undergoes 50 exchange events, the mass of the protein increases by 50 Da. The method does not rely on a measurement of signal intensity, as do most other spectroscopic techniques, but merely records a small perturbation in mass.

There are two general procedures to label proteins with deuterium: pulse labeling and continuous labeling. The pulse-labeling protocol utilizes a quench-flow apparatus to rapidly deuterate a protein to uncover protein conformational changes at short times immediately preceding ESI-MS analysis [59–61]. A combination of monitoring the charge-state distribution of the protein with its deuterium uptake enables the identification of folding intermediates, the existence of which continues to be of interest to the protein folding community.

Continuous labeling is by far the more common form of deuterium incorporation and more widely used. The protein is typically diluted 20–40 times into a D_2O-containing buffer and allowed to exchange for certain periods of time, in the range of seconds to hours. In the continuous labeling experiment, multiple populations of protein states can be sampled owing to natural fluctuations in structure, caused by either environmental conditions (pH, temperature) or a response to a ligand-binding event. Once an appropriate time of exchange has been reached, the reaction is quenched by dropping the pH to 2–3 and the temperature to 0°C. At this point, the experimenter has two options for mass analysis: inject the full-length protein to determine its extent of deuterium incorporation or digest the deuterated protein into peptides and determine their extents of exchange (Figure 6.1). Full-length protein analysis reveals the number of amide sites accessible for exchange. A comparison of two different states of a protein affords information regarding each individual condition by comparing the difference in deuterium uptake. Increased levels of deuterium uptake (as compared to a standard) indicate a more unstructured or open conformation resulting from fewer and/or decreased strengths of amide hydrogen bonds throughout the protein. Decreased levels of deuterium uptake (as compared to a standard) indicate a more compact, closed and/or more stable conformation resulting from more and/or increased strengths of amide hydrogen bonds throughout the protein. The determination of deuterium levels of full-length proteins can be very rapid, sometimes less than 1 minute after quench; however, it does not disclose where deuterium differences are taking place.

A common approach to localize the extent of deuteration throughout the protein is to perform a digestion with a protease at acidic pH [62]. A commonly used, stable protease at acidic pH is pepsin; however, there are new proteases on the market such as type XIII protease from *Aspergillus saitoi* that also work at pH 2. Although pepsin is considered a nonspecific protease, meaning that it does not consistently cleave peptide bonds at the same amino acid residue(s), it does reproducibly cleave a given protein in

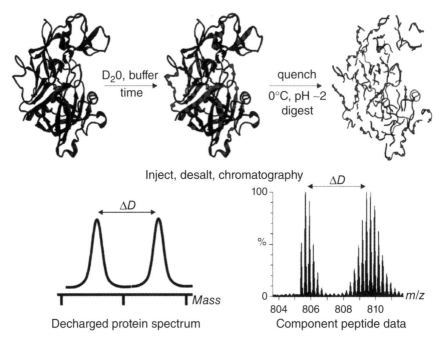

Figure 6.1. Schematic of the two continuous labeling H/D exchange approaches for proteins. The protein of interest is diluted in deuterated buffer with the appropriate formulation components and allowed to exchange for discrete periods of time. Amide hydrogens (represented as gray dots) take up deuterium at varying exchange times. The reaction is quenched by dropping the pH to around 2 and the temperature to near 0°C. One can chose to analyze the intact protein or to localize differences using an acidic protease such as pepsin. The total mass change (ΔD) of the protein is representative of the total exchangeable amide sites for that particular condition, whereas the mass change of a component peptide (also ΔD) is representative of the total exchangeable amide sites in that particular region of the protein. Low ΔD values indicate decreased H/D exchange rates from strong hydrogen bonds and high ΔD values indicate increased H/D exchange rates from weaker hydrogen bonds.

the same place under the same conditions. The identifications of peptides from peptic digests are verified using standard proteomic protocols (i.e., accurate mass measurement, sequencing of product-ion spectra, and database searching).

A major concern in H/D exchange experiments is the retention of the deuterium label for as long as possible. The loss of deuterium with time is referred to as back-exchange. Back-exchange occurs when the protein and/or peptides are subjected to protiated solvents in, for example, a liquid chromatography (LC) step before mass analysis. In a controlled experiment, where the LC solvents are kept near 0°C and pH 2, the back-exchange rate is on the order of one deuterium per minute. This can amount to 1–5% of the extent of deuteration on the protein, although some have reported as high as 20% in ESI-MS and greater than 50% in MALDI-MS [63]. A common way to correct for back-exchange was introduced by Zhang and Smith [62]. The correction for back-exchange is widely used, but it is often difficult to produce a fully deuterated control without elevating the temperature or using denaturants.

The use of automation in sample handling and sample introduction into the instrument, as well as automated data analysis are becoming more common. Completely automated H/D exchange-MS systems have been reported and are commercially available [64–68]. Automated H/D exchange-MS systems may benefit from sample-temperature stability, accurate quench conditions, and a reproducible injection technique. An increasing number of semiautomated software systems were reported as well [69–71]. Software systems gather deuterated peptide isotope patterns from instrument software and calculate their centroids, an average measure of deuterium uptake per peptide per unit time. Most of the software packages allow one to correct for back-exchange using appropriate controls, create deuterium uptake plots, and export data to spreadsheet programs for additional data workup.

Several recent applications of H/D exchange combined with mass spectrometry have delivered useful information regarding the characterization of biopharmaceuticals. The largest segment of biopharmaceuticals in development is recombinant immunoglobulin gamma (IgG) monoclonal antibodies (mAbs). Mass spectrometry is the primary characterization technique to verify the primary structure of mAbs, as well as their inherent posttranslational modifications and degradation products. In recent years academic laboratories and biopharmaceutical companies have sought to exploit the advantages of H/D exchange to this important class of therapeutics. Houde et al. [72] demonstrated that IgG1 mAbs undergo differential deuterium exchange as a result of de-N-glycosylation. The removal of N-linked glycans from the consensus glycosylation site in the Fc region of the mAb revealed several regions around the site that had more deuterium than in the native (reference) state, an indication of less protection from exchange. On the other hand, several regions were also identified as having less deuterium exchange, or more protection from exchange, as compared to the reference molecule. These regions were localized using proteolytic digestion with pepsin and rapid chromatography techniques with an ultrahigh pressure liquid chromatography (UPLC) system.

Burkitt et al. [73] studied the effects of oxidation of IgG1 monoclonal antibodies by H/D exchange. The antibody samples were oxidatively stressed under accelerated conditions by the addition of hydrogen peroxide to a concentration near 100 mM. The oxidation was allowed to proceed for 30 minutes at 37°C and then the samples were dialyzed back into their respective formulation buffer. This procedure afforded 90% oxidation at Met253 and 70% oxidation at Met430. Immobilized pepsin was utilized to digest the antibodies after deuterium exchange to locate the regions affected by oxidation. The interface between the C_H2 and C_H3 domains was identified as being a hot spot in terms of observed differences in deuterium uptake. In particular, residues 247–253 were most affected by oxidation and resulted in increased levels of deuterium uptake, indicative of decreased levels of amide hydrogen bonding in the area of the mAb. Interestingly, Met253 resides on this particular peptic peptide and was significantly oxidized during the experiments.

A more thorough investigation of IgG1 mAbs was later conducted by Houde et al. [74] and included the effects of receptor binding, varying levels of glycosylation variants, and the formation of degradation products. The work demonstrated H/D exchange's capabilities to localize conformational changes as a result of methionine oxidation and varying extents of fucosylation and galactosylation in an IgG1 mAb. Most of the changes were observed in the C_H2 domain, N-terminal to the consensus glycosylation site of the antibody. As the level of galactosylation increased, the relative level of deuterium uptake decreased in this region, indicating an increased structural rigidity. This change in conformation altered the ability of FcγRIIIa receptor binding in a positive manner. The levels of fucosylation did not affect the conformation of the IgG1 mAb to any noticeable difference. However, the fucosylation level did alter the extent of FcγRIIIa binding, with a complete removal of fucose generating an approximately 50-fold increase in binding.

Another class of biopharmaceuticals includes small recombinant proteins such as interferons. Bobst, Kaltashov, and coworkers [75, 76] used interferon-β1a as a model system to assess H/D exchange's ability to monitor higher order structure and protein degradation. This particular interferon analogue contains an unpaired Cys at residue 17. Two mass spectrometry-based methods, and data from orthogonal biophysical assays such as CD spectroscopy, fluorescence emission, and size exclusion chromatography, afforded a direct comparison of native IFN-β1a to a chemically modified form (alkylation using N-ethylmaleimide, NEM). The first, termed native electrospray, showed a partial unfolding due to the presence of higher charge states in the electrospray mass spectrum. Although the results of this assay were definitive in assigning a difference between the two proteins, it, like the biophysical assays, did not locate where the structural change took place on the protein. The H/D exchange experiments clearly showed three major regions of change upon alkylation, one near the alkylation site and two distal from the modification. Those distal to the modification (in terms of amino acid sequence) were actually close in terms of tertiary structure. The largest change was α-helix C from residues 89 to 104. This particular region interacts with α-helix A which contains the modification site. The alkylation of Cys17 disrupted the side chain contacts of α-helix A to α-helix C, thus affording large changes in the extents of deuterium uptake in this particular region.

Interferon was also used as a system to assess biopharmaceutical comparability, including the development of new plotting techniques to display complex H/D exchange data sets in a concise and accurate manner. Comparability assessments are core character-ization activities in the biotechnology industry in that various analytical- and biological-based assays can determine if two or more biopharmaceutical products are deemed comparable without the need for clinical data. Houde et al. [77] demonstrated that H/D exchange has the ability to become a technique for routine use in biopharma-ceutical comparability assessments. A thorough discussion regarding the experimental uncertainty in H/D exchange data and the establishment of reliable difference limits to increase the confidence of comparability assessments was presented. Several graphical formats of comparability were suggested as a standard means of communicating data. The most useful, visually, was a mirror plot (or butterfly plot) of a control versus a test sample. A simple difference plot from the data clearly shows regions of similarity or difference. This difference plot, along with displayed confidence limits, affords visible and quantifiable evidence of comparability.

6.4   INTRODUCTION TO COVALENT LABELING TECHNIQUES

Protein footprinting via covalent labeling methods can be used to characterize biophar-maceuticals and, in particular, protein–protein and protein–ligand interactions. Covalent labeling methods utilize the formation of covalent bonds between a reagent and an amino acid side chain or between two side chains to probe macromolecules. The use of covalent labeling methods coupled with MS has been valuable for studying the structure of macromolecular complexes. This approach uses reagents that can either react nonspecifically with many amino acids or react specifically with certain amino acids. This section focuses on irreversible covalent labeling approaches. H/D exchange, discussed earlier, is a subset of many types of chemical labeling methods. The covalent labeling strategies discussed here permanently modify the side chain of amino acids thus these methods provide different information than H/D exchange and can be used in conjunction with H/D exchange to provide a greater level of information. The use of covalent labels is advantageous because of the irreversible nature of the label. These methods are not subject to the same limitation as H/D exchange where back-exchange and label scrambling are major concerns. Unlike H/D exchange, covalent labels may perturb protein structure, thus it is important to maintain tight control over the amount of labeling. Here, we will focus on three types of covalent labeling, hydroxyl radical labeling, chemical cross-linking, and specific amino acid labeling.

6.5   OVERVIEW AND APPLICATIONS OF HYDROXYL RADICAL FOOTPRINTING TO MASS SPECTROMETRY

In recent years, hydroxyl radicals (*OH) have been widely used as a probe for oxidative labeling of macromolecules [78–81]. Unlike specific amino acid labeling, *OH react with low specificity making it a more general method for covalent labeling. Hydroxyl radical footprinting was first developed by Tullius and Dombroski [82] to map DNA–protein interactions. Chance and coworkers [83] were the first to couple oxidative labeling with MS. The coupling of these two methods has greatly increased the utility of hydroxyl labeling for the study of macromolecules.

In proteins, *OH reacts with side chains of amino acids and in some cases can cause backbone bond cleavages. The radicals incorporate oxygen into amino acid side chains resulting in adducts of +16 Da. Other types of modifications can occur with varying mass changes [84]. In theory all amino acid side chains can react with *OH; however, only 14 amino acids are usefully modified in footprinting experiments [84]. The sulfur-containing amino acids cysteine and methionine are highly reactive with *OH as well as the aromatic residues tryptophan, tyrosine, phenylalanine, and histidine. In the case of DNA, *OH attacks the deoxyribose sugar on the surface eventually causing the backbone to break.

There are multiple methods to generate *OH for covalent labeling. These methods are described in detail by Xu and Chance [84]. Three radical generating methods, Fenton chemistry, synchrotron radiolysis, and laser photolysis, will be described briefly here. In Fenton chemistry methods, *OH is generated via the reduction of hydrogen peroxide

(H_2O_2) by Fe(II) ions. In their study, Tullius and Dombroski [82] used Fenton chemistry to generate $^{\bullet}OH$ for mapping DNA–protein interactions. Fenton-like reactions can also be carried out using other transition metals such as Cu(II), Co(II), Ni(II), and Mn(II). Fenton chemistry methods generate $^{\bullet}OH$ for footprinting with reaction times of minutes or longer [85]. The timescale of the reaction is a disadvantage when the interactions that are being studied are very fast.

The use of synchrotron X-rays for the radiolysis of water to generate $^{\bullet}OH$ was first described by Chance and coworkers [83]. In this reaction, ionized water molecules react with water to produce $^{\bullet}OH$. This method of generating hydroxyl radical has been used to study DNA–protein interactions, RNA folding, and protein–protein interactions. Synchrotron radiolysis of water generates $^{\bullet}OH$ on the millisecond timescale, faster than Fenton chemistry. This method is advantageous because no additional reagents have to be added to the sample to produce radicals. A major limitation of this method is the lack of accessibility to a synchrotron by everyday users.

A third method for radical generation is laser photolysis of H_2O_2. In this method, H_2O_2 directly absorbs photons, leading to its cleavage. Photolysis has been achieved using UV irradiation [86], a YAG laser [87], and an excimer laser [88]. These methods are more accessible to various labs because they require moderately expensive sources for photolysis. A disadvantage of these methods, however, is that they require an additional reagent, H_2O_2.

As discussed previously, covalent labeling methods are susceptible to label-induced conformational changes in the molecule of interest. This phenomenon has been observed in hydroxyl labeling where excess $^{\bullet}OH$ exposure leads to oxidation-induced altered protein conformations, thereby creating experimental artifacts [88–90]. Although these changes are considered small compared to other covalent methods owing to the small size of the incorporated oxygen [91], steps should be taken to minimize these structural changes.

Oxidation-induced conformational changes are well correlated with the time of exposure to $^{\bullet}OH$; therefore, it is important to minimize $^{\bullet}OH$ exposure time [88]. Fenton chemistry, synchrotron radiolysis, and UV irradiation labeling methods are on a timescale range of milliseconds to minutes. Studies have shown that protein unfolding can occur within this timescale, even as fast as 3 µs [92–94]. One method that aims to limit $^{\bullet}OH$ exposure to shorter than the fastest possible conformational change is fast photochemical oxidation of proteins (FPOP) developed by Hambly and Gross [80, 88]. FPOP reduces oxidation-induced conformational changes in two ways. First, this approach uses a pulsed laser directed at an optical window in a protein flow tube. The continuous flow of the protein ensures single exposure conditions. Second, a radical scavenger is used to reduce the labeling time to microseconds. The FPOP labeling method has been shown to be faster than protein unfolding [95]. This method reduces the number of experimental artifacts that result from oxidation-induced unfolding.

Similar to H/D exchange, hydroxyl radical labeling has been used to determine interaction sites primarily through the comparison of solvent accessibilities of an apo and holo state of a macromolecule. Hydroxyl radical labeling has been used to study the structure and interactions of varying macromolecules including DNA [85], RNA [96], and various protein systems [78, 81]. The method has also been extended to measure the thermodynamic stability of proteins. This technique, stability of proteins from rates of oxidation (SPROX), utilizes H_2O_2, to specifically oxidize methionine, in the presence of increasing concentrations of a chemical denaturant [97]. The folding free energy can be evaluated based on the dependence of the oxidation reaction rate on the denaturant concentration. The method is similar to the H/D exchange-based method SUPREX.

A particularly interesting application of hydroxyl radical labeling is in the study of integral membrane proteins [79, 98–100]. Recently, Gupta et al. [98] have used $^\bullet$OH to probe conformational changes that occur during the gating of the potassium channel Kir-Bac3.1. Although high-resolution structures are available for some eukaryotic channels, they all represent the closed state. No structural information is available for the open conformational state of channels. KirBac3.1 was labeled via synchrotron X-ray radiolysis of water in its stabilized open or closed states. The differential labeling between the two states was quantitated by LC/MS/MS. Structural changes between the two states were identified using $^\bullet$OH labeling. The data provided a more detailed model of the major conformational changes that occur during channel gating where other medium resolution methods such as cryo-electron microscopy could not provide such details [98]. The interpretation of results for this study was aided by the availability of a high-resolution structure for the closed state of the channel. This demonstrates the efficacy of combining protein footprinting data with high-resolution structural methods to provide a more detailed picture of protein interactions.

A recent application of FPOP for epitope mapping of biopharmaceuticals was demonstrated by Jones et al. [101]. In this study a model antibody–antigen complex of thrombin to antithrombin antibody was utilized to compare epitope-mapping data determined from X-ray crystallography, H/D exchange, and hydroxyl radical footprinting. Extents of oxidative modification in thrombin alone versus thrombin in the presence of the antibody have both increased and decreased levels, suggesting that portions of thrombin become inaccessible to modification, while other portions become more accessible to modification. The region identified by H/D exchange as the potential epitope was corroborated by the FPOP data, specifically monitored by approximately 14 residues that were modified.

6.6   OVERVIEW AND APPLICATIONS OF CHEMICAL CROSS-LINKING TO MASS SPECTROMETRY

Chemical cross-linking involves the formation of covalent bonds between pairs of functional groups within a protein and protein complex, or between a protein and its ligand, which includes peptides, small molecules, and nucleic acids. Cross-links can be either intermolecular or intramolecular contacts. Intermolecular cross-links occur between different molecules and can provide information about binding interactions between molecules. Intramolecular cross-links occur within parts of a molecule and can provide insight into the fold of that molecule. Both inter- and intramolecular cross-links can provide a greater level of information by defining through-space distance constraints on macromolecular interactions [102]. Coupled with molecular modeling methods, which results in a number of possible structures, distance restraints obtained by chemical cross-linking can be useful in reducing the number of predicted structures [102]. Cross-linking is also useful in studying transient, intermediate complexes [103, 104]. Chemical cross-links can trap intermediate assemblies or specific conformations to shed light on assembly and protein folding pathways.

Cross-linking combined with MS has several advantages. The size of the cross-linked complex is not limited with a bottom-up approach for MS analysis, MS requires small amount of protein (femtomoles) compared to other structural methods, and membrane proteins and protein mixtures are amenable to analysis [105]. Cross-linked samples have been analyzed using both bottom-up [106] and top-down [107, 108] MS approaches. Identification of cross-links by MS can be difficult because of the low abundance of the cross-linked species in the peptide digest milieu. The use of top-down MS eases this complication with gas phase isolation of the cross-linked species prior to fragmentation.

Commercially available cross-linkers are based on organic chemical reactions to target specific functional groups. Cross-linkers can be homobifunctional, heterobifunctional, or trifunctional. Homobifunctional cross-linkers have identical functional groups at both reactive sites while heterobifunctional cross-linkers contain two different reactive groups. Trifunctional cross-linkers have a third functional group that can either be used for cross-linking another species or for affinity purification.

One of the most widely used class of cross-linking reagents are amine reactive [105]. *N*-Hydroxysuccinimide (NHS) esters, imidoesters, and carbodiimides are all used to modify primary amines. Another class of cross-linkers target SH groups of cysteines via a maleimide functional group. There is also a class of photoreactive cross-linkers that include aryl azides, diazarines, and benzophenones. These nonspecific cross-linkers react with target molecules after exposure to UV light.

The length of the spacer regions of cross-linkers provides the basis for distance constraint information derived from cross-linking experiments. Commercially available cross-linkers can have a spacer region between functional groups of varying lengths. Some cross-linkers, however, are zero-length cross-linkers. They create a bond between two proteins without a linker. Carbodiimides, which react with carboxylic acids, are a widely used type of zero-length cross-linkers.

As mentioned earlier, a problem with the bottom-up approach to identifying cross-links is the low abundance of cross-linked species in the protein digest. To address this problem, different types of cross-linkers have been developed. The trifunctional cross-linkers contain a third functional group that can be used for affinity purification. Affinity cross-linkers generally contain a biotin group that allows for purification of the cross-linked species on avidin beads [109]. This purification step increases the identification of cross-linked species. Another method to increase cross-linker identification is the use of isotopically labeled cross-linkers. A mixture of labeled and unlabeled cross-links provides a distinctive isotopic pattern in the MS analysis that is readily identifiable [110].

Cross-linking has been applied to living cell systems to obtain information on the way cellular processes are organized. The coupling of *in vivo* cross-linking of cellular proteins is hampered owing to the increased degree of complexity of digested samples where the cross-linked sample is only a fraction of the total peptide levels [111]. The use of immunoaffinity chromatography after *in vivo* cross-linking has been successful in identifying protein–protein interactions. Vasilescu et al. [112] used formaldehyde, which rapidly permeates cell membranes, to generate protein–protein cross-links *in vivo*. Immunoaffinity chromatography was used to purify the proteins cross-linked to the tagged protein of interest. After SDS-PAGE separation the proteins were analyzed by MS and a novel interaction partner was identified [112].

A recent development in cross-linking living cells that does not require purification of the cross-linked samples is the first example of a nontargeted *in vivo* cross-linking strategy. This method employs a novel class of cross-linkers called protein interaction reporters (PIRs) [113]. In these PIR compounds, two specific mass spectrometry-cleavable bonds, a mass-encoded reporter tag, and an affinity tag are incorporated into the spacer chain of a traditional cross-linker. The cross-linked cells are analyzed by a novel two-stage MS strategy [113]. The presence of two highly specific MS-cleavable bonds allows for the detection of the cross-linked peptide or peptide complex during a low-energy precursor scan. The released intact peptide masses are then identified in a subsequent high-energy scan. Zhang et al. [113] successfully used this method to identify a set of protein–protein interactions in the microbial system *Shewanella oneidensis*.

The use of antibodies as biopharmaceuticals has increased the need for characterization of antibody interactions. Epitope mapping is valuable in gaining information on how an antibody interacts with its antigen. The coupling of chemical cross-linking with MS has been useful in epitope mapping. Pimenova et al. [114] has used chemical cross-linking to characterize the epitope on the bovine prion protein bPrP(25–41). The cross-linking pattern of bPrP alone was compared to the pattern of the antigen when bound to its antibody. Two isotope-labeled amine-reactive cross-linkers with varying spacer arms were used to cross-link the samples. The cross-linked peptides were analyzed using Fourier transform MS. In this case, the amino acid sequence of the antibody is unknown so cross-links between the antigen and antibody could not be determined using proteomic methods. The epitope was determined by identifying the mono-linked peptides from bPrP alone and in the presence of antibody. The mono-links not observed in the antibody-bound samples were presumed to be involved in binding to the antibody. The antibody epitope on bPrP(25–41) was identified by this method using only 200 pmol of antigen [114]. This technique provides a powerful tool for identifying epitopes without consuming a large amount of sample. This further provides evidence that chemical cross-linking can be useful in characterizing protein therapeutics.

6.7 OVERVIEW AND APPLICATIONS OF SPECIFIC AMINO ACID LABELING TO MASS SPECTROMETRY

A third covalent labeling method is specific amino acid labeling. Unlike H/D exchange and hydroxyl radical labeling, this footprinting method is not general. Rather, the method specifically targets certain amino acids that are important in complex formation of the system being studied. For example, arginine is the second most prevalent amino acid at protein–protein interfaces due to its involvement in electrostatic interactions [115]. The labeling of this specific amino acid in determining protein–protein interactions is very useful.

The challenges of specific amino acid labeling are similar to those faced in other covalent labeling methods. Amino acid-specific labels are relatively large, thereby increasing the possibility that they will alter protein structure. In addition, these labels can alter the side chain charge or hydrophobicity, further increasing the likelihood of structural perturbation. The prospect of structural changes can be reduced by limiting the extent of modification. These challenges as well as those regarding determining the extent of modification when the label alters the ionization efficiency of the measured peptide are reviewed by Mendoza and Vachet [116].

Specific amino acid labeling relies on reagents that use organic reactions. Of the 20 amino acids, only 8 have been used to monitor protein structure [116]. These eight residues include arginine, aspartate, glutamate, cysteine, histidine, lysine, tryptophan, and tyrosine. The properties of these amino acids make them useful depending on the information that is being determined. To probe protein structure, charged residues such as aspartate and glutamate are advantageous because they are commonly found on the surface of proteins. Due to their ability to form salt bridges, however, they are often found at protein–protein interfaces. This also makes them appropriate for analyzing protein–ligand and protein–protein interactions.

Specific amino acid labeling is especially suited for mapping metal-binding sites. Aspartate and glutamate are found in calcium-binding proteins where they are the predominant ligands for Ca(II) [117]. Histidine, a common ligand in metalloproteins that bind transition metals, is modified using diethylpyrocarbonate (DEPC). DEPC modification of histidine was used to assess the Cu(II) binding sites of the prion protein [118]. The alkylation of cysteines using iodoacetamide [119] and NEM [120] has also been used to study the binding sites of metalloproteins.

Specific amino acid labeling has been used to study some biotherapeutics. Kalkum et al. [121] used DEPC to modify histidine residues on insulin. DEPC modifies histidine with a carboethoxy group. At low DEPC concentrations a single modification is observed. Increased concentration of DEPC can lead to two modifications on histidine imidazole ring. In their study, Kalkum et al. [121] incubated histidine with various concentrations of DEPC. The extent of modification of histidine residues in insulin was determined by MALDI-MS. They found that the reactivity agreed well with surface accessibility calculations of imidazole nitrogen atoms thus providing information about the protein tertiary structure.

Specific amino acid modification has also been used for epitope mapping. The first application of specific amino acid labeling to map conformationally dependent epitopes was reported by Fiedler et al [122]. They used arginine, lysine, and tyrosine modification to map the epitope of hen egg white lysozyme that is recognized by an IgM-type antibody. Arginine modification was also used to map the epitope of the E2 protein, a hepatitis C virus envelope glycoprotein [123]. In both studies arginine was modified using 1,2-cyclohexanedione (CHD) which results in a mass increase of 112 Da. This method relies on the differential labeling of arginine in the antigen compared with the immune complex. In the E2 protein study, the modified samples were digested with protease and analyzed by LC-MS/MS. A single arginine residue, Arg587, was modified in E2 alone but not modified in the complex. Two other arginine residues, Arg630 and Arg651, were modified in the complex but to a lesser extent than in E2 alone. These data indicate these three arginine residues are either involved in the antibody-binding site or are part of a region that undergoes a conformational change upon antibody binding [123].

Other applications of specific amino acid labeling coupled with MS to characterize protein therapeutics include the mapping of the heparin-binding region of bovine seminal plasma using arginine modification [124] and disulfide mapping of antibodies using NEM alkylation [125, 126].

6.8 CONCLUSIONS

Mass spectrometry-based techniques to confirm the primary structure of biopharmaceuticals will continue to afford valuable information regarding amino acid sequence fidelity, posttranslational modifications, and degradation pathways. In recent years, there is an increasing focus to apply chemical labeling approaches to discern secondary and tertiary structure properties of proteins, which will expand on traditional biophysical techniques for higher order structure analysis. H/D exchange labels proteins in solution in a formulation of choice which makes the technique highly valuable to the biopharmaceutical industry. With the advent of automated sample-handling systems, high-resolution separations, and automated software analysis, H/D exchange is becoming more commonplace in routine biopharmaceutical characterization/comparability workflows. Other chemical labeling approaches, to date, are not as common in the biopharmaceutical industry, but their application from academic labs is rapidly increasing. A major advantage for covalent labeling approach is just that, the modification is a permanent entity of the peptide/protein until MS analysis. A downfall to covalent labeling is the inability to automate sample preparation, work up data, and display large data sets in a concise manner. These methods can be used in a stand-alone fashion or in combination with one another to obtain a higher sensitivity and resolution of the structural properties of biopharmaceuticals.

REFERENCES

1. Zhang, Z., Pan, H., and Chen, X. (2008) Mass spectrometry for structural characterization of therapeutic antibodies. *Mass Spectrom. Rev.*, **28**(1), 147–176.

2. Domon, B. and Aebersold, R. (2006) Mass spectrometry and protein analysis. *Science*, **312**, 212–217.

3. Chen, G., et al. (2011) Characterization of protein therapeutics by mass spectrometry: recent developments and future directions. *Drug Discov. Today*, **16**(1/2), 58–64.

4. Fenn, J.B., et al. (1989) Electrospray ionization for mass spectrometry of large biomolecules. *Science*, **246**, 64–71.

5. Fenn, J.B. (2003) Electrospray wings for molecular elephants—nobel lecture. *Angew. Chem. Int. Ed.*, **42**, 3871–3894.

6. Smith, R.D. and Loo, J.A. (1990) New developments in biochemical mass spectrometry: electrospray ionization. *Anal. Chem.*, **62**, 882–899.

7. Collette, C. and Pauw, E.D. (1998) Calibration of the internal energy distribution of ions produced by electrospray. *Rapid Commun. Mass Spectrom.*, **12**, 165–170.

8. Loo, J.A. (1997) Studying noncovalent protein complexes by electrospray ionization mass spectrometry. *Mass Spectrom. Rev.*, **16**, 1–23.

9. Schwartz, J.C. and Jardine, I. (1996) Quadrupole ion trap mass spectrometry. *Methods Enzymol.*, **270**, 552–586.

10. Jonscher, K.R. and Yates, J.R. (1997) The quadrupole ion trap mass spectrometer— a small solution to a big challenge. *Anal. Biochem.*, **244**(1), 1–15.

11. Fitzgerald, M.C., et al. (1996) Probing the oligomeric structure of an enzyme by electrospray ionization time-of-flight mass spectrometry. *Proc. Natl. Acad. Sci.*, **93**(14), 6851–6856.

12. Marshall, A.G., Hendrickson, C.L., and Jackson, G.S. (1998) Fourier transform ion cyclotron resonance mass spectrometry: a primer. *Mass Spectrom. Rev.*, **17**(1), 1–35.

13. Loo, J.A., Edmonds, C.G., and Smith, R.D. (1990) Primary sequence information from intact proteins by electrospray ionization tandem mass spectrometry. *Science*, **248**, 201–204.

14. Borchers, C., Parker, C.E., Deterding, L.J., and Tomer, K.B. (1999) Preliminary comparison of precursor scans and liquid chromatography-tandem mass spectrometry on a hybrid quadrupole time-of-flight mass spectrometer. *J. Chromatogr. A*, **854**, 119–130.

15. Chernushevich, I.V., Loboda, A.V., and Thomson, B.A. (2001) An introduction to quadrupole-time-of-flight mass spectrometry. *J. Mass Spectrom.*, **36**, 849–865.

16. Morris, H.R., et al. (1996) High sensitivity collisionally-activated decomposition tandem mass spectrometry on a novel quadrupole/orthogonal-acceleration time-of-flight mass spectrometer. *Rapid Commun. Mass Spectrom.*, **10**(8), 889–896.

17. Heeren, R.M.A., Kleinnijenhuis, A.J., McDonnell, L.A., and Mize, T.H. (2004) A mini-review of mass spectrometry using high-performance FTICR-MS methods. *Anal. Bioanal. Chem.*, **378**(4), 1048–1058.

18. Pauling, L. and Corey, R.B. (1950) Two hydrogen-bonded spiral configurations of the polypeptide chain. *J. Am. Chem. Soc.*, **72**, 5349.

19. Pauling, L. and Corey, R.B. (1951) Atomic coordinates and structure factors for two helical configurations of polypeptide chains. *Proc. Natl. Acad. Sci.*, **37**, 235–240.

20. Pauling, L. and Corey, R.B. (1951) The structure of synthetic polypeptides. *Proc. Natl. Acad. Sci.*, **37**, 241–250.

21. Pauling, L. and Corey, R.B. (1951) The pleated sheet, a new layer configuration of polypeptide chains. *Proc. Natl. Acad. Sci.*, **37**, 251–256.

22. Pauling, L., Corey, R.B., and Branson, H.R. (1951) The structure of proteins: two hydrogen-bonded helical configurations of the polypeptide chain. *Proc. Natl. Acad. Sci.*, **37**, 205–211.

23. Hvidt, A. and Linderstrøm-Lang, K. (1954) Exchange of hydrogen atoms in insulin with deuterium atoms in aqueous solutions. *Biochim. Biophys. Acta*, **14**(4), 574–575.

24. Hvidt, A. and Linderstrøm-Lang, K. (1954) The kinetics of the deuterium exchange of insulin with D_2O. An amendment. *Biochim. Biophys. Acta*, **16**(1), 168–169.

25. Linderstrøm-Lang, K. (1955) The pH-dependence of the deuterium exchange of insulin. *Biochim. Biophys. Acta*, **18**(2), 308.

26. Benson, E.E. and Linderstrøm-Lang, K. (1959) Deuterium exchange between myoglobin and water. *Biochim. Biophys. Acta*, **32**, 579–581.

27. Hvidt, A. and Nielsen, S.O. (1966) Hydrogen exchange in proteins. *Adv. Protein Chem.*, **21**, 287–386.

28. Englander, S.W., Mayne, L., Bai, Y., and Sosnick, T.R. (1997) Hydrogen exchange: the modern legacy of Linderstrom-Lang. *Protein Sci.*, **6**, 1101–1109.

29. Bai, Y., Sosnick, T.R., Mayne, L., and Englander, S.W. (1995) Protein folding intermediates: native-state hydrogen exchange. *Science*, **269**, 192–197.

30. Clarke, J., Itzhaki, L.S., and Fersht, A.R. (1997) Hydrogen exchange at equilibrium: a short cut for analysing protein-folding pathways? *Trends Biochem. Sci.*, **22**, 284–287.

31. Dempsey, C.E. (2001) Hydrogen exchange in peptides and proteins using NMR spectroscopy. *Prog. Nucl. Magn. Reson. Spectrosc.*, **39**, 135–170.

32. Englander, S.W. (2006) Hydrogen exchange mass spectrometry: a historical perspective. *J. Am. Soc. Mass Spectrom.*, **17**(11), 1481–1489.

33. Englander, S.W., Downer, N.W., and Teitelba, H. (1972) Hydrogen-exchange. *Annu. Rev. Biochem.*, **41**, 903–924.

34. Englander, S.W. and Kallenbach, N.R. (1984) Hydrogen exchange and structural dynamics of proteins and nucleic acids. *Q. Rev. Biophys.*, **16**, 521–655.

35. Klotz, I.M. and Frank, B.H. (1965) Deuterium-hydrogen exchange in amide N-H groups. *J. Am. Chem. Soc.*, **87**(12), 2721–2728.

36. Maier, C.S. and Deinzer, M.L. (2005) Protein conformations, interactions, and H/D exchange. *Methods Enzymol.*, **402**, 312–360.

37. Woodward, C., Simon, I., and Tuchsen, E. (1982) Hydrogen exchange and the dynamic structure of proteins. *Mol. Cell. Biochem.*, **48**, 135–160.

38. Leichtling, B.H. and Klotz, I.M. (1966) Catalysis of hydrogen-deuterium exchange in polypeptides. *Biochemistry*, **5**(12), 4026–4037.

39. Hilton, B.D. and Woodward, C.K. (1978) Nuclear magnetic resonance measurement of hydrogen exchange kinetics of single protons in basic pancreatic trypsin inhibitor. *Biochemistry*, **17**(16), 3325–3332.

40. Roder, H., Wagner, G., and Wuthrich, K. (1985) Individual amide proton exchange rates in thermally unfolded basic pancreatic trypsin inhibitor. *Biochemistry*, **24**, 7407–7411.

41. Bai, Y., Milne, J.S., Mayne, L., and Englander, S.W. (1993) Primary structure effects on peptide group hydrogen exchange. *Proteins Struct. Funct. Genet.*, **17**, 75–86.

42. Wales, T.E. and Engen, J.R. (2006) Hydrogen exchange mass spectrometry for the analysis of protein dynamics. *Mass Spectrom. Rev.*, **25**(1), 158–170.

43. Miranker, A., et al. (1993) Detection of transient protein folding populations by mass spectrometry. *Science*, **262**(5135), 896–900.

44. Katta, V. and Chait, B.T. (1991) Conformational changes in proteins probed by hydrogen-exchange electrospray-ionization mass spectrometry. *Rapid Commun. Mass Spectrom.*, **5**(4), 214–217.

45. Engen, J.R. and Smith, D.L. (2001) Investigating protein structure and dynamics by hydrogen exchange MS. *Anal. Chem.*, **73**(9), 256A–265A.

46. Ferraro, D.M., Lazo, N.D., and Robertson, A.D. (2004) EX1 hydrogen exchange and protein folding. *Biochemistry*, **43**, 587–594.

47. Miller, D.W. and Dill, K.A. (1995) A statistical mechanical model for hydrogen exchange in globular proteins. *Protein Sci.*, **4**, 1860–1873.

48. Weis, D.D., et al. (2006) Identification and characterization of EX1 kinetics in H/D exchange mass spectrometry by peak width analysis. *J. Am. Soc. Mass Spectrom.*, **17**, 1498–1509.

49. Rosenberg, A. and Chakravarti, K. (1968) Studies of hydrogen exchange in proteins. I. The exchange kinetics of bovine carbonic anhydrase. *J. Biol. Chem.*, **243**, 5193–5201.

50. Roder, H., Wagner, G., and Wuthrich, K. (1985) Amide proton exchange in proteins by EX1 kinetics: studies of the basic pancreatic trypsin inhibitor at variable pH2 and temperature. *Biochemistry*, **24**, 7396–7407.

51. Sivaraman, T. and Robertson, A.D. (2001) Kinetics of conformational fluctuations by EX1 hydrogen exchange in native proteins. *Methods. Mol. Biol.*, **168**, 193–214.

52. Wales, T.E. and Engen, J.R. (2006) Partial unfolding of diverse SH3 domains on a wide time scale. *J. Mol. Biol.*, **357**, 1592–1604.

53. Arrington, C.B. and Robertson, A.D. (2000) Microsecond to minute dynamics revealed by EX1-type hydrogen exchange at nearly every backbone hydrogen bond in a native protein. *J. Mol. Biol.*, **296**, 1307–1317.

54. Miranker, A., et al. (1993) Detection of transient protein folding populations by mass spectrometry. *Science*, **262**, 896–900.

55. Hamuro, Y., Weber, P.C., and Griffin, P.R. (2005) *High-throughput analysis of protein structure by hydrogen/deuterium exchange mass spectrometry*. In: Figeys, D. editor. *Industrial Proteomics*. Wiley. Chapter 5.

56. Hoofnagle, A.N., Resing, K.A., and Ahn, N.G. (2003) Protein analysis by hydrogen exchange mass spectrometry. *Annu. Rev. Biophys. Biomol. Struct.*, **32**, 1–25.

57. Smith, D.L. and Dharmasiri, K. (1998) *Protein-Ligand Binding Studied by Amide Hydrogen Exchange and Mass Spectrometry*. Springer.

58. Tsutsui, Y. and Wintrode, P.L. (2007) Hydrogen/deuterium exchange mass spectrometry: a powerful tool for probing protein structure, dynamics and interactions. *Curr. Med. Chem.*, **14**, 2344–2358.

59. Hossain, B.M. and Konermann, L. (2006) Pulsed hydrogen/deuterium exchange MS/MS for studying the relationship between noncovalent protein complexes in solution and in the gas phase after electrospray ionization. *Anal. Chem.*, **78**, 1613–1619.

60. Pan, J., Wilson, D.J., and Konermann, L. (2005) Pulsed hydrogen exchange and electrospray charge-state distribution as complementary probes of protein structure in kinetic experiments: implications for ubiquitin folding. *Biochemistry*, **44**(24), 8627–8633.

61. Yang, H. and Smith, D.L. (1997) Kinetics of cytochrome c folding examined by hydrogen exchange and mass spectrometry. *Biochemistry*, **36**, 14992–14999.

62. Zhang, Z. and Smith, D.L. (1993) Determination of amide hydrogen exchange by mass spectrometry: a new tool for protein structure elucidation. *Protein Sci.*, **2**, 522–531.

63. Kipping, M. and Schierhorn, A. (2003) Improving hydrogen/deuterium exchange mass spectrometry by reduction of the back-exchange effect. *J. Mass Spectrom.*, **38**(3), 271–276.

64. Woods, V.L. (1997) Method for characterization of the fine structure of protein binding sites. USA patent 5,658,739.

65. Woods, V.L. (2001) Methods for the high-resolution identification of solvent-accessible amide hydrogens in polypeptides or proteins and for characterization of the fine structure of protein binding sites. USA patent 6,291,189.

66. Woods, V.L. (2001) Methods for identifying hot-spot residues of binding proteins and small compounds that bind to the same. USA patent 6,599,707.

67. Woods, V.L. (2001) Method for characterization of the fine structure of protein binding sites. USA patent 6,331,400.

68. Chalmers, M.J., et al. (2006) Probing protein-ligand interactions by automated hydrogen/deuterium exchange mass spectrometry. *Anal. Chem.*, **78**(4), 1005–1014.

69. Hotchko, M., Anand, G.S., Komives, E.A., and Eyck, L.F.T. (2006) Automated extraction of backbone deuteration levels from amide H/^{2}H mass spectrometry experiments. *Protein Sci.*, **15**, 583–601.

70. Weis, D.D., Engen, J.R., and Kass, I.J. (2006) Semi-automated data processing of hydrogen exchange mass spectra using *HX-Express*. *J. Am. Soc. Mass Spectrom.*, **17**(12), 1700–1703.

71. Pascal, B.D., et al. (2007) The deuterator: software for the determination of backbone amide deuterium levels from H/D exchange MS data. *BMC Bioinform.*, **8**, 156–167.

72. Houde, D., et al. (2009) Characterization of IgG1 conformation and conformational dynamics by hydrogen/deuterium exchange mass spectrometry. *Anal. Chem.*, **81**, 2644–2651.

73. Burkitt, W., Domann, P., and O'Connor, G. (2010) Conformation changes in oxidatively stressed monoclonal antibodies studied by hydrogen exchange mass spectrometry. *Protein Sci.*, **19**, 826–835.

74. Houde, D., Peng, Y., Berkowitz, S.A., and Engen, J.R. (2010) Post-translational modifications differentially affect IgG1 conformation and receptor binding. *Mol. Cell. Proteomics*, **9**(8), 1716–1728.

75. Bobst, C.E., et al. (2008) Detection and characterization of altered conformations of protein pharmaceuticals using complementary mass spectrometry-based approaches. *Anal. Chem.*, **80**, 7473–7481.

76. Kaltashov, I.A., et al. (2010) Conformation and dynamics of biopharmaceuticals: transition of mass spectrometry-based tools from academe to industry. *J. Am. Soc. Mass Spectrom.*, **21**, 323–337.

77. Houde, D., Berkowitz, S.A., and Engen, J.R. (2010) The utility of hydrogen/deuterium exchange mass spectrometry in biopharmaceutical comparability studies. *J. Pharm. Sci.*, **100**(6), 2071–2086.

78. Zhang, H., et al. (2010) Fast photochemical oxidation of proteins for comparing structures of protein-ligand complexes: the calmodulin-peptide model system. *Anal. Chem.*, **83**(1), 311–318.

79. Pan, Y., Brown, L., and Konermann, L. (2009) Mapping the structure of an integral membrane protein under semi-denaturing conditions by laser-induced oxidative labeling and mass spectrometry. *J. Mol. Biol.*, **394**(5), 968–981.

80. Hambly, D. and Gross, M. (2007) Laser flash photochemical oxidation to locate heme binding and conformational changes in myoglobin. *Int. J. Mass Spectrom.*, **259**(1–3), 124–129.

81. Sharp, J.S., Sullivan, D.M., Cavanagh, J., and Tomer, K.B. (2006) Measurement of multisite oxidation kinetics reveals an active site conformational change in Spo0F as a result of protein oxidation. *Biochemistry*, **45**(20), 6260–6266.

82. Tullius, T.D. and Dombroski, B.A. (1986) Hydroxyl radical "footprinting": high-resolution information about DNA-protein contacts and application to lambda repressor and Cro protein. *Proc. Natl. Acad. Sci. U.S.A.*, **83**(15), 5469–5473.

83. Maleknia, S.D., Brenowitz, M., and Chance, M.R. (1999) Millisecond radiolytic modification of peptides by synchrotron X-rays identified by mass spectrometry. *Anal. Chem.*, **71**(18), 3965–3973.

84. Xu, G. and Chance, M.R. (2007) Hydroxyl radical-mediated modification of proteins as probes for structural proteomics. *Chem. Rev.*, **107**(8), 3514–3543.

85. Brenowitz, M., Chance, M.R., Dhavan, G., and Takamoto, K. (2002) Probing the structural dynamics of nucleic acids by quantitative time-resolved and equilibrium hydroxyl radical "footprinting". *Curr. Opin. Struct. Biol.*, **12**(5), 648–653.

86. Sharp, J.S., Becker, J.M., and Hettich, R.L. (2004) Analysis of protein solvent accessible surfaces by photochemical oxidation and mass spectrometry. *Anal. Chem.*, **76**(3), 672–683.

87. Aye, T.T., Low, T.Y., and Sze, S.K. (2005) Nanosecond laser-induced photochemical oxidation method for protein surface mapping with mass spectrometry. *Anal. Chem.*, **77**(18), 5814–5822.

88. Hambly, D.M. and Gross, M.L. (2005) Laser flash photolysis of hydrogen peroxide to oxidize protein solvent-accessible residues on the microsecond timescale. *J. Am. Soc. Mass. Spectrom.*, **16**(12), 2057–2063.

89. Sharp, J.S. and Tomer, K.B. (2007) Analysis of the oxidative damage-induced conformational changes of apo- and holocalmodulin by dose-dependent protein oxidative surface mapping. *Biophys. J.*, **92**(5), 1682–1692.

90. Shum, W.K., Maleknia, S.D., and Downard, K.M. (2005) Onset of oxidative damage in alpha-crystallin by radical probe mass spectrometry. *Anal. Biochem.*, **344**(2), 247–256.

91. Konermann, L., Stocks, B.B., Pan, Y., and Tong, X. (2010) Mass spectrometry combined with oxidative labeling for exploring protein structure and folding. *Mass Spectrom. Rev.*, **29**(4), 651–667.

92. Chung, H.S., Ganim, Z., Jones, K.C., and Tokmakoff, A. (2007) Transient 2D IR spectroscopy of ubiquitin unfolding dynamics. *Proc. Natl. Acad. Sci. U.S.A.*, **104**(36), 14237–14242.

93. Chung, H.S., et al. (2005) Conformational changes during the nanosecond-to-millisecond unfolding of ubiquitin. *Proc. Natl. Acad. Sci. U.S.A.*, **102**(3), 612–617.

94. Naganathan, A.N. and Munoz, V. (2005) Scaling of folding times with protein size. *J. Am. Chem. Soc.*, **127**(2), 480–481.

95. Gau, B.C., Sharp, J.S., Rempel, D.L., and Gross, M.L. (2009) Fast photochemical oxidation of protein footprints faster than protein unfolding. *Anal. Chem.*, **81**(16), 6563–6571.

96. Brenowitz, M., Erie, D.A., and Chance, M.R. (2005) Catching RNA polymerase in the act of binding: intermediates in transcription illuminated by synchrotron footprinting. *Proc. Natl. Acad. Sci. U.S.A.*, **102**(13), 4659–4660.

97. West, G.M., Tang, L., and Fitzgerald, M.C. (2008) Thermodynamic analysis of protein stability and ligand binding using a chemical modification- and mass-spectrometry based strategy. *Anal. Chem.*, **80**(11), 4175–4185.

98. Gupta, S., et al. (2010) Conformational changes during the gating of a potassium channel revealed by structural mass spectrometry. *Structure*, **18**(7), 839–846.

99. Pan, Y., Brown, L., and Konermann, L. (2010) Site-directed mutagenesis combined with oxidative methionine labeling for probing structural transitions of a membrane protein by mass spectrometry. *J. Am. Soc. Mass Spectrom.*, **21**(11), 1947–1956.

100. Pan, Y., Stocks, B.B., Brown, L., and Konermann, L. (2009) Structural characterization of an integral membrane protein in its natural lipid environment by oxidative methionine labeling and mass spectrometry. *Anal. Chem.*, **81**(1), 28–35.

101. Jones, L.M., Sperry, J.B., Carroll, J.A., and Gross, M.L. (2011) Fast photochemical oxidation of proteins for epitope mapping. *Anal. Chem.*, **83**(20), 7657–7661.

102. Back, J.W., de Jong, L., Muijsers, A.O., and de Koster, C.G. (2003) Chemical cross-linking and mass spectrometry for protein structural modeling. *J. Mol. Biol.*, **331**(2), 303–313.

103. Kurucz, E., et al. (2002) Assembly of the drosophila 26 S proteasome is accompanied by extensive subunit rearrangements. *Biochem. J.*, **365**(Pt 2), 527–536.

104. Baker, A., et al. (2000) Biochemical and molecular approaches to understanding protein import into peroxisomes. *Biochem. Soc. Trans.*, **28**(4), 499–504.

105. Sinz, A. (2006) Chemical cross-linking and mass spectrometry to map three-dimensional protein structures and protein-protein interactions. *Mass Spectrom. Rev.*, **25**(4), 663–682.

106. Young, M.M., et al. (2000) High throughput protein fold identification by using experimental constraints derived from intramolecular cross-links and mass spectrometry. *Proc. Natl. Acad. Sci. U.S.A.*, **97**(11), 5802–5806.

107. Kruppa, G.H., Schoeniger, J., and Young, M.M. (2003) A top down approach to protein structural studies using chemical cross-linking and Fourier transform mass spectrometry. *Rapid Commun. Mass Spectrom.*, **17**(2), 155–162.

108. Novak, P., Young, M.M., Schoeniger, J.S., and Kruppa, G.H. (2003) A top-down approach to protein structure studies using chemical cross-linking and Fourier transform mass spectrometry. *Eur. J. Mass Spectrom. (Chichester, Eng.)*. **9**(6), 623–631.

109. Alley, S.C., Ishmael, F.T., Jones, A.D., and Benkovic, S.J. (2000) Mapping protein-protein interactions in the bacteriophage T4 DNA polymerase holoenzyme using a novel tri-functional photo-cross-linking and affinity reagent. *J. Am. Chem. Soc.* **122**(25), 6126–6127.

110. Muller, D.R., et al. (2001) Isotope tagged cross linking reagents. A new tool in mass spectrometric protein interaction analysis. *Anal. Chem.*, **73**(9), 1927–1934.

111. Rinner, O., et al. (2008) Identification of cross-linked peptides from large sequence databases. *Nat. Methods*, **5**(4), 315–318.

112. Vasilescu, J., Guo, X., and Kast, J. (2004) Identification of protein-protein interactions using in vivo cross-linking and mass spectrometry. *Proteomics*, **4**(12), 3845–3854.

113. Zhang, H., et al. (2009) Identification of protein-protein interactions and topologies in living cells with chemical cross-linking and mass spectrometry. *Mol. Cell. Proteomics*, **8**(3), 409–420.

114. Pimenova, T., et al. (2008) Epitope mapping on bovine prion protein using chemical cross-linking and mass spectrometry. *J. Mass Spectrom.*, **43**(2), 185–195.

115. Bogan, A.A. and Thorn, K.S. (1998) Anatomy of hot spots in protein interfaces. *J. Mol. Biol.*, **280**(1), 1–9.

116. Mendoza, V.L. and Vachet, R.W. (2009) Probing protein structure by amino acid-specific covalent labeling and mass spectrometry. *Mass Spectrom. Rev.*, **28**(5), 785–815.

117. Pidcock, E. and Moore, G.R. (2001) Structural characteristics of protein binding sites for calcium and lanthanide ions. *J. Biol. Inorg. Chem.*, **6**(5–6), 479–489.

118. Qin, K., Yang, Y., Mastrangelo, P., and Westaway, D. (2002) Mapping Cu(II) binding sites in prion proteins by diethyl pyrocarbonate modification and matrix-assisted laser desorption ionization-time of flight (MALDI-TOF) mass spectrometric footprinting. *J. Biol. Chem.*, **277**, 1981–1990.

119. Gonzalez de Peredo, A., et al. (1999) Identification of the two zinc-bound cysteines in the ferric uptake regulation protein from Escherichia coli: chemical modification and mass spectrometry analysis. *Biochemistry*, **38**(26), 8582–8589.

120. Apuy, J.L., et al. (2001) Ratiometric pulsed alkylation/mass spectrometry of the cysteine pairs in individual zinc fingers of MRE-binding transcription factor-1 (MTF-1) as a probe of zinc chelate stability. *Biochemistry*, **40**(50), 15164–15175.

121. Kalkum, M., Przybylski, M., and Glocker, M.O. (1998) Structure characterization of functional histidine residues and carbethoxylated derivatives in peptides and proteins by mass spectrometry. *Bioconjug. Chem.*, **9**(2), 226–235.

122. Fiedler, W., et al. (1998) Molecular characterization of a conformational epitope of hen egg white lysozyme by differential chemical modification of immune complexes and mass spectrometric peptide mapping. *Bioconjug. Chem.*, **9**(2), 236–241.

123. Iacob, R.E., et al. (2008) Structural elucidation of critical residues involved in binding of human monoclonal antibodies to hepatitis C virus E2 envelope glycoprotein. *Biochim. Biophys. Acta Proteins Proteomics*, **1784**(3), 530–542.

124. Calvete, J.J., Campanero-Rhodes, M.A., Raida, M., and Sanz, L. (1999) Characterisation of the conformational and quaternary structure-dependent heparin-binding region of bovine seminal plasma protein PDC-109. *FEBS Lett.*, **444**(2–3), 260–264.

125. Zhang, W., Marzilli, L.A., Rouse, J.C., and Czupryn, M.J. (2002) Complete disulfide bond assignment of a recombinant immunoglobulin G4 monoclonal antibody. *Anal. Biochem.*, **311**(1), 1–9.

126. Yen, T.Y., Yan, H., and Macher, B.A. (2002) Characterizing closely spaced, complex disulfide bond patterns in peptides and proteins by liquid chromatography/electrospray ionization tandem mass spectrometry. *J. Mass Spectrom.*, **37**(1), 15–30.

7

APPLICATION OF BIOPHYSICAL AND HIGH-THROUGHPUT METHODS IN THE PREFORMULATION OF THERAPEUTIC PROTEINS —FACTS AND FICTIONS

Ahmad M. Abdul-Fattah and Hanns-Christian Mahler

Pharmaceutical Development and Supplies, Pharma Technical Development Biologics EU, F. Hoffmann-La Roche Ltd, Basel, Switzerland

Biophysical Methods for Biotherapeutics: Discovery and Development Applications, First Edition. Edited by Tapan K. Das.
© 2014 John Wiley & Sons, Inc. Published 2014 by John Wiley & Sons, Inc.

7.1 INTRODUCTION

To date, over 75 therapeutic protein drugs, also known as biopharmaceuticals, are approved worldwide, and more than 500 other therapeutic proteins are currently in development [1, 2]. A key success factor for the commercialization and the quality and speed of the development program is the pharmaceutical product design, providing, for example, stability, usability, and convenience. The overall formulation strategy of a company and timing of introduction of a commercial formulation during clinical development drive a number of key decisions. A formulation planned for commercialization is typically introduced latest for use in clinical phase 3 studies, in order to minimize changes to the product between phase 3 and launch, thus reducing the potential risk of any comparability studies between pivotal clinical trials and launch. Interestingly, a number of companies in the pharmaceutical industry seem to be adopting the use of commercial formulations in earlier clinical studies, as early as clinical phase 1 trials. However, a number of considerations challenge a very early investment into huge formulation development studies (and hence incur higher costs of development).

First, protein degradation mechanisms are diverse and complex [3, 4]. Depending on the conditions encountered, a protein is subject to degradation in a variety of pathways. Chemical degradation pathways include oxidation, deamidation, isomerization, fragmentation, disulfide shuffling, and reactions with formulation components such as reducing sugars to form adducts of Maillard reactions (e.g., glycation). Proteins are structurally defined by primary (amino acid sequence), secondary, tertiary (folding to structural elements such as α-helix, β-sheet, or random coil and higher order structure), and even quaternary structure. Physical instabilities include structural alterations, such as denaturation (unfolding), adsorption, aggregation, or precipitation.

Second, comprehensive understanding of all degradation pathways may not be possible in early stages of development. The availability of adequate and representative analytical methods may pose a major challenge during early stages of development. For example, new impurities and product heterogeneities may be found upon optimization or introduction of new analytical methods. Also, analytical method development to study specific degradation pathways may not be complete. In essence, if one is not able to study specific pathways such as oxidation due to challenges in analytical method availability, it is of course not possible to optimize a formulation against that attribute.

Third, changes in fermentation and purification processes limit availability of representative developmental material in very early stages, and hence the depth and level of formulation studies—thereby impacting a solid decision-making related to formulations and product design in early stages. Furthermore, any changes in the fermentation and purification process may potentially alter the stability profile of the protein, making it challenging to determine formulations intended for commercialization early on.

Fourth, with clinical and nonclinical studies progressing along pharmaceutical development, the doses intended for commercialization of novel biopharmaceuticals are typically not known early in development. This makes informed product design, that is, planning of dose strength (X mg/mL) and volume to be filled in container and/or administered to the patient, a challenge if not impossible in early stages.

Fifth, a huge number of R&D projects are terminated during development for various reasons including insufficient therapeutic efficacy, poor bioavailability, safety concerns (toxicology, immunogenicity), changed business cases, or high development costs [5].

Sixth, throughout development, a therapeutic protein is subjected to environmental stresses that may result in alterations in protein structure that may jeopardize stability, safety, and/or efficacy of the drug product (DP) such as freeze-thaw stress, temperature excursions, and agitation. Incomplete knowledge of required process development activities may also limit the usefulness of formulation parameters or excipients that are added in early stages.

Finally, a key question is how to supply for preclinical (first-in-human enabling) studies, including toxicological studies. In a number of cases, the formulation used for toxicological studies may be identical to the phase 1 formulation, while in other cases it may not be. Toxicological studies typically require higher protein concentrations in the formulation, given the limitations in dosing volume for most test animals. Also, some formulation excipients widely used for formulations may result in adverse events in animals. As an example, polysorbates may lead to hypersensitivity reactions in dogs, potentially due to their histamine-releasing properties [6] or their hemolytic potential [7].

The development of a formulation during early stages of product development depends on the company-specific business strategies and prior knowledge gained about the molecule. For example, monoclonal antibodies (mAbs) may typically drive the use of platform formulations in several companies for early stages—formulations that are proven suitable for the specific collection of antibodies encountered in that group and company. With the huge variety of antibodies tested and studied, significant knowledge about the relevance and impact of specific excipients and their concentration and stress conditions has been generated. Also freeze-dried formulations have been chosen as platform formulations for antibodies by some groups in the past. Interestingly, frozen formulations are also used in early clinical studies by some groups, despite the number of challenges encountered, such as freeze-thaw stress, supply chain logistics, requirement of thawing on (clinical) site, and, most importantly, potential container closure integrity issues (proximity to glass transition temperature of rubber stoppers). In any case, various forms of platform formulations have been, are, and will be used for antibodies and also other precedential classes of therapeutic proteins—mostly without extensive preformulation activities [8]. In other cases, formulation development may include staged studies of the various formulation parameters (e.g., pH and buffer, surfactant and excipient) early in development. Considering that protein is available only in small amounts for development purposes in early stages of R&D and other factors discussed previously, it may be advisable to limit the formulation space that is planned to be explored during subsequent studies. This is where "preformulation" and the use of biophysical methods come into play.

Preformulation is defined differently from one source to another. Volkin et al. define preformulation as a process in which a bulk drug material is characterized sufficiently such that it can be converted to a pharmaceutically acceptable drug substance [9]. Kamerzell et al. described preformulation as "to determine the protein's physicochemical properties and pathways of instability, allowing for the design of formulations containing various excipients to ensure protein stability under defined storage conditions" [10]. We suggest to define protein preformulation as R&D studies that aim at predicting what excipient(s), and solution conditions that may help stabilize (or destabilize) the protein. The formulation strategies for a given therapeutic protein may thus determine the timing and extent of development activities as well the number of potential future "changes" for a given formulation during the overall clinical development.

This chapter will discuss different preformulation strategies including application of biophysical and high-throughput (HT) methods that may be used for protein therapeutics and the pros and cons associated with these strategies. Preformulation can come into play for the "developability" assessment and molecule optimization prior to the selection of a molecule for further clinical development (before clinical lead selection). The analysis and identification of degradation hot spots for chemical degradation reactions and potentially also for physical instability reactions (such as aggregation) can make use of primary sequence analyses and molecular modeling efforts [11–14]. High-throughput formulation (HTF) approaches can be classified under preformulation category. The idea of HT testing is to accelerate development timelines while minimizing the amount of protein needed for the studies and being able to study a wider range of formulation excipients and solution conditions for stability or other endpoints considered relevant. Preformulation studies have been performed evaluating surrogate or single "endpoint" testing, trying to narrow the later formulation development test space. For example, the measurement of melting temperature (Tm), second virial coefficient (B22, A2), hydrodynamic radius, and spectroscopic features (using extrinsic dye binding fluorescence, circular dichroism (CD), etc.) has been assessed in this context. When trying to use "surrogate" endpoints, some formulation scientists try to skip typical development studies by solely assessing these endpoints, for example, studying Tm as a function of pH or different excipients.

Lastly, typical protein characterization studies such as structural analyses (by CD, Fourier-transformed infrared spectroscopy (FTIR), etc.) and melting temperature can form the basis for formulation development studies in later clinical stages. The idea of performing some characterization beforehand is to guide the formulation scientist not necessarily for the selection of excipients, but, for example, for the selection of stress conditions. The analysis of the Tm of a protein would be helpful for other purposes such as evaluating unsuitable thermal conditions for accelerating testing. For example, 40°C accelerated stress testing data of an enzyme with a Tm of ca. 38°C should be interpreted with caution or preferentially not be studied at all, as the enzyme will denature significantly. Thus, totally different degradation routes and rates will be the outcome of such stress testing that cannot aid in predicting reactions at intended storage condition, such as refrigerated (2–8°C).

7.2 CONSIDERATIONS FOR A SUCCESSFUL PROTEIN DRUG PRODUCT

In order to understand typical challenges of a protein formulation, we first try to explain and highlight challenges that are to be overcome for a successful protein product. The

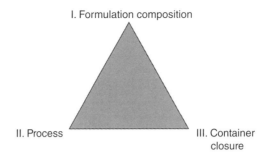

Figure 7.1. Pillars for successful formulation development.

properties of a DP are impacted by the interactions between formulation composition and/or DP manufacturing unit operations (process) and/or container closure/primary packaging/administration device (Figure 7.1). Therefore, successful formulation development of a robust DP involves optimization of all three aspects. The main phases in the life cycle of a DP include DP manufacturing, storage, and end use (Figure 7.2). During each phase, the DP may encounter several stresses, some of which may be common among the different phases. Major stress conditions a DP may encounter in different phases of the life cycle are summarized in Table 7.1. Successful formulations of therapeutic protein should be able to provide sufficient protein stabilization against various stress conditions during manufacturing unit operations, as well as during shelf life at intended storage conditions and during end use by patients and clinicians. In addition, they need to provide patient convenience, for example, ease-of-use as well as need to ensure acceptable local tolerability, ensure complete delivery of DP dose, and bioavailability.

7.2.1 Formulation Composition

In general, the technical target product profile (tTPP) will dictate the choice of the dosage form (i.e., liquid solution for parenteral administration and freeze-dried concentrate for reconstitution), as well as the intended dose, route of administration, desired presentation (e.g., vials, prefilled syringe, autoinjector), dosing frequency, and sometimes even excipients used. Accordingly, this narrows down the choice of formulation ingredients. (Pre)formulation screening activities may cover optimizing the following aspects: protein concentration (DP strength), pH, osmolality (preferably, close to isotonicity), and excipient (type and levels). Formulation excipients typically include a choice from nonionic surfactants (e.g., polysorbate 20, polysorbate 80), sugars (e.g., sucrose, trehalose),

Figure 7.2. Life cycle of a drug product.

TABLE 7.1. Stresses encountered in different phases during the lifecycle of a drug product (DP) and risk category (R, relevant; SR, sometimes relevant; N/A, not applicable)

Stress	Compounding and manufacturing	Transportation	Storage	End use
Elevated temperature	SR	R	R	SR
Relative humidity[a]	SR	SR	SR	SR
Freeze-thaw	R	SR	SR	SR
Mixing	R	N/A	N/A	SR
Filtration	R	N/A	N/A	SR
Oxygen[b]	SR	N/A	SR	SR
Air–liquid interfacial stress (e.g., shaking, pumping)	SR	SR	SR	SR
Metal ions	R	N/A	SR	SR
Pressure fluctuations	SR	R	N/A	N/A
Light	SR	SR	SR	SR
Solid–liquid interfacial stress (e.g., adsorption to tubings, organic leachables)	R	SR	R	SR

[a]This is a critical parameter if DPs are in the solid state (e.g., lyophilized, spray dried).
[b]Sources may be air oxygen, peroxides as contaminants, and vaporized hydrogen peroxide (VHP) residuals.

amino acids, buffers, salts, and other excipients as needed (e.g., antioxidants, preservatives). They are added to either stabilize the protein, aid in manufacture, control or target delivery to the target or to provide pH or tonicity control [10, 15].

It is essential to establish buffering agents and buffer strength for the formulation to ensure maintenance of the pH at the desired target and within the pH range. It is known that different buffer ions can have specific ion effects on the stability of proteins—either stabilizing or destabilizing in nature. Amino acids (e.g., histidine, arginine, or glycine) are often used and can provide buffering capacity and sometimes also antioxidant protection to the protein. Sugars (e.g., sucrose, trehalose) are commonly used for lyophilized and liquid products to provide sufficient stabilization while ensuring adequate tonicity. Obviously, glucose or fructose can lead to protein glycation at lysine residues via the Maillard reaction [16], rarely used nowadays. The sugar alcohol mannitol shows little ability to stabilize proteins, although it is an agent providing cake elegance as a bulking agent [17]. The effect of salts on protein stability is well studied and it is known they can significantly affect protein integrity and stability. Gokarn et al. demonstrated that anions, but not cations, directly interact and associate with the protein surface and that ion binding can modulate protein aggregation (following temperature and agitation stress) by neutralizing protein charge following the Hofmeister series [18–20]. Surfactants such as polysorbate 20 or 80 are widely used in protein formulations to protect the protein from interfacial stresses. Other excipients (preservatives, chelators, specific ligands) may also play a role depending on the desired product and the protein to be formulated. For more insight and references to excipients and excipient–protein interactions, the reader is referred to read the review by Kamerzell et al. [10].

The choice of excipients that can be tested in the development of a parenteral protein formulation is somewhat limited, based on TPP and on regulatory and toxicological considerations. The applicability of an excipient planned for parenteral administration generally depends on the intended concentration, volume, type of administration, patient population, indication, and other factors. Excipients that are listed on the FDA inactive ingredient list for parenteral use may in cases not be used for specific routes of administrations. For example, intrathecally administered excipients may require further attention. Some excipients listed in the FDA excipient list may not be usable or of limited use for other routes. Additionally, the selection of excipients for formulation testing may further be narrowed down as a result of stability- or process-related challenges observed with some of these. Impurities or degradation products in excipients such as trace metal ions and hydroperoxides are known to potentially impact protein stability. Lam et al. showed that chloride-induced corrosion of stainless steel at acidic pH and elevated temperatures caused the formation of free ions, catalyzing methionine oxidation in a monoclonal antibody formulation [21]. Polysorbates are examples of excipients that may contain significant peroxide and can undergo degradation (oxidation and hydrolysis) in aqueous protein formulations [22]. When using phosphate-buffered solutions, it is known they can undergo drastic pH changes upon freezing. The pH of Tris and several other buffers can change with temperature [23]. Another challenge that was observed for frozen bulk formulations is the potential of certain excipients to undergo crystallization in frozen state above the glass transition temperature and their potential to generate protein degradation, such as aggregation [24]. Another example that may limit the choice of an excipient is its potential negative impact on local tolerance. Citrate buffers are believed to play a role in pain perception upon subcutaneous injection in humans [25].

The critical choice of the excipients to be studied during preformulation and formulation studies is one of the major responsibilities of the formulation scientist. Testing excipients that finally are not going to make it into a formulation due to other factors, such as toxicological concerns, would be a waste of protein and labor.

In later stages of development, formulation robustness studies on an optimized formulation are often performed to assess the impact of variations on formulation ingredients due to composition variations encountered during different manufacturing operations such as weighing and dilution. Extreme variations in formulation composition can help assess edge of failure and can help support the assessment of impact of specification ranges.

7.2.2 Testing under Different Stress Conditions

Most protein formulations to date are still intended for long-term storage under refrigerated conditions (2–8°C). Nevertheless, these products typically need to withstand higher temperatures than 8°C in order to be able to manufacture and market them. Stability studies for protein formulations need to study both real-time intended storage conditions (e.g., 2–8°C) and accelerated conditions (such as 25°C/60% relative humidity (RH), 30°C/65% RH, 40°C/75% RH [26]. During the development of protein formulations, the use of accelerated conditions is of utmost importance in order to be able to quickly differentiate between "good" and "bad" formulations at the given stress condition and parameters and analytics. Unfortunately, the degradation mechanism at accelerated conditions may not always predict well for the intended storage conditions.

The critical choice of the stress conditions to be studied during preformulation and formulation studies is also one of the most important responsibilities of the formulation scientist. Finally, testing stress conditions which are irrelevant for final DP manufacturing, handling, shipment, or storage may lead to inconsistent or inadequate conclusions from the development studies. Among the stress conditions that proteins may encounter, the following are most noteworthy in the context of protein (pre)formulation development and are often applied when assessing different excipients and excipient levels:

Temperature

Relative humidity (RH) (e.g., in case of lyophilized DPs or in some cases other dosage forms such as prefilled syringes)

Interfacial stresses:

- Freezing and thawing
- Shaking, stirring (e.g., as encountered during mixing and shipping of liquid goods)

Additionally, further stress conditions and studies are required after the design of the product formulation in order to assess and determine potential precautions and parameters for manufacturing and handling [27–30]:

Photostability

Oxygen, oxidants (e.g., air oxygen, peroxides as contaminants, vaporized hydrogen peroxide (VHP) residuals in filling isolators)

It is widely known that temperature can cause protein unfolding, that is, denaturation, and protein degradation. It can also impact the degradation rate of various chemical pathways, such as succinimide formation and deamidation [31] as well as physical pathways such as aggregation [32].

RH can impact the stability of dried dosage forms, as water would typically accelerate degradation. The impact of humidity on the stability of lyophilisates is thus of utmost importance. The susceptibility of lyophilisates toward water drives various measures typically applied during DP development and manufacture, including adequate drying of rubber stoppers prior use [33–35]. Both temperature and humidity potentially play a role in the drying of protein on filling lines (e.g., filling needles and tubings) and in needles of prefilled syringes or autoinjectors, after removal of protective measures such as the rigid needle shield (RNS) or cap or if water permeability across the RNS is significant (unpublished; Müller, Robert).

Given that protein bulk is typically frozen—both to provide physicochemical and microbiological stability over storage time—freezing and thawing unit operations are often encountered in the manufacture of protein bulks and DPs. The potential impact of

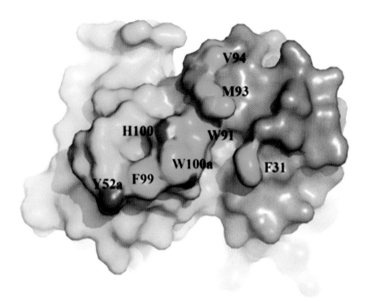

Figure 3.3. Antigen recognition surface of CNTO607 and its interaction with IL-13. The hydrophobic ridge in the CNTO607 antigen-binding site. VL and VH are colored cyan and magenta. The hydrophobic ridge is colored green. N, S, and O atoms in the ridge are colored blue, yellow, and red. Adapted from Reference 13.

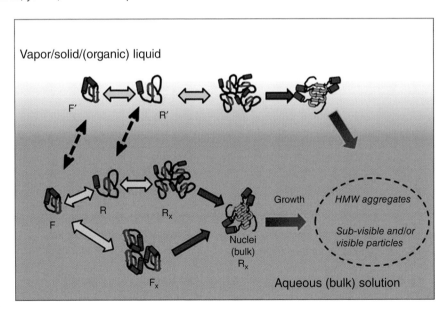

Figure 5.1. Schematic overview of the multistage nature of aggregate nucleation, including multiple steps that potentially involve conformational changes.

Biophysical Methods for Biotherapeutics: Discovery and Development Applications, First Edition. Edited by Tapan K. Das.
© 2014 John Wiley & Sons, Inc. Published 2014 by John Wiley & Sons, Inc.

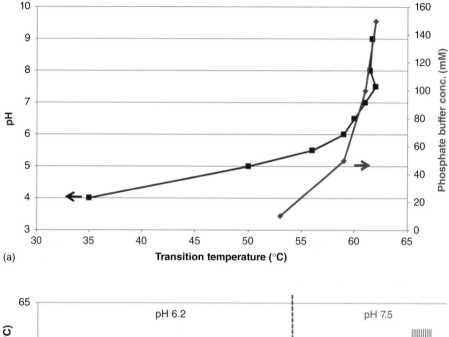

(a)

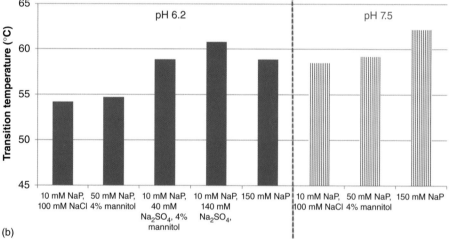

(b)

Figure 7.8. Screening different protein formulations based on thermal transition (melting) temperatures at selected pH values. (a) Transition temperature as a function of pH (4–9) and buffer concentration (10–150 mM) from a pH and buffer screening study. (b) Protein transition temperature during a formulation screening study. Solid bars are formulations at pH 6.2, striped bars are formulations at pH 7.5. Figures modified from Reference 78.

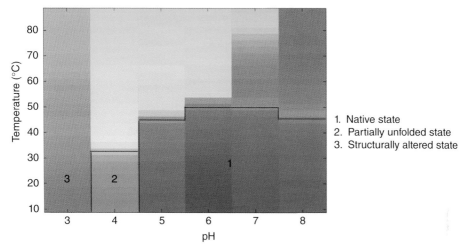

Figure 7.7. Empirical phase diagram of a recombinant protein. The diagram was prepared from CD at 217 nm, intrinsic fluorescence (peak position), static light scattering, and ANS fluorescence data as a function of pH (3–8) and temperature (10–87.5°C). Unpublished data, figure used with kind permission and courtesy of Sangeeta Joshi and Russ Middaugh.

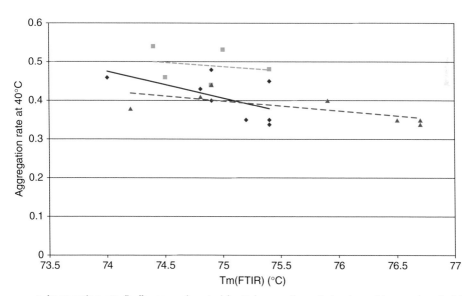

◆ Aggregation rate (buffer screening study) ■ Aggregation rate (amino acid screening study)
▲ Aggregation rate (stabilizer screening study)

Figure 7.10. Correlation between physical aggregation rate constants (assuming first-order kinetics) to Tm(FTIR) values in buffer, amino acid, and stabilizer (sugar) screening studies. Figure modified from Reference 81.

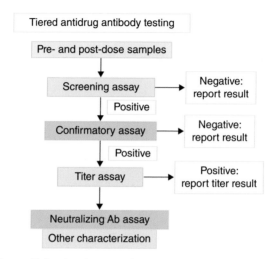

Figure 8.1. Tiered approach to antidrug antibody testing.

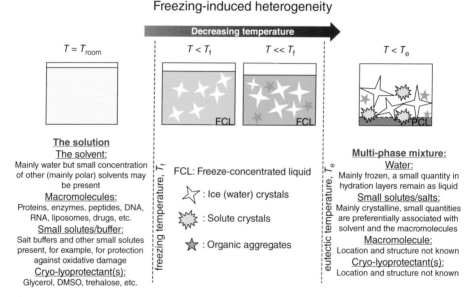

Figure 10.1. Transition of a "simple" solution to a multi-phase mixture during freezing. Note that the actual freezing-induced heterogeneity is much more complicated than what is shown here. In many cases, there are no clear boundaries among different phases but there are heterogeneous regions composed of different constituents (e.g., proteins entrapped in ice; see Reference 3). The amount of remaining water at temperatures below the eutectic temperature is a complex function of the thermal history of the specimen as well as its specific constituents.

freeze/thaw operations and parameters on protein quality is known and well researched [36–38]. As mentioned before, surfactants are often used to stabilize proteins against freezing/thawing stresses.

Mechanical stress, such as shaking and stirring, generates excess air/liquid interfaces. This stress condition has been extensively studied in recent years [39, 40]. Given that DP solutions in primary packaging without headspace remain stable despite extensive shaking, the primary cause of degradation is considered to be air–liquid interfaces. Extensive stirring, however, may create different interfaces than solely shaking (e.g., Teflon–water interfaces) or effects such as cavitation that may play a role in protein aggregation and particulate formation during stirring [40]. Again, surfactants are often used to stabilize proteins against these kinds of interfacial stresses.

Photostability testing is often done only with the chosen formulation and considering the light- and UV-induced degradation pathways, most protein products are not considered photostable [29], especially when testing against ICH conditions. However, secondary packaging typically provides adequate protection from light during product storage. Still, product stability against (lower levels of) visible and UV light needs to be considered, as the respective manufacturing conditions may sometimes be performed without protection from light, including fill/finish operations and labeling. A thorough understanding and control of light conditions during processing and an understanding of related (potential) product impact are important.

Oxygen and oxidants are widely known to impact protein quality. Oxidation can occur, for example, at methionine and tryptophan residues that are exposed to the protein surface. Sources of oxidants in the final product can be contaminants in excipients (e.g., contaminants in polysorbates) and contaminants from processing (e.g., VHP residuals in filling isolators), just to name a few [30]. Peroxides in polysorbates are known to lead to protein oxidation in various cases [41]. Measures to limit peroxides in excipient raw material (e.g., cold storage of polysorbate raw material in brown glass and under oxygen protection) and studies to elaborate on the sensitivity of protein degradation with oxidant contaminants are considered important during the course of development.

In summary, the choice of adequate stress conditions for preformulation and formulation studies is of significant relevance. A combination of different conditions, including temperature, humidity, and interfacial stresses relevant for bulk and DP processing (e.g., shaking and freeze/thaw operations), is relevant to identify a sufficiently stable and robust formulation. The choice of single stress conditions for preformulation and formulation studies may lead to misleading conclusions.

7.2.3 Primary Packaging/Container Closure

Components of primary packaging, an aspect we consider to be too much overlooked at least until late stages of formulation development, play an important role in safety and efficacy of the DP, manufacturing processes (such as filling and freeze-drying), and establishing the shelf life of the DP. A nice overview of primary packaging is described by Bhambhani et al. [42]. Although we consider secondary packaging to be just as important to provide protection (e.g., from direct light), our discussion will focus on primary packaging. Primary packaging refers to the components in direct contact with the DP such as vials, stoppers, ampoules, dual chamber syringes, prefillable syringes, and cartridges. Container closure configurations depend on dosage form (i.e., dry powders versus liquid). Other considerations include fill volume and end use.

Significant problems are reported in literature and experienced in industry as a result of interaction of primary packaging with liquid formulations. Extractables and leachables are the most common problems from container closures in direct contact with DP. Most commonly reported extractables and leachables include phthalates (e.g., di(2-ethylhexyl)-phthalate), metal ions (e.g., Zn, Fe, Ba, Ca, Al, and Ni), fatty acids, cyclic esters, silicone oil, organic solvents, nitrosamines, vulcanizing agents, antioxidants, polycyclic aromatic hydrocarbons, antistatic agents, cleaning agents, and chelating agents [43–46].

Leachables from containers into the liquid formulation during storage can interact with excipients and/or proteins catalyzing many degradation reactions such as oxidation, aggregation, and particulate formation. Sulfate ions from parenteral formulations can react with barium ions leaching from the glass surface resulting in the slow crystallization of barium sulfate over storage time [45, 47] (also Wigand Weirich, published data). Benzoic acid used as a preservative in a liquid parenteral formulation reacted with aluminum leaching from the inner surface of glass vials resulting in the slow crystallization of aluminum benzoate over storage time (Wigand Weirich, published data). Similar incidents have also been observed with calcium leachable from glass surface (Wigand Weirich, unpublished data).

Tungsten, a component of prefillable syringes, has been reported to potentially impact stability of some proteins [48, 49]. However, many therapeutic proteins are not impacted adversely by tungsten. The source of tungsten in prefillable syringes is tungsten oxide vapor deposits in the syringe funnel area from the application of high temperatures during the shaping process on tungsten pins to produce the channel through which the needle is mounted [49]. Additionally, tungsten oxides are deposited in the bore and funnel area of the syringe as a result of oxidation [50]. In addition to protein concentration and ionic strength, formulation conditions that induce protein aggregation by tungsten include acidic pH [50]. The latter results in the formation of soluble tungsten polyanions that are most likely responsible for tungsten-induced protein aggregation [51–53]. Acidified sodium tungstate (Na_2WO_4) was found to be the most potent species under acidic pH conditions. Incompatibility with tungsten has been assessed by performing stability studies in actual prefilled syringes (although batch-to-batch heterogeneity of tungsten deposits poses a challenge), as well as spiking studies with soluble tungsten species such as tungsten (IV) oxide, tungsten (VI) oxide, sodium tungstate dihydrate, sodium polytungstate, ammonium tungstate, or tungsten metal powder [48].

Silicone oil from siliconized glass containers and/or siliconized rubber closures in vials and prefillable syringes has been reported to interact with proteins in solution potentially impacting stability. The number of reported incompatibilities with biological products at relevant concentrations, however, is low and occurs under nonoptimized formulation conditions at accelerated storage temperatures, for example, Abatacept [50, 54], as well as under certain stress conditions (agitation). Similar to tungsten studies, silicone oil incompatibilities have been studied by performing stability studies in actual containers (batch-to-batch heterogeneity of silicone oil may pose a challenge), as well as spiking studies with irrelevant high amounts of silicone oil. However, one additional challenge is interference of silicone oil droplets with subvisible particle characterization methods used to characterize proteinaceous subvisible particles, once the silicone oil droplets detach from the inner wall or stopper of the container closure [50].

Acrylic acid, a component of acrylic adhesive used to attach the needle to the barrel or prefillable syringes, has been identified as a leachable from certain types of syringes in one study leading to interaction with lysine side chains, N-terminus, and histidine side chains of a IgG2 by Michael reaction [55].

Incompatibilities with Tungsten and/or silicone oil can be avoided by several measures such as using prefillable syringes with lower and more controlled levels of tungsten or silicone (as required) or trying to avoid these components for very sensitive proteins (e.g., replacing tungsten pins by other materials as transition metals). These measures, however, may create significant issues as materials that replace tungsten or silicone may also create significant incompatibilities with proteins. Other challenges in addition to leachables/extractables include adsorption and glass delamination. Interferon was observed to undergo considerable adsorption onto siliconized glass barrels resulting in a decrease and a large variation in protein concentration between different syringes (unpublished data). Similar findings on adsorption of recombinant human interferon-β and increase in adsorption propensity as a function of ionic strength are also reported [56]. In some type I pharmaceutical glass vials, delamination has been identified at the vial bottom and shoulder. A consequence of glass delamination was the appearance of shiny needle-shaped particles in the range of 50–200 µm in the liquid [57]. It was speculated that extensive flaming during the conversion process at these sites promotes evaporation of alkali and borate species and formation of enriched silica layers. Contact with alkaline pH formulations and silica complexing agents as citrate, phosphate, and acetate buffers as well as glutaric acid and EDTA can increase the rate of glass corrosion and subsequent detachment of flakes following hydration of these layers after long incubation times during storage [57,58]. Other factors that affect delamination are ionic strength, terminal sterilization process, storage time, type of glass (e.g., expansion 33 glass), and so on [57]. Accelerated testing and systematic monitoring are encouraged to test if this phenomenon will occur or not in actual product [57]. If the container/optimized DP combination are found to generate glass flakes, one can test other type I glasses or containers or use coated containers (e.g., Schott Type I plus) [58].

Although the risk of interaction between primary container components and dry powdered formulations are low, similar adverse effects to those described previously can occur during pre-lyophilization and post-reconstitution. Additionally, the choice of container closure may have long-term adverse effects on DP stability. Moisture and oxygen ingress during product storage is known to occur, especially under conditions of accelerated storage stability per ICH guidelines (30°C/70% RH, 30°C/75% RH, 40°C/75% RH). This is not surprising given that stoppers are amorphous materials by nature that under normal (or accelerated stability) storage conditions are stored significantly above their glass transition temperature thereby rendering them permeable to small atmospheric molecules such as water vapor and oxygen. Similarly, moisture and other volatile leachables present in stoppers may ingress into the headspace and thereby into the hygroscopic lyophilized cake, causing a negative impact on DP quality during long-term storage. Vial breakage during manufacturing of mannitol may lead to increased glass breakage upon accidental or intended freezing (e.g., during

Figure 7.3. Glass fogging phenomenon observed post lyophilization.

freeze-drying) [59, 60]. Cosmetic defects of lyophilized product may also occur (Figure 7.3). These phenomena can be described as "fogging" and "clouding" (a form of white "haze" along the inner walls of a vial). Depending on the formulation properties (e.g., surface tension), type of vial used, and inner glass vial surface energy (e.g., hydrophilicity), DP solution may creep up the inside walls of the vial already during filling and will be dried at the inner vial walls during lyophilization [61].

In summary, given the potential interactions of primary packaging components and formulation, there is a necessity to consider relevant primary packaging early in the selection and development of protein formulations. Preformulation activities that may take place should be adequately designed to consider such interactions.

7.3 PROTEIN PREFORMULATION STRATEGIES

A successful protein preformulation strategy is one that leads to a formulation robust enough to achieve stability in all aspects described previously or that sufficiently enables formulation development for a therapeutic protein to be used in clinical studies. Different strategies employed for conducting preformulation studies are described in this section, and a general outline of the different strategies is presented in Figure 7.4. In general, preformulation screening strategies can be clustered into the following four main buckets:

1. developability studies (for site mutations on proteins) and molecule candidate selection;
2. HT screening studies;
3. studies relying on a limited number of endpoints (surrogate measures); or
4. traditional biophysical and/or biochemical characterization.

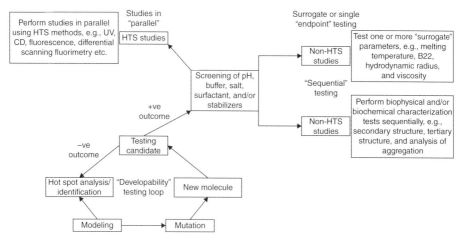

Figure 7.4. The complex loop of different preformulation/biophysical approaches preceding formulation development.

7.3.1 Developability Assessment and Molecule Candidate Selection

The discussion so far has focused on a given molecule, studying various excipients or concentrations in development or preformulation studies. However, major consideration should be given to the initial choice of molecule for clinical development in the first place. Traditionally, therapeutic proteins were chosen for preclinical and clinical development based on the analysis of their potency (e.g., binding properties), preliminary or predicted pharmacokinetics, and an early assessment of the drug substance process (bioprocess), especially the stability and productivity of the cell line. Whereas all these aspects are key for a successful protein therapeutic, the analysis and optimization of stability of a therapeutic protein has received insufficient attention. The primary sequence and structure of a protein itself would determine whether a protein is susceptible to specific types of degradation and the short- and/or long-term impact of this degradation reaction on the molecule. As an example, a monoclonal antibody with a (surface-exposed) deamidation hot spot residue in its binding region would likely show reduced binding upon modification of this residue [62]. A mAb with a surface-exposed methionine in its binding region may likely show decreased binding upon oxidation. Rather than solely trying to minimize the degradation by modulating formulation parameters (e.g., in case of oxidation, the use of antioxidants, scavengers, or chelators or use of nitrogen flushing and overlay during DP processing), it may be worthwhile to explore whether a version (mutant) of the antibody not containing this degradation residue would still be a good drug candidate (fulfilling all of the above-mentioned requirements including potency and cell line productivity) while at the same time not being susceptible to degradation at this critical residue.

A number of potential "degradation hot spots" can be analyzed nowadays by means of software and modeling, and examples for this strategy are given in literature [63,64].

For appropriate in silico analyses, for example, toward likelihood of protein aggregation for a given protein molecule, a number of (software) tools are being researched nowadays, including Aggrescan, Page, Pasta, Salsa, Tango, Zyggregator, Amylpred, Waltz, and others. An essential step in the developability assessment is to study mutants (variants) of the respective drug molecule, based on the conclusion drawn and recommendations provided by software analyses. This requires to put sufficient additional efforts into generating and producing these mutants and to subsequently study their stability behavior, at least in small scale.

Trout et al. are researching in silico modeling and prediction as potential tools for candidate screening during discovery for developability by applying the spatial aggregation propensity (SAP) tool and the developability index (DI) tool, screening mAbs for their aggregation propensity using SEC [13, 14]. The SAP tool aims to predict aggregation propensity via hydrophobic and electrostatic interactions, considering the protein's net charge and SAP of the modeled protein, building on its primary sequence. It has been found that net charge derived from sequence-based calculations and experimental determinations can vary significantly [65]. However, the assessment of net charge and potential aggregation propensity is considered a rather relative ranking tool, not being able to provide accurate predictions on an absolute scale. Kayser et al. [66] performed a study to rank the stability of seven different mAb constructs from different classes and correlate two parameters to long-term physical stability (as measured by SEC). One parameter was SAP—a measure of the degree of hydrophobicity of a molecule—and the second was Thioflavin T (ThT) binding constant to aggregated protein ($1/K$) from a fluorescence dye binding assay. The authors argue that both parameters could be used to rapidly screen proteins for their propensity to aggregate, especially if coupled with SEC in detecting aggregation. However, further analysis of the data showed poor correlation between parameter studies (Figure 7.5). Furthermore, the rate of physical aggregation was not accounted for in this assessment but rather absolute values of monomers at the end of the storage stability studies (at different temperatures), assuming percentage of bound ThT is proportional to monomer loss.

A common challenge for these tools is to identify an adequate analytical endpoint for correlation. While this is more easily achievable with chemical degradation routes, it remains a challenge for protein aggregation, as the mechanisms as well as analytical methods for aggregation remain broad and diverse [32] and thus the structural regions in a protein that are aggregation prone are often not known or may differ significantly between molecules and aggregation reaction. Lauer et al. have mainly used SEC at accelerated temperature to study aggregation propensity and to validate their SAP model. While SEC still remains the workhorse in industry, it is capable to provide results for soluble aggregates only [13].

Another aspect that is considered relevant for the choice of a molecule for clinical development, apart from optimization of clinical lead molecules for degradation and stability, is the analysis of viscoelastic and stability properties. Viscosity has been a major issue for (certain) high-concentration formulations and products, leading to viscosity values that may create significant issues during processing or handling [67]. For proteins intended for high-concentration formulations, the comparative analysis of concentration/viscosity curves (at given or various temperatures) may help to stratify and identify

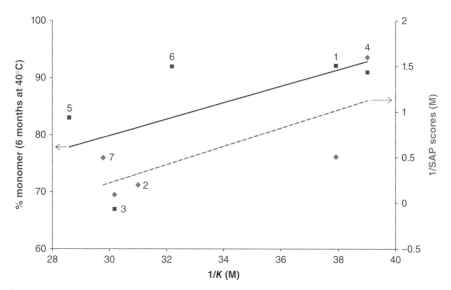

Figure 7.5. Correlation of physical stability of different mAbs (numbered 1–7 in the figure) after storage for 6 months at 40°C to ThT binding constants (K) from equilibrium ThT fluorescence binding assays and SAP scores (SAP scoring was performed only on mAbs of the same class). Figure modified from Reference 66.

molecules which show abnormal viscosity behavior. However, it is noteworthy that all potential hot spots may not actually manifest themselves. For example, a degradation hot spot that is structurally not accessible is less likely to show degradation reactions.

In summary, if proteins or their mutants with comparable cell line, binding, and kinetic properties are found, comparative stability and concentration/viscosity studies of the drug candidates may help to identify differences and select the most ideal drug, in connection to the TPP. The selection of "more stable" or "less viscous" molecules for development and, thus ranking of clinical candidate proteins based on early studies, is certainly worth pursuing. Despite current limitations in understanding of degradation mechanisms and availability of analytical tools for stability prediction, developability is worth conducting. Further insight into the mechanism of degradation likely will improve the algorithms used for predictions.

7.3.2 High-Throughput Formulation Development

In order to study a huge variety of formulation parameters with low amounts of protein, HT approaches have been developed and applied for protein formulations. To date, formulation development is often largely empirical, with poor prediction of stability and performance on long-term storage, while building on a strong knowledge base. Early stage formulation development typically explores HT approaches for potential savings in material and time and potentially enabling a broader design space for

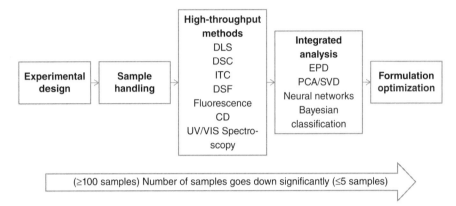

Figure 7.6. As described by Kamerzell et al., the general flow diagram illustrates HTS steps for protein formulation development. DLS, dynamic light scattering, DSC, differential scanning calorimetry; ITC, isothermal titration calorimetry; DSF, differential scanning fluorescence; CD, circular dichroism; UV/Vis, ultraviolet/visible; EPD, empirical phase diagram; PCA/SVD, principle component analysis/singular value decomposition. Initially, a large number of prospective experiments are designed. Robotics can then be used for liquid sample preparation prior to the use of high-throughput analytical methods for physicochemical characterization. The high-throughput methods are equipped with 96–364 micro-well plates. The large amounts of data are rapidly analyzed by advanced numerical methods using pre-programmed criteria. Figure modified from Reference 10.

formulation parameter screening, while still coping with limitations in available protein and (more) stringent timelines early in development. The application of HT and automation approaches in development has increased steadily since the launch of the robotic liquid handling systems in the mid-1980s. HT approaches for protein formulation development are being marketed by various vendors (Tecan Ltd, SimBioSys Inc., Chem-Bridge, In Vivo Sciences LLC, etc.). Specific designs of HT formulation development also have been successfully applied for salmon calcitonin by Arvinte and coworkers [68], Li et al. [69], and Bambhani et al. [70] for monoclonal antibodies.

An outline for a potential strategy and approach for HT formulation development is provided in Figure 7.6. HT formulation development would aim at

1. automated formulation (sample) preparation, considering relevant and representative components (e.g., excipients, primary packaging materials) and adequate microbiological quality of the preparation;
2. material handling during HTS processing and stress conditions, including application of (relevant and hopefully, a variety of) stress conditions for forced degradation;
3. minimizing sample demands and analyses time when using appropriate analytical methods (potentially also using or exploring "surrogates" for analyses); and finally,
4. automatized data analysis, for large body of analytical data generated.

Automation of the preparation of formulation samples provides significant opportunity for improvement, compared to manual preparation. Formulation manufacturing can typically be standardized in automated preparation, thus leading to less error in preparation and potentially saving valuable labor that is well spent for a different activity. However, the preparation of formulation samples needs to consider various elements that require adequate design. As an example, formulation sample preparation should be performed aseptically, in order to avoid contamination from bacteria and/or particulates. Microbiological contaminations could lead to loss of a sample. Particulate contamination could be misleading for the formulation scientists, as often the data on particulates in a formulation are one of the formulation decision criteria to consider. Also, use and storage time of excipients and stock solutions require some thought, as stock solution stability (microbiologically and physicochemically) needs to be considered. As an example, given the fact that polysorbates degrade in solution, stock solutions of polysorbate used for formulation sample preparation purposes should typically be freshly prepared and not used and stored for weeks at uncontrolled ambient conditions. The design space of the excipients (type and concentration) to be tested for a given protein molecule also requires a robust setup. It is interesting to note that some HT formulation studies use some excipients that may be of limited suitability for protein formulations [71]. As an example, the authors observed the use of lactose or glucose in various HT formulation experiments [71, 72], although it is known that reducing sugars may covalently react with primary amino groups in the protein, leading to protein glycation [16] and are thus not desirable. Also, various buffers tested in HT formats (such as HEPES, MOPS) are not practical for use in a protein parenteral formulation [73].

Formulation sample preparation needs to consider the type of primary packaging to be used. As subsequent analyses are often based on 96-well plate reader or comparable formats in order to automatize and ensure HT, the formulation samples are often prepared directly in multi-well plates. As the incubation of samples typically includes elevated temperature, a tight seal needs to be warranted for the well plates, to ensure that no loss of water due to, for example, evaporation occurs during incubation. A huge number of HTS studies in the field are using other types of primary packaging components than the ones to be used in clinical studies. For example, plastic material (e.g., well plates, HPLC tubes, Eppendorf caps) has been applied. It is noteworthy that due to water permeability of (most) plastic components, observations were made of poorly reproducible increase in protein concentrations, limiting the use of the data to make conclusions on a future formulation (Mahler, HC, published data). Thus, the integrity of the primary storage components is key. Leachables from the plastic material may also interact differently with solution components (excipients and proteins alike) compared to glass components, again limiting the value of the generated HTS data.

Obviously, the applicability of stress conditions also needs to be considered. For example, if using only few microliters in multi-well plates, the samples cannot easily be subjected to mechanical stress such as shaking and freezing/thawing processes which will of course be significantly different from DP or drug substance containers. This is a major challenge for HT studies. If protein degradation at test conditions cannot be extrapolated to the intended storage conditions (2–8°C), the results obtained at accelerated conditions may be misleading. An HT formulation experiment based only on using temperature stress as differentiating factor may also be misleading, as it would neglect the impact of, for example, interfacial stresses on the protein and thus may underestimate the need for certain stabilizers such as surfactants. This is specifically important as often data from Tm measurements may suggest to exclude polysorbates due to their known effect of lowering the Tm (melting temperature) or Tonset (onset temperature of unfolding) [70].

Analytical methods that are often used in the context of HT formulation testing are spectroscopic, light scattering, and imaging techniques (see Table 7.2). These

TABLE 7.2. Some biophysical techniques commonly applied to formulation development in high-throughput approaches

Analytical/biophysical technique	Primary output	Application
Differential scanning calorimetry (DSC)	Protein unfolding/melting profile, Tm, Tonset	Conformational stability (as function of temperature)
Differential scanning fluorimetry (DSF)	Protein unfolding/melting profile, Tonset	Conformational stability (as function of temperature)
Circular dichroism (CD)	Secondary structure Protein unfolding/melting profile, Tonset	Conformational stability (as function of temperature)
Intrinsic fluorescence	Protein tertiary structure	Conformation
Extrinsic fluorescence	Protein surface hydrophobicity, solution polarity	Protein aggregate detection, solution viscosity
Static light scattering (SLS)	Scattering intensity	Protein aggregate and particle detection
Multiangle light scattering (MALS)	Scattering particle size	Protein aggregate and particle characterization
Dynamic light scattering (DLS)	Protein hydrodynamic size distribution, protein diffusion, diffusion of external tracer	Protein characterization, aggregate/particle detection, protein interaction/aggregation propensity, solution viscosity
Analytical ultracentrifugation (AUC)	Hydrodynamic properties based on centrifugal force or phase equilibrium, protein sedimentation velocity	Size and shape analyses of protein monomers and aggregates
High/ultra performance liquid chromatography	Protein separation, protein self- or cross-interaction	Protein species characterization (size, charge, etc.) Protein aggregation propensity

Source: Table adapted/reconstructed from References 10 and 74.

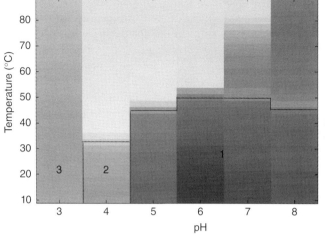

Figure 7.7. Empirical phase diagram of a recombinant protein. The diagram was prepared from CD at 217 nm, intrinsic fluorescence (peak position), static light scattering, and ANS fluorescence data as a function of pH (3–8) and temperature (10–87.5°C). Unpublished data, figure used with kind permission and courtesy of Sangeeta Joshi and Russ Middaugh. (See color insert)

include optical density (e.g., absorption at 350 nm as a surrogate for solution turbidity), differential scanning calorimetry (DSC), CD, dynamic and static light scattering (DLS and SLS), and intrinsic and extrinsic fluorescence. Commonly used fluorescence dyes include Sypro Orange, Nile Red, 4-(dicyanovinyl)juloidine (DCVJ), 4,4'-bis(1-anilonaphthalene-8-sulfonate) (bis-ANS), and ThT. Differential scanning fluorimetry (DSF) characterizes the shift in melting temperature of proteins using extrinsic probes, in combination with RT PCR instruments. A list of HT biophysical techniques commonly applied is given in Table 7.2. These analyses can often be based on 96-well plate or comparable formats. Bhambhani et al. [70] have assessed the formulation design and excipient selection in a HT mode using an IgG1 monoclonal antibody, using a variety of structural and conformational analyses. The overall data set has been summarized by the authors using an "empirical phase diagram" (EPD) approach, in order to make the results of individual methods visually accessible (see Figure 7.7).

Generally speaking, it is tempting to use single or a combination of analytical tools that provide "surrogate stability endpoints" for the HT experiments, such as the above-mentioned assays. The idea behind using very accelerated stresses combined with surrogate analytical endpoints is to accelerate overall time required for the HT approach. Conventional formulation development often includes testing of analytical endpoints for purity and content derived from the testing control strategy (e.g., SE-HPLC, IE-HPLC, and UV280) and potentially optimized for the purpose of analyses for development in combination with long-term stability testing, requiring weeks or sometimes months of incubation. Further discussion on analytical surrogate endpoints is included in Section 7.3.3.

Typically, the volume of analytical data generated by automated sample preparation and analyses is significant. Thus, an important aspect of HT formulation development is the need to also implement automated and software-guided data analysis. Therefore, data analyses for formulation development purposes require some thorough upfront design and establishing related criteria: the absence or presence of (impurity) peaks in a SE-HPLC may, for example, be a guiding principle, as well as monomer peak percent. For methods that are not so well established like SE-HPLC, knowledge about method performance (e.g., precision, reproducibility, LoD, LoQ) and the potential for a method interface to the analyses software are often key factors.

There are a variety of other possible approaches towards HT and miniaturization such as trying to apply elements of a more "conventional environment." For example, the "downscaling" of stress conditions such as stirring was performed as part of mAb preformulation development by Fraunhofer [75]. They developed a stir-bar-based model and tested aggregation as a function of stirring stress at low concentration (1 mg/mL) to study the effectiveness and concentration of excipients (surfactants) as stabilizers against this stress. The data would have to be extrapolated to protein concentrations as needed in order to fulfill the TPP, for antibodies typically in the range of 20–200 mg/mL. Another interesting approach for HT development was studied by Gibson et al. [76]. In order to determine "relative solubility" of a series of monoclonal antibodies in different formulations, they studied polyethylene glycol (PEG)-induced precipitation. Varying concentrations of PEG were added to the mAb formulations that lead to protein precipitation. The data were linearly fitted and extrapolated to 0% PEG concentration in order to calculate "protein solubility." This approach was also applied later for another mAb [77, 78]. The obtained solubility values were quite low (9.7–73.1 mg/mL) and did not reflect actual protein solubility limits, given that the concentration achieved by regular UF-DS well exceeded 170 mg/mL. As mentioned by the authors later, the drawback of this approach is that it does not take into account any interaction between PEG and the proteins as well as impact of PEG on the dielectric constant of the solution. However, still, the relative estimation of the solubility might be feasible using this approach.

In summary, HT and miniaturization approaches may provide significant opportunities, especially in early stages of protein formulation development. The formulation design space may be explored much wider compared to conventional formulation development approaches, with less material consumption and potentially, less time. However, a number of key challenges can be encountered when embarking on HT and miniaturization approaches for protein formulations. The challenge lies in linking the experimental data to stability at real time and intended storage conditions. Even if not desiring to predict shelf life based on these studies, degradation pathways may differ significantly that may mislead the formulation scientist. HT and downscaled formulation development studies thus have to be set up and designed with great care. The authors believe that results from HT should not be considered a replacement of current stability testing in the configurations planned to be clinically or commercially used, but potentially can serve to broaden the design space of early formulation scouting. Results of the HT can then guide the formulation scientist to fine-tune the formulation toward its use in clinical stages.

7.3.3 Surrogate Parameters from Biophysical Studies during Formulation Screening

A multitude of studies in the literature aim at using surrogate endpoints to study the impact of excipients on proteins in order to avoid long-term stability studies. The main assumption in these studies is that thermal and/or colloidal protein stability is/are predictive of long-term storage stability and hence can be used in preformulation screening studies and potentially in combination with HT approaches. Surrogate endpoints used for liquid formulation development include

protein melting temperature (Tm, Tonset), obtained from measurements with methods such as
- Differential scanning calorimetry, DSC;
- Fluorescence or differential scanning fluorimetry, DSF;
- Circular dichroism, CD;
- Fourier-transformed infrared spectroscopy, FTIR;

second virial coefficient (B_{22}), net charge, diffusion coefficient, isoelectric point, hydrophobicity measures, etc.

turbidity, optical density
- $A_{350-450}$.

The following sections provide a brief overview and discussion on currently studied surrogate endpoints.

7.3.3.1 *Protein Melting Temperature (Tm).* The protein melting temperature (Tm), temperature midpoint at which thermal protein unfolding occurs, has been researched widely to develop formulations and attempts were made to predict long-term refrigerated storage based on related studies. Although DSC is an informative technique about the melting profile of biomolecules, it only probes thermodynamic stability under conditions far from storage conditions. Additionally, the usability of Tm is questionable, given that unfolding of most proteins under experimental conditions is not reversible. While it is common to use DSC to screen pH and buffer (see example in Figure 7.8) [79], preformulation screening using thermal transition temperature as a surrogate parameter to select a commercial protein formulation in the same example has not accounted for stresses such as freeze-thaw, temperature deviations during transport, interaction with barium leachable from glass surfaces, chemical degradation, and prediction of long-term storage stability. In a study by Matheus et al. [80], perturbation of secondary structure (by FTIR) as a function of temperature was evaluated on four different proteins at different concentrations. Signal–temperature curves were generated from which protein melting point—Tm(FTIR)—was derived. Results of Tm(FTIR) were compared against Tm values obtained from DSC studies—Tm(DSC)—on the same proteins at lower concentrations. Highlights from the study are presented in Figure 7.9. Tm(FTIR) values were comparable with Tm(DSC) values only with β-sheet proteins (i.e., IgG1). However, that was not the case with α-helical proteins in another study with recombinant

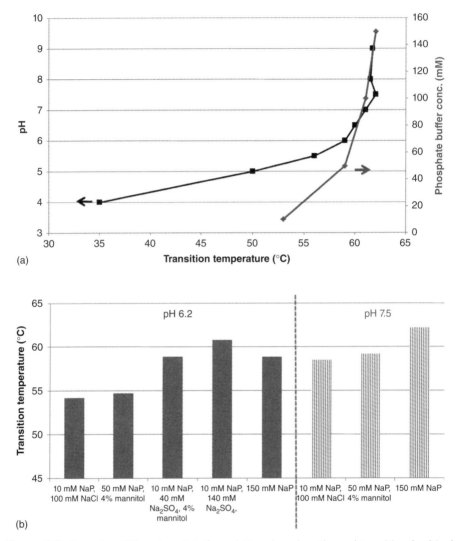

Figure 7.8. Screening different protein formulations based on thermal transition (melting) temperatures at selected pH values. (a) Transition temperature as a function of pH (4–9) and buffer concentration (10–150 mM) from a pH and buffer screening study. (b) Protein transition temperature during a formulation screening study. Solid bars are formulations at pH 6.2, striped bars are formulations at pH 7.5. Figures modified from Reference 78. (See color insert)

human interferon-γ [81]. Furthermore with α-helical proteins, Tm(FTIR) values were consistently higher at lower protein concentrations. The choice of Tm value as a "surrogate parameter" obtained from two different methods, as seen from this study, posed as a challenge. Besides the type of protein, dilution may also be another variable resulting in inconsistencies between methods (dilution is required with DSC, but not with FTIR). Diluting the sample can make the data irrelevant when studying concentration-dependent

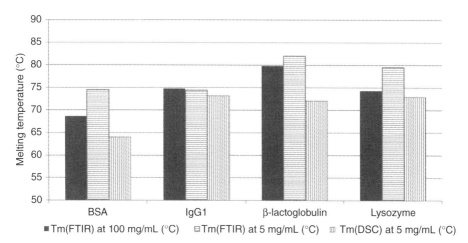

Figure 7.9. A comparison of Tm(FTIR) values at high and low protein concentrations and Tm(DSC) values at low protein concentrations. Figure reconstructed from Reference 79.

instability reactions such as aggregation and association reactions. This is a significant challenge especially when screening studies are performed for formulation development of proteins at high concentration. The predictability of Tm(FTIR) as a surrogate parameter to determine the stability of IgG1 was further investigated in a follow-up study [82]. Three main screening studies were conducted on IgG1 after determination of optimum pH for stability, namely buffer, stabilizer, and amino acid screening studies. Physical stability of the prepared formulations was determined at 40°C and Tm(FTIR) measurements on each formulation were taken, as previously described [80]. Using the published data, we calculated the aggregation rate constants (assuming first-order kinetics, where R^2 values were all >0.98) and plotted them against Tm(FTIR) values. As can be seen from Figure 7.10, there were poor correlation coefficients between aggregation rate constants (k) and Tm(FTIR) in both buffer screening and amino acid screening studies. However, there was a fair correlation ($R^2 > 0.5$) between k and Tm(FTIR) values for excipient screening studies. Chemical instability was not assessed in these studies, and it may very well be that degradation pathways in buffer screening and amino acid screening studies differ from those in the presence of excipients (trehalose, sucrose, and mannitol).

Stress conditions other than temperature are also overlooked when focusing only on Tm. As an example, when studying the potential use of polysorbates using Tm analyses by µDSC or FTIR, it is often found that polysorbates can slightly decrease Tm. At the same time, surfactants such as polysorbates are often critical components in protein formulations to protect against interfacial stresses such as shaking [40]. The sole analysis of Tm values may lead to inappropriately excluding surfactants in a formulation. Of course, Tm is a relevant parameter to research during characterization (see Section 7.3.4); however, the formulation scientist needs to maintain focus on the scope and applicability of Tm.

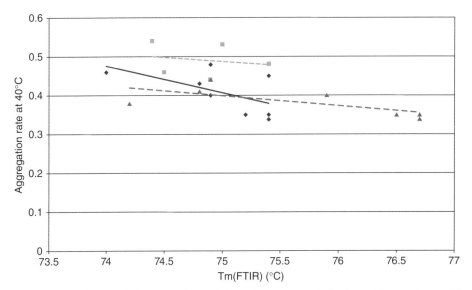

Figure 7.10. Correlation between physical aggregation rate constants (assuming first-order kinetics) to Tm(FTIR) values in buffer, amino acid, and stabilizer (sugar) screening studies. Figure modified from Reference 81. (See color insert)

Intrinsic fluorescence makes use of intrinsic chromophores in the protein structure such as tryptophan, tyrosine, and phenylalanine. As side chains become more surface exposed, the emission maxima shifts to higher wavelengths. Changes in protein (un)folding are often directly reflected by changes in intrinsic fluorescence. Unfortunately, the method cannot resolve the local conformational changes around these residues in a given protein. DSF characterizes the shift in melting temperature of proteins using extrinsic probes. Sypro orange has high binding affinity to hydrophobic sites in the protein structure. It is hypothesized that the dye will bind to hydrophobic patches of the protein that are exposed during protein unfolding, leading to enhanced fluorescence emission. Significant concerns exist on the potential of the applied dyes to induce structural changes in the protein (potentially leading to measurement artifacts) as well as on the interaction of dyes such as Sypro orange with organic solvents and polysorbates. Measurements of extrinsic fluorescence in the presence of polysorbates should be performed with care. Considering that polysorbates are a typical component of most protein formulations, extrinsic fluorescence may only have limited applicability [74, 83–86].

In summary, extensive caution should be taken with the implementation and use of analytical and biophysical techniques that are considered to "predict" protein stability. The use of such endpoints during formulation development can be a challenge if the predictive nature of the individual technique is not well understood, as to what it actually translates into the actual stability profile of a protein, at real-time or accelerated and stress studies.

7.3.3.2 Second Virial Coefficient (B_{22} or A_2). The second virial coefficient (B_{22} or A_2) is a thermodynamic parameter that reflects nonideality of protein solutions at higher concentrations, depending on the geometry and interaction potentials of two interacting molecules in solution. It has received significant attention by various research groups. It serves to study whether two proteins attract or repel each other in a given environment (protein–protein interactions, PPI). Of course, this depends on the protein and its structure, as well as on solution parameters such as pH, ionic strengths, and type of ions [87,88]. B_{22} has been traditionally used to study conditions where protein would crystallize [89]. It has also been used to predict aggregation in protein formulations and to rank formulation excipients or conditions for their potential to stabilize a given protein [19]. B_{22} can be derived from analyses using, for example, confined membrane electrophoresis and SLS via self-interaction chromatography or indirectly via DLS [90–92]. B_{22} values can be determined through the diffusion interaction parameters, K_D, and the sedimentation interaction parameters, K_S. While K_S requires significant analytical efforts to measure (using analytical ultra centrifugation), the diffusion interaction parameter (K_D) can be empirically determined by measuring diffusion coefficient, D, as a function of protein concentration using DLS. K_D has been applied as a potentially predictive tool to study protein aggregation propensity (as a result of PPI) and to compare and rank formulations or molecules [93]. Also, K_D can provide information on the viscoelastic behavior, as this is also governed by PPI [94]. Obviously, B_{22} is limited to assessing a (two-state) aggregation model, which is not able to judge or rank other pathways of degradation [87,88,95,96].

7.3.3.3 Optical Density, Turbidity, Light Scattering. Optical density measured in the visible or near-UV light wavelengths (such as 350 nm) is often used as a surrogate endpoint to study turbidity despite its limitations (no information on particle sizing and quantity). Light blockage at these wavelengths is often determined by protein aggregates, but may also have other causes (e.g., optical property of the protein solution and precursor to liquid–liquid phase separation). Measurements at 350 nm can make use of typical photometers or equipment that is capable to downscale sample requirements (e.g., nanodrops). SLS measures the amount of light scattered by particles in solution. Coupled with detection at different angles (MALS, MALLS), it can nicely be combined with size-exclusion chromatography or field-flow fractionation (FFF) to further characterize protein species separated by size, for example, to estimate molecular mass. DLS measures time-dependent intensity fluctuations of scattered light caused by Brownian motion of particles/diffusing species in solution. Data from DLS experiments are fitted using autocorrelation functions to obtain diffusion coefficients and can potentially yield information on size (hydrodynamic radii), distribution, geometry, and diffusion patterns. DLS has been used to study the impact of formulation parameters during (pre)formulation on the generation of protein aggregation or particles [97–100]. Accompanying temperature gradients to DLS experiments can be used to study formation of larger aggregates or even particles in solutions of different protein formulations, and hence help in ranking them. Typically, one would study the increase in DLS signal as a function of increase in temperature (suggesting the formulation of larger aggregates or particles in a protein formulation). This may also be linked to the protein Tm. For more information and discussion on the application of light scattering or turbidity measurements, we refer the reader to References 32 and 95–98.

7.3.3.4 Other Surrogate Parameters.

Recent studies have pointed to the utility of surrogate parameters for (pre)formulation studies for lyophilization formulation development such as secondary structure and molecular mobility [101–103]. Furthermore, a few studies have surfaced on composition heterogeneity as a surrogate parameter, but this may be of more interest to specific formulations that are stabilizer rich (e.g., typically found in formulations of different vaccines) [103–105]. Stabilization mechanism in the solid state is explained by either the glass dynamics theory or by the water substitution hypothesis, while stabilization in the liquid state is explained by the solute exclusion mechanism. Additionally, degradation mechanisms in the liquid state are much more complex than in the solid state [103].

The biophysical and biochemical characterization of a therapeutic protein has also been considered "preformulation" by a number of scientists [106]. These methods include structural analyses (e.g., FTIR, CD), calorimetric analyses (e.g., uDSC), and the analyses of charge and charge distribution (e.g., calculated and measured isoelectric point, B_{22}, surface charge analyses and modeling) and may all be valuable to guide the development (or use of a platform formulation) of a formulation scientist. Thus, they serve to study the protein of choice that is intended for development and to understand its behavior. However, details about these methods are not provided further in this chapter, since they are covered in details elsewhere in this book.

7.4 CONCLUSIONS

Understanding key protein and solution properties can guide formulation scientists to design clever studies for development and help them to avoid pitfalls and obvious mistakes. The overall formulation strategy of a company and timing of introduction of a commercial formulation during clinical development drive a number of key decisions. Several key challenges need to be overcome for a successful protein product, as the properties of a DP are impacted by the interactions between formulation composition, DP manufacturing unit operations (process), and container closure/primary packaging. Preformulation is defined differently between practitioner groups. Protein preformulation may be defined as R&D studies that aim at predicting what excipient(s) and solution conditions may help stabilize (or destabilize) the protein. A number of examples for preformulation activities are available in literature and can be clustered into (1) "developability" assessment and molecule optimization prior to the selection of a molecule for further clinical development (prior to clinical lead selection), (2) HT formulation approaches to accelerate development timelines while minimizing the amount of protein needed for the studies and being able to study a wider range of formulation excipients for stability or other endpoints considered relevant, (3) evaluation of surrogate or single "endpoint" testing, and (4) protein characterization studies, such as structural analyses and melting temperature, that form the basis for later formulation development studies. Whatever the definition is, it is of utmost importance to consider the setup of any preformulation/biophysical study being closely connected to the goal of achieving a successful commercial formulation and DP. If, for example, primary packaging/formulation interactions are not studied given that the formulation composition was solely derived from HT formulation studies or surrogate studies not using intended primary packaging (e.g., multiwall plates), the development scientist runs a significant risk of finding these only at a late time point during development. In any case, early stage studies need to be conducted with minimal amount of protein and resource, given the risk of attrition. Platform formulations in early stage development help to tackle these challenges. The formulation(s) and dosage form(s) used in early stages of development (phase 0, phase 1, and potentially phase 2) need to consider the overall TPP and impact on later stage development. For example, if a liquid prefilled syringe product is required for launch, the selection of freeze-dried dosage forms in early stages may pose significant challenges in later development for bridging and comparability. Therefore a platform lyophilized formulation could be avoided if a liquid alternative is available.

REFERENCES

1. Maynard, H.D. (2013) Proteins in a Pill. *Nature Chemistry*, **5**(7), 557–558.
2. Pavlou, A.K. and Belsey, M.J. (2005) The therapeutic antibodies market to 2008. *Eur. J. Pharm. Biopharm.*, **59**(3), 389–396.
3. Manning, M.C., et al. (2010) Stability of protein pharmaceuticals: an update. *Pharm. Res.*, **27**(4), 544–575.
4. Manning, M.C., Patel, K., and Borchardt, R.T. (1989) Stability of protein pharmaceuticals. *Pharm. Res.*, **6**(11), 903–918.
5. Zurdo, J., et al. (2011) Improving the developability of biopharmaceuticals. *Innov. Pharm. Technol.*, **37**, 34–40.
6. Masini, E., et al. (1985) Histamine-releasing properties of polysorbate 80 in vitro and in vivo: correlation with its hypotensive action in the dog. *Agents Actions*, **16**(6), 470–477.
7. Krantz, J.C., Jr., et al. (1948) Sugar alcohols; pharmacodynamic studies of polyxyalkylene derivatives of hexitol anhydride partial fatty acid esters. *J. Pharm. Exp. Ther.*, **93**(2), 188–195.
8. Warne, N.W. (2010) Formulation development of phase 1–2 biopharmaceuticals: an efficient and timely approach. In: Jameel, F.H. and Hershenson, S. editors. *Formulation and Process Development Strategies for Manufacturing Biopharmaceuticals*. John Wiley & Sons, Inc, pp. 147–159.
9. Volkin, D., Sanyal, G., Burke, C., and Middaugh, R. (2002) Preformulation studies as an essential guide to formulation development and manufacture of protein pharmaceuticals. *Pharma. Biotechnol.*, **14**, 1–46.
10. Kamerzell, T.J., et al. (2011) Protein-excipient interactions: mechanisms and biophysical characterization applied to protein formulation development. *Adv. Drug Deliv. Rev.*, **63**(13), 1118–1159.

11. Geiger, T. and Clarke, S. (1987) Deamidation, isomerization, and racemization at asparaginyl and aspartyl residues in peptides. Succinimide-linked reactions that contribute to protein degradation. *J. Biol. Chem.*, **262**(2), 785–794.

12. Wu, X.-H., et al. (2012) Identifying the hotspots on the top faces of WD40-repeat proteins from their primary sequences by beta-bulges and DHSW tetrads. *PloS One*, **7**(8), e43005.

13. Lauer, T.M., et al. (2012) Developability index: a rapid in silico tool for the screening of antibody aggregation propensity. *J. Pharm. Sci.*, **101**(1), 102–115.

14. Agrawal, N.J., et al. (2011) Aggregation in protein-based biotherapeutics: computational studies and tools to identify aggregation-prone regions. *J. Pharm. Sci.*, **100**(12), 5081–5095.

15. Mahler, H.-C. and Mekking, A. (2009) Formulation development of therapeutic proteins. In: Mahler, H.-C., Borchard, G., and Luessen, H. editors. *Protein Pharmaceuticals*. Editio Cantor Verlag, pp. 144–162.

16. Fischer, S., Hoernschemeyer, J., and Mahler, H.C. (2008) Glycation during storage and administration of monoclonal antibody formulations. *Eur. J. Pharm. Biopharm.*, **70**(1), 42–50.

17. Pikal, M.J., et al. (1991) The effects of formulation variables on the stability of freeze-dried human growth-hormone. *Pharm. Res.*, **8**(4), 427–436.

18. Gokarn, Y.R., et al. (2011) Effective charge measurements reveal selective and preferential accumulation of anions, but not cations, at the protein surface in dilute salt solutions. *Protein Sci.*, **20**(3), 580–587.

19. Saluja, A., et al. (2010) Diffusion and sedimentation interaction parameters for measuring the second virial coefficient and their utility as predictors of protein aggregation. *Biophys. J.*, **99**(8), 2657–2665.

20. Saluja, A., et al. (2009) Anion binding mediated precipitation of a peptibody. *Pharm. Res.*, **26**(1), 152–160.

21. Lam, X.M., Yang, J.Y., and Cleland, J.L. (1997) Antioxidants for prevention of methionine oxidation in recombinant monoclonal antibody HER2. *J. Pharm. Sci.*, **86**(11), 1250–1255.

22. Kishore, R.S., et al. (2011) Degradation of polysorbates 20 and 80: studies on thermal autoxidation and hydrolysis. *J. Pharm. Sci.*, **100**(2), 721–731.

23. Stoll, V.S. and Blanchard, J.S. (2009) Buffers: principles and practice. *Methods Enzymol.*, **463**, 43–56.

24. Singh, S.K., et al. (2011) Frozen state storage instability of a monoclonal antibody: aggregation as a consequence of trehalose crystallization and protein unfolding. *Pharm. Res.*, **28**(4), 873–885.

25. Yu, A.W., et al. (1998) Pain perception following subcutaneous injections of citrate-buffered and phosphate-buffered epoetin alpha. *Int. J. Artif. Organs*, **21**(6), 341–343.

26. Ganan Jimenez, A. and Brake, B. European Medicines Agency, Human Medicines Development and Evaluation. (2011). ICH Q5C: Stability testing of Biotechnological / Biological products. In: ICGH CGC ASEAN training. Kuala Lumpur. 30–31 May 2011.

27. Lasch, P., et al. (2001) Hydrogen peroxide-induced structural alterations of RNAse A. *J. Biol. Chem.*, **276**(12), 9492–9502.

28. Aman, W. and Thoma, K. (2003) ICH guideline for photostability testing: aspects and directions for use. *Pharmazie*, **58**(12), 877–880.

29. Mason, B.D., Schoneich, C., and Kerwin, B.A. (2012) Effect of pH and light on aggregation and conformation of an IgG1 mAb. *Mol. Pharm.*, **9**(4), 774–790.

30. Yin, J., et al. (2005) Effects of excipients on the hydrogen peroxide-induced oxidation of methionine residues in granulocyte colony-stimulating factor. *Pharm. Res.*, **22**(1), 141–147.

31. Wakankar, A.A. and Borchardt, R.T. (2006) Formulation considerations for proteins susceptible to asparagine deamidation and aspartate isomerization. *J. Pharm. Sci.*, **95**(11), 2321–2336.

32. Mahler, H.-C., et al. (2009) Protein aggregation: pathways, induction factors and analysis. *J. Pharm. Sci.*, **98**(9), 2909–2934.

33. Pikal, M.J. and Shah, S. (1992) Moisture transfer from stopper to product and resulting stability implications. *Dev. Biol. Stand.*, **74**, 165–177; discussion 177–179.

34. Pikal, M.J., Dellerman, K., and Roy, M.L. (1992) Formulation and stability of freeze-dried proteins: effects of moisture and oxygen on the stability of freeze-dried formulations of human growth hormone. *Dev. Biol. Stand.*, **74**, 21–37; discussion 37–38.

35. Roy, M.L., et al. (1992) The effects of formulation and moisture on the stability of a freeze-dried monoclonal antibody-vinca conjugate: a test of the WLF glass transition theory. *Dev. Biol. Stand.*, **74**, 323–339; discussion 340.

36. Zhang, A., et al. (2011) A new approach to explore the impact of freeze-thaw cycling on protein structure: hydrogen/deuterium exchange mass spectrometry (HX-MS). *Pharm. Res.*, **28**(5), 1179–1193.

37. Padala, C., et al. (2010) Impact of uncontrolled vs controlled rate freeze-thaw technologies on process performance and product quality. *PDA J. Pharm. Sci. Technol.*, **64**(4), 290–298.

38. Mitchell, B.L., et al. (2005) Impact of freeze-thaw cycles and storage time on plasma samples used in mass spectrometry based biomarker discovery projects. *Cancer Inform.*, **1**, 98–104.

39. Mahler, H.C., et al. (2005) Induction and analysis of aggregates in a liquid IgG1-antibody formulation. *Eur. J. Pharm. Biopharm.*, **59**(3), 407–417.

40. Kiese, S., et al. (2008) Shaken, not stirred: mechanical stress testing of an IgG1 antibody. *J. Pharm. Sci.*, **97**(10), 4347–4366.

41. Lam, X.M., et al. (2011) Site-specific tryptophan oxidation induced by autocatalytic reaction of polysorbate 20 in protein formulation. *Pharm. Res.*, **28**(10), 2543–2555.

42. Bhambhani, A. and Medi, B.M. (2010) Selection of containers/closures for use in lyophilization applications: possibilities and limitations. *Am. Pharm. Rev.*, **13**(4). http://www.americanpharmaceuticalreview.com/Featured-Articles/36725-Selection-of-Containers-Closures-for-Use-in-Lyophilization-Applications-Possibilities-and-Limitations/

43. Markovic, I. (2007) Evaluation of safety and quality impact of extractable and leachable substances in therapeutic biologic protein products: a risk-based perspective. *Expert Opin. Drug Saf.*, **6**(5), 487–491.

44. Kushwaha, P. and Madan, A. (2008) Extractables and leachables: an overview of emerging challenges. *Pharm. Technol.*, **32**(8). http://www.pharmtech.com/pharmtech/article/articleDetail.jsp?id=583058

45. Markovic, I. (2012) Keynote address: extractables and leachables in prefilled syringes. In: *European Pre-Filled Syringes Conference*, London, United Kingdom.

46. Rupertus, V. and Rothhaar, U. (2012) PDA Europe thanks universe of pre-filled syringes platinum. Two ways to minimize the delamination risk of glass containers. PDA Lett. 42–43.

47. Boddapati, S., et al. (1980) Identification of subvisible barium sulfate crystals in parenteral solutions. *J. Pharm. Sci.*, **69**(5), 608–610.

48. Seidl, A., et al. (2012) Tungsten-induced denaturation and aggregation of epoetin alfa during primary packaging as a cause of immunogenicity. *Pharm. Res.*, **29**(6), 1454–1467.

49. Vilivalam, V., Waxman, L., and Steeley, T. (2011) Protein sensitivity to tungsten. *Pharm. Technol.*, November, s28–s32.

50. Adler, M. (2012) Challenges in the development of pre-filled syringes for biologics from a formulation scientist's point of view. *Am. Pharm. Rev.*, **15**(1), 96, 98–101.

51. Jiang, Y., et al. (2009) Tungsten-induced protein aggregation: solution behavior. *J. Pharm. Sci.*, **98**(12), 4695–4710.

52. Liu, W., et al. (2010) Root cause analysis of tungsten-induced protein aggregation in pre-filled syringes. *PDA J. Pharm. Sci. Technol.*, **64**(1), 11–19.

53. Bee, J.S., et al. (2009) Precipitation of a monoclonal antibody by soluble tungsten. *J. Pharm. Sci.*, **98**(9), 3290–3301.

54. Majumdar, S., et al. (2011) Evaluation of the effect of syringe surfaces on protein formulations. *J. Pharm. Sci.*, **100**(7), 2563–2573.

55. Liu, D., et al. (2012) Interactions between therapeutic proteins and acrylic acid leachable. *PDA J. Pharm. Sci. Technol.*, **66**(1), 12–19.

56. van Beers, M.M.C., et al. (2011) Oxidized and aggregated recombinant human interferon beta is immunogenic in human interferon beta transgenic mice. *Pharm. Res.*, **28**(10), 2393–2402.

57. Guadagnino, E. and Zuccato, D. (2012) Delamination propensity of pharmaceutical glass containers by accelerated testing with different extraction media. *PDA J. Pharm. Sci. Technol.*, **66**(2), 116–125.

58. Walther, M., et al. (2002) Pharmaceutical vials with extremely high chemical inertness. *PDA J. Pharm. Sci. Technol.*, **56**(3), 124–129.

59. Jiang, G., et al. (2007) Mechanistic studies of glass vial breakage for frozen formulations. I. Vial breakage caused by crystallizable excipient mannitol. *PDA J. Pharm. Sci. Technol.*, **61**, 441–451.

60. Williams, N.A. and Guglielmo, J. (1993) Thermal mechanical analysis of frozen solutions of mannitol and some related stereoisomers: evidence of expansion during warming and correlation with vial breakage during lyophilization. *J. Parenter. Sci. Technol.*, **47**, 119–123.

61. Bauer Dauphin, I. and Mahler, H.C. (2010) A method for avoiding glass fogging of pharmaceutical containers during freeze drying. *PCT Int. Appl.*, 26. http://www.google.com/patents/WO2010115728A2?cl=en

62. Harris, R.J., et al. (2001) Identification of multiple sources of charge heterogeneity in a recombinant antibody. *J. Chromatogr. B, Biomed. Sci. Appl.*, **752**(2), 233–245.

63. Buck, P.M., et al. (2012) Computational methods to predict therapeutic protein aggregation. *Methods Mol. Biol.*, **899**, 425–451.

64. Pawar, A.P., et al. (2005) Prediction of "aggregation-prone" and "aggregation-susceptible" regions in proteins associated with neurodegenerative diseases. *J. Mol. Biol.*, **350**(2), 379–392.

65. Lehermayr, C., et al. (2011) Assessment of net charge and protein-protein interactions of different monoclonal antibodies. *J. Pharm. Sci.*, **100**(7), 2551–2562.

66. Shire, S.J., Shahrokh, Z., and Liu, J. (2004) Challenges in the development of high protein concentration formulations. *J. Pharm. Sci.*, **93**(6), 1390–1402.

67. Kayser, V., et al. (2012) A screening tool for therapeutic monoclonal antibodies: identifying the most stable protein and its best formulation based on thioflavin T binding. *Biotechnol. J.*, **7**(1), 127–132.

68. Capelle, M.A., Gurny, R., and Arvinte, T. (2009) A high throughput protein formulation platform: case study of salmon calcitonin. *Pharm. Res.*, **26**(1), 118–128.

69. Li, Y., Mach, H., and Blue, J.T. (2011) High throughput formulation screening for global aggregation behaviors of three monoclonal antibodies. *J. Pharm. Sci.*, **100**(6), 2120–2135.

70. Bhambhani, A., et al. (2012) Formulation design and high-throughput excipient selection based on structural integrity and conformational stability of dilute and highly concentrated IgG1 monoclonal antibody solutions. *J. Pharm. Sci.*, **101**(3), 1120–1135.

71. Capelle, M.A., Gurny, R., and Arvinte, T. (2007) High throughput screening of protein formulation stability: practical considerations. *Eur. J. Pharm. Biopharm.*, **65**(2), 131–148.

72. Kim, A. (2009) Rational high-throughput screening for formulations that physically stabilize recombinant proteins. In: *Chemical Engineering*. Carnegie-Mellon Univ, Pittsburgh, PA. p. 264. http://udini.proquest.com/view/rational-high-throughput-screening-pqid: 1877818241/

73. Capelle, M. and Arvinte, T. (2008) High-throughput formulation screening of therapeutic proteins. *Drug Discov. Today Technol.*, **5**(2–3), e71–e79.

74. Ingham, K.C. and Bolotin, C. (1978) Intrinsic and extrinsic fluorescence probes of subunit interactions in ovine lutropin. *Arch. Biochem. Biophys.*, **191**(1), 134–145.

75. Fraunhofer, W. (2006) Innovative and Practical Approaches to Understand and Handle Physical Instability & Aggregation of Protein Pharmaceuticals. In: IBC's 10th International Conference Well Characterized Biologicals.

76. Gibson, T.J., et al. (2011) Application of a high-throughput screening procedure with PEG-induced precipitation to compare relative protein solubility during formulation development with IgG1 monoclonal antibodies. *J. Pharm. Sci.*, **100**(3), 1009–1021.

77. Matheus, S. (2006) Development of high concentration cetuximab formulations using ultra-filtration and precipitation. In: *Fakultat fur Chemie und Pharmazie*, Ludwig-Maximilians-Universitat Munchen, München, Germany.

78. Middaugh, C.R., et al. (1979) Determination of the apparent thermodynamic activities of saturated protein solutions. *J. Biol. Chem.*, **254**(2), 367–370. http://www.jbc.org/content/254/2/367

79. Papadimitriou, A. (2001) *New pharmaceutical composition*. In: Hoffmann, F. editor. *P.I. Appl.* La Roche A.-G., Switzerland. 64 pp.

80. Matheus, S., Friess, W., and Mahler, H.-C. (2006) FTIR and nDSC as analytical tools for high-concentration protein formulations. *Pharm. Res.*, **23**(6), 1350–1363.

81. Kendrick, B.S., et al. (1998) Aggregation of recombinant human interferon gamma: kinetics and structural transitions. *J. Pharm. Sci.*, **87**(9), 1069–1076.

82. Matheus, S, Mahler, H., and Friess, W. (2006) A critical evaluation of Tm(FTIR) measurements of high-concentration IgG1 antibody formulations as a formulation development tool. *Pharm. Res.*, **23**(7), 1617–1627.

83. Uniewicz, K.A., et al. (2010) Differential scanning fluorimetry measurement of protein stability changes upon binding to glycosaminoglycans: a screening test for binding specificity. *Anal. Chem.*, **82**(9), 3796–3802.

84. Niesen, F.H., Berglund, H., and Vedadi, M. (2007) The use of differential scanning fluorimetry to detect ligand interactions that promote protein stability. *Nat. Protoc.*, **2**(9), 2212–2221.

85. Bhattacharya, M., et al. (2011) pH-Induced conformational isomerization of bovine serum albumin studied by extrinsic and intrinsic protein fluorescence. *J. Fluoresc.*, **21**(3), 1083–1090.

86. Samra, H.S. and He, F. (2012) Advancements in high throughput biophysical technologies: applications for characterization and screening during early formulation development of monoclonal antibodies. *Mol. Pharm.*, **9**(4), 696–707.

87. Bajaj, H., et al. (2006) Protein structural conformation and not second virial coefficient relates to long-term irreversible aggregation of a monoclonal antibody and ovalbumin in solution. *Pharm. Res.*, **23**(6), 1382–1394.

88. Bajaj, H., Sharma, V.K., and Kalonia, D.S. (2004) Determination of second virial coefficient of proteins using a dual-detector cell for simultaneous measurement of scattered light intensity and concentration in SEC-HPLC. *Biophys. J.*, **87**(6), 4048–4055.

89. Deszczynski, M., Harding, S.E., and Winzor, D.J. (2006) Negative second virial coefficients as predictors of protein crystal growth: evidence from sedimentation equilibrium studies that refutes the designation of those light scattering parameters as osmotic virial coefficients. *Biophys. Chem.*, **120**(2), 106–113.

90. Moody, T.P. and Shepard, H.K. (2004) Nonequilibrium thermodynamics of membrane-confined electrophoresis. *Biophys. Chem.*, **108**(1–3), 51–76.

91. Yadav, S., et al. (2011) Use of dynamic light scattering to determine second virial coefficient in a semidilute concentration regime. *Anal. Biochem.*, **411**(2), 292–296.

92. Einaga, Y., Abe, F., and Yamakawa, H. (1992) Light scattering method of determining the second virial coefficient for simple molecules and oligomers. *J. Phys. Chem.*, **96**(10), 3948–3953.

93. Yadav, S., Shire, S.J., and Kalonia, D.S. (2010) Factors affecting the viscosity in high concentration solutions of different monoclonal antibodies. *J. Pharm. Sci.*, **99**(12), 4812–4829.

94. Connolly, B.D., et al. (2012) Weak interactions govern the viscosity of concentrated antibody solutions: high-throughput analysis using the diffusion interaction parameter. *Biophys. J.*, **103**(1), 69–78.

95. Saito, S., et al. (2012) Behavior of monoclonal antibodies: relation between the second virial coefficient (B (2)) at low concentrations and aggregation propensity and viscosity at high concentrations. *Pharm. Res.*, **29**(2), 397–410.

96. Yadav, S., et al. (2011) Establishing a link between amino acid sequences and self-associating and viscoelastic behavior of two closely related monoclonal antibodies. *Pharm. Res.*, **28**(7), 1750–1764.

97. Shiba, K., et al. (2010) Polydispersity as a parameter for indicating the thermal stability of proteins by dynamic light scattering. *Anal. Sci. Int. J. Jpn Soc. Anal. Chem.*, **26**(6), 659–663.

98. Chayen, N., et al. (2004) Size and shape determination of proteins in solution by a non-invasive depolarized dynamic light scattering instrument. *Ann. N. Y. Acad. Sci.*, **1027**, 20–27.

99. Arzenšek, D. (2010) *Dynamic Light Scattering and Application to Proteins in Solutions*. Department of Physics, Faculty of Mathematics and Physics, University of Ljubljana, Ljubljana.

100. Li, S.X., Xing, D., and Li, J.F. (2004) Dynamic light scattering application to study protein interactions in electrolyte solutions. *J. Biol. Phys.*, **30**(4), 313–324.

101. Murphy, B.M., et al. (2012) Structure, stability, and mobility of a lyophilized IgG1 monoclonal antibody as determined using second-derivative infrared spectroscopy (Vol. **101**, p. 81, 2012). *J. Pharm. Sci.*, **101**(5), 1941–1946.

102. Abdul-Fattah, A.M., Kalcinia, D.S., and Pikal, M.I. (2007) The challenge of drying method selection for protein pharmaceuticals: product quality implications. *J. Pharm. Sci.*, **96**(8), 1886–1916.

103. Truong, V. and Abdul-Fattah, A. (2010) The impact of formulation and drying processes on the characteristics and performance of biopharmaceutical powders. In: Jameel, F. and Hershenson, S. editors. *Formulation and Process Development Strategies for Manufacturing Biopharmaceuticals.* John Wiley & Sons, Inc., Hoboken, NJ, pp. 565–585.

104. Abdul-Fattah, A.M., et al. (2007) Drying-induced variations in physico-chemical properties of amorphous pharmaceuticals and their impact on stability (I): stability of a monoclonal antibody. *J. Pharm. Sci.*, **96**(8), 1983–2008.

105. Abdul-Fattah, A.M., et al. (2007) Drying-induced variations in physico-chemical properties of amorphous pharmaceuticals and their impact on stability II: stability of a vaccine. *Pharm. Res.*, **24**(4), 715–727.

106. Bedu-Addo, F.K., et al. (2004) Use of biophysical characterization in preformulation development of a heavy-chain fragment of botulinum serotype B: evaluation of suitable purification process conditions. *Pharm. Res.*, **21**(8), 1353–1361.

8

BIOANALYTICAL METHODS AND IMMUNOGENICITY ASSAYS

Bonita Rup, Corinna Krinos-Fiorotti, Boris Gorovits, and Hendrik Neubert

Department of Pharmacokinetics, Pharmacodynamics, and Metabolism, Pfizer, Andover, MA, USA

Biophysical Methods for Biotherapeutics: Discovery and Development Applications, First Edition. Edited by Tapan K. Das.
© 2014 John Wiley & Sons, Inc. Published 2014 by John Wiley & Sons, Inc.

8.1 INTRODUCTION

Bioanalytical methods are required to characterize pharmacokinetics (PK), pharmaco-dynamics (PD), and immunogenicity of a biotherapeutic candidate in nonclinical and clinical studies. PK, PD, and immunogenicity assays are utilized throughout the discovery and development stages with increased attention to assay reliability and limitations, assay quality, and regulatory requirements being applied as the candidate advances. The structural complexity and heterogeneity that is common for most biotherapeutics can be characterized by a diverse set of biophysical methods described in earlier chapters. Ultimately it will be important to understand the impact that the biophysical characteristics have on PK, PD, and immunogenicity of the biotherapeutic candidate during the stages in which the candidate structure and manufacturing process are being designed and controlled. Because the biophysical properties also affect performance of the bioanalytical and immunogenicity assays, the potential for biophysical characteristics to change during the product life cycle must be a consideration in strategies to develop, maintain, and interpret the results of these assays. Rapid developments in the biotechnology discovery area are resulting in the generation of a greater number of candidates across a wider range of modalities, in turn resulting in greater challenges for biophysical, bioanalytical, and immunogenicity assay development.

8.1.1 Biotherapeutic Modalities

As shown in Table 8.1, today the field of biotherapeutics covers a diverse range of protein-based modalities, including recombinant versions of natural human proteins,

TABLE 8.1. Diversity of biotherapeutic protein modalities

Modality	Sources	Examples
Peptides	Natural, synthetic, recombinant	Insulins, incretin mimetics
Recombinant proteins	Human or animal derived, recombinant cytokines, growth factors, enzymes and zymogens, and others	Interferons, human growth hormone, bone morphogenetic protein-2, tissue plasminogen activator, factor VIII
Monoclonal antibodies	Chimeric, humanized, human	Rituximab, adalimumab
Fc fusion proteins	Fusions with peptides, soluble receptors, recombinant proteins	Enbrel (etanercept)
Antibody conjugates	Antibody–toxin conjugates, antibody–drug conjugates, pegylated proteins	Mylotarg (gemtuzumab ozogamicin), Cimzia (certolizumab pegol)
Antibody fragments	Fabs, single-chain Fvs, Nanobodies, camel Vhh-based single chains	Reopro (abciximab), Lucentis (ranibizumab), Cimzia (certolizumab pegol)
Novel scaffolds	Domains of fibronectin, albumen, transferrin, Staph -protein A	

synthetic peptides, monoclonal antibodies (MAbs) and fragments, and conjugates and fusion proteins that may be based on antibody Fc regions or on novel scaffolds [1–5].

Nonprotein-based biotherapeutics, including oligonucleotide-based therapies, are also entering development more frequently; multiple reviews have been written on bioanalytical strategies for these therapies [6–9]. However, nonprotein therapies will not be covered in this chapter.

One of the important distinguishing aspects of biotherapeutic candidates is the characteristic heterogeneity. A large protein may contain multiple isoforms that differ in size, charge, presence, and type. These isoforms may differ in their PK, PD/efficacy, and immunogenicity profiles and also in terms of their performance in assays. In addition, product-related and process-related impurities, though usually present in very low quantities, can also affect assay performance.

8.1.2 Application of Bioanalytical and Immunogenicity Assays in Discovery and Development of Biotherapeutics: Stage-Specific Requirements

In the discovery stage, selecting the best target and drug candidates for the disease area are the primary drivers. PK and PD properties are important aspects of biotherapeutic candidate selection. Because many biotherapeutics have species-specific activity, *in vitro* assays may be used to characterize cross-species biological activity in order to select appropriate nonclinical species for *in vivo* studies. It is useful to compare the activity using cell lines or primary cells derived from the intended test species with measurements such as potency, receptor occupancy or target-binding/inhibition potential, affinity, and selectivity. Early characterization of PK may be performed in nonclinical species to select among candidates and to optimize the design (e.g., dose and frequency) of nonclinical efficacy and safety studies. Therefore PK assays should usually be available at this stage. The ligand-binding assay (LBA) format has been most commonly used, and therefore assay reagent development may be the critical activity that drives timelines. This represents a special challenge when there are multiple candidates from which to select, since generation of reagents for candidates that are unlikely to be selected for advancement would be costly. Alternatively target-derived reagents, radiolabeled drug, generic LBAs (e.g., assays that detect human immunoglobulin Fc used to measure the humanized MAb in rodents or nonhuman primates), or more recently mass spectrometry-based assays have been employed to overcome the need for early generation of candidate-specific LBA reagents to enable early PK assessments.

Increasingly, biotherapeutics are being developed for subcutaneous or other non-intravenous routes of administration [10]. In some cases initial development may proceed with an intravenous formulation with a decision to develop a subcutaneous formulation occurring later in development after some confidence in the mechanism has been developed. Regardless of when the decision on route of administration is made, nonclinical studies are usually conducted to ensure adequate bioavailability of the biotherapeutic candidate by the non-intravenous route. Because higher concentration drug product is usually required for subcutaneous administration, it may be important to evaluate the subcutaneous formulated lot for any impact on assay performance when conducting the bioavailability studies.

Assays for PD markers or target could also be implemented in early stages to characterize PK/PD relationships. Understanding whether the PK and target assay designs favor detection of total, bound, or free drug and total, bound, or free target is important in modeling these relationships.

Assessment of anti-drug antibody development is performed in early stages to aid in the interpretation of the nonclinical studies, not for the purpose of predicting human anti-drug antibody development. In particular, such assessments can help determine if the presence of anti-drug antibodies has reduced exposure to active drug, causing poor outcomes in PK and efficacy models. Assays for anti-drug antibodies are not always needed in early studies, especially if the drug exposure is short (e.g., <7–14 days) or if the drug concentration and/or activity profiles show no evidence of anti-drug antibody effects. In addition, changes in the PK or PD activity assays consistent with the timing of anti-drug antibody formation are often used to infer that anti-drug antibodies were present, even if not directly measured.

As the candidate advances, nonclinical single- and repeat-dose toxicology studies are conducted. In these studies, it is important to demonstrate that exposure to active drug has been maintained throughout the study and to characterize the exposure–response relationship of any toxicological findings. Furthermore, toxicity is frequently associated with exaggerated pharmacology, therefore measurements of PD activity may be useful. Development of anti-drug antibodies can affect the exposure or safety findings, especially in repeat-dose studies or studies of candidates with prolonged exposure or delayed pharmacological activity. Assays to characterize toxicokinetics, PD activity, and immunogenicity are therefore regularly employed in these studies. Generally, anti-drug antibody assays are needed if adequate exposure to active biotherapeutic cannot be assured based on PK and PD assay measurements or if there is a need to establish that PK, PD, or safety findings are anti-drug antibody related [11]. Because anti-drug antibody development against human proteins in animals is usually not predictive of the incidence of anti-drug antibody in humans, generally anti-drug antibody assay results are only needed for interpretation of the nonclinical study results. For IND-enabling studies that require compliance with the Good Laboratory Practices (GLPs) Regulations, the PK and anti-drug antibody assays should be validated [12, 13]. Early clinical studies are usually designed to characterize the drug PK, PD, and safety profile. The first in human (FIH) studies may be conducted in either normal human volunteers or the intended patient population. As the project stage approaches IND submission for FIH studies, assay requirements need to be established. Selection of assays and optimization strategy require consideration of study-specific requirements including whether there will be sample matrix-specific effects and differences in the analyte concentration. For example, some populations (e.g., rheumatoid arthritis) may be characterized by the presence of auto- or cross-reactive antibodies that may affect the performance of the assays. In particular, for immunogenicity assessments, generation of an immunogenicity risk and mitigation plan is a regulatory expectation. The risk assessment conducted at this stage is intended to determine what types of assays and sample analysis strategy to employ. Products with higher perceived risk may require more aggressive monitoring for development or consequences of anti-drug antibody formation. An immunogenicity risk assessment is usually expected to monitor for any development and consequences of anti-drug antibodies.

Additional nonclinical studies may be conducted in parallel with the clinical studies, for example, longer term (chronic) toxicology and reproductive toxicology studies, and studies to support manufacturing changes are often not conducted until the FIH studies become available to demonstrate the viability of the candidate for further development and support refinement of the intended clinical dosing regimen and indication.

Assays used in early clinical studies are usually carried over into pivotal studies and post-marketing studies although modification may be necessary, for example, adjustment of sensitivity, dynamic range, and optimization of background. Changes in the manufacturing process may lead to the need for qualification of different lots of reference materials. Changes in route of administration may trigger the need for adjustment of the assay ranges.

8.2 ASSAYS TO ASSESS PK

Analysis of drug concentration for characterization of PK is an essential element of the drug development process. PK evaluation is crucial when determining appropriate exposure to the drug. PK data are also utilized to construct PK–PD models that can link drug exposure to the pharmacological effect [14, 15].

A careful evaluation of a biotherapeutic candidate project-specific analytical strategy should be initiated early on in order to ensure a successful transition through the project development milestones. Commonly the type of the analyte as well as the analytical approach in testing should be reviewed. Ideally the analyte type should be determined based on the PK–PD evaluation of the biotherapeutic candidate, for example, using information about biotherapeutic mechanism of action (MOA) and biotherapeutic binding partners *in vivo*. Typically, multiple isoforms of the biotherapeutic may be present in circulation, including intact biotherapeutic and products of partial metabolism (catabolism). In many cases, the biotherapeutic will be present as free, partially free, bound, and partially bound to the target. This is particularly exemplified by MAb-based biotherapeutics [16].

Often, availability of assay reagents and laboratory capabilities will play a substantial role in the selection of the assay analytical platform and the assay design. Yet, requirements generated based on PK–PD data review should be used as primary input in decision process. The assay format together with the specificity of the critical reagents used in the method will determine the nature of the analyte that can be detected. For example, assays using the biological target as a capture reagent will generally detect free or partially free biotherapeutic. At the same time, the specificity of an anti-biotherapeutic antibody used as a capture reagent will determine whether the assay measures free or total (free and bound) biotherapeutic concentration. Here, use of an anti-idiotypic anti-biotherapeutic reagent will allow for detection of free or partially free biotherapeutic while use of a general scaffold-specific anti-biotherapeutic antibody (e.g., antibody to immunoglobulin Fc) will likely result in measuring of total (free and bound) biotherapeutic concentration.

One has to remember that project requirements may change as the compound matures though the development cycle, resulting in a change in the required PK evaluation, for example, a shift from the need to measure total versus free biotherapeutic. Such changes may require development of a new PK assay, often associated with a need for assay reagents with different specificity.

It is often assumed that biotherapeutics substantially lose their biological activity after initial degradation *in vivo*. Depending on the modality of the biotherapeutic compound, it could be important to understand whether products of its degradation are themselves stable *in vivo* and continue to exhibit any biological activity or response in the PK assays. It could be possible that minor proteolytic modifications of a protein-based therapeutic may result in the generation of a stable yet inactive fragment with a PK profile similar to that for the original full-length biotherapeutic. Although not typically required, evaluation of specificity of the assay with regard to detecting only the active and/or intact form of the biotherapeutic may provide helpful information. As an example, assays using the liquid chromatography-mass spectrometry (LCMS) platform may be designed to allow for specific measurement of the peptides found both within the active site and within other domains of the biotherapeutic. To design LBA methods with similar capability will require specific and often proprietary, custom-made reagents.

Information about potential degradation products for a biotherapeutic compound may be obtained from the *in vitro* stability program required as part of the typical biotherapeutic drug product release program as well as from the matrix stability studies conducted as part of the PK assay development and validation process, as described later. Finally, parallelism assessment, requiring analysis of incurred study sample performance, will yield important information about potential biotherapeutic processing *in vivo*.

LBAs, the most common analytical platforms for PK support of biotherapeutics, include various types of solid- and solution-phase enzyme-linked immunoassays (ELISAs) and other methodologies that vary depending on the type of protein–protein complex separation and detection mechanism. Formats include, but are not limited to, sandwich, competition, direct, and indirect binding. The electrochemiluminescence plate-based platform (e.g., Meso Scale Technology (MSD)), fluorescence bead-based platform (e.g., Gyrolab® by Gyros), microparticle-based immunoassays with single-molecule counting technology (Erenna® by Singulex), and a combination of standard immunoassay and real-time PCR detection (e.g., Imperacer® by Chimera Biotec) are innovative assay technologies that can be used to build PK assays. LBA platforms are commonly used at all stages of the drug development process, including discovery, candidate selection, nonclinical and clinical evaluation, and post-marketing monitoring.

Successful development and implementation of LBA methods hinges on the availability of specific and high-quality assay reagents. LBA methodology is based on a series of specific and unique biochemical interactions between the analyte and assay reagents resulting in selection and detection of the biotherapeutic or a fragment of it rather than other matrix components. Therefore, characteristics of critical reagents that provide the necessary selectivity and sensitivity are imperative to the assay performance.

Recently, wider application of LCMS platforms to PK evaluation of biotherapeutics has been reported. The LCMS analytical platform presents an attractive alternative to the LBA-based methodology [17, 18]. Analytical options for LCMS evaluation of biotherapeutics largely depend on the molecular weight. Methods used in support of peptides or low molecular weight proteins are often similar to those applied for small-molecule biotherapeutics. For example, peptide analysis can be based on an initial sample pretreatment using a solid-phase extraction (SPE) protocol. LCMS application has been successfully used for oligonucleotide-based compounds [8]. A more typical biotherapeutic with a molecular weight of 50–150 kDa will require a more complex sample pretreatment procedure, often comprised of immunoaffinity (IA) separation followed by extraction and digestion steps [17]. An IA purification step requires the availability of a specific reagent, often an antibody to the biotherapeutic antibody. Other bioanalytical platforms include enzyme activity measurements (e.g., for enzyme homologues), hybridization (oligonucleotides), radioimmunoprecipitation, and surface-plasmon resonance-based platforms.

Critical reagents commonly include monoclonal and polyclonal antibodies (MAbs and PAbs), antibody fragments, binding partners or targets for the biotherapeutic, fusion proteins, or otherwise engineered proteins, enzymes, chemically synthesized molecules, and others. Conjugates of the same reagents with small organic molecules (e.g., biotin or fluorescent dyes) or enzymes are commonly used. More complex biological systems, such as cells, tissues, cell organelles, or membranes, can be used as assay system components. Other assay components may also be viewed as critical elements of the method and should be evaluated with care. These may include cell lines, complex biologics, or mixtures, for example, blocking agents, biological matrices, solid-phase components of the assay, or other reagents.

Critical reagents should be viewed as absolutely essential components of any LBA. Biophysical characteristics of critical reagents together with the nature of the analytical platform chosen will dictate fundamental assay parameters such as sensitivity, dynamic range, reproducibility, and robustness. A change in the batch or lot of a critical reagent will often require re-qualification or revalidation of the assay. On the opposite side, investigations initiated to understand the reasons for assay failure will commonly include evaluation of the quality and stability of reagents used in the method. The best practices for critical reagent generation, characterization, resupply, and overall life cycle management have been reviewed [19, 20].

Assay reference material, generally prepared from the drug substance material, is a critical component of the PK assay. Availability and quality of the reference material lot used during assay development and later stages are absolutely crucial for the success of bioanalytical support in general. The best effort should be made to information on characteristics of the reference material available, including the presence and the nature of aggregated components, the nature and extent of glycosylation, any variations in the degree of the active versus inactive form (e.g., in cases when protein requires a refolding step during purification procedure), and presence of a co-purified material (e.g., host cell proteins). Commonly, characterization of the reference material is expected as a part of material release [21]. Earlier lots of material may not receive the level of characterization applicable for later, Good Manufacturing Practice (GMP) grades. However, a timely and all inclusive communication between the biotherapeutic manufacturing and formulation group and the bioanalytical lab is imperative for the success of the bioanalytical work in general and in particular during assay development, validation, and implementation phases.

8.2.1 PK Assay Development Considerations

Typically the steps for developing a LBA-based PK assay are [22]

- Identification of assay reagents: Selection of capture and detector reagents is usually driven by the nature of analyte required for PK evaluation.
- Identification of the assay format.
- Identification of assay matrix, sample preparation procedure.
- Establishing the minimal required dilution (MRD) parameter.
- Establishing assay specificity and selectivity characteristics.
- Establishing the standard calibrator range and selection of the model for the standard curve, for example, 4-parameter versus 5-parameter (4PL vs. 5PL) logistic nonlinear regression model.
- Evaluation of the assay range of quantification.
- Evaluation of assay precision and accuracy.
- Initiation of the sample stability evaluation.
- Evaluation of assay dilutional linearity.
- Evaluation of assay robustness and ruggedness.

Assay development generally starts with the selection of the assay format and assay reagents. The assay detection system should be carefully considered in order to provide an acceptable signal to noise ratio. The sources and the nature (lot number, batch number) of the assay critical reagents, including reference material that was used to prepare assay standards and quality control (QC) solutions, should be carefully documented.

Assay parameters, presented below, should be considered to simulate assay use during assay validation and study support. These include, but are not limited to, the positioning of the calibrator standards, QC samples, study samples on the plate, the number of replicate wells used, the number of QC samples per run, and other sources of variability. QC samples are expected to be positioned evenly across the assay quantification range and generally include a low QC (approximately three times the lower limit of quantification (LLOQ)), midpoint QC (approximately midpoint of the calibration curve), and high QC (approximately 75% of the upper level of quantification (ULOQ)). The number of calibrator standards may be greater during the assay development phase in order to enable a better understanding of the concentration–response curve performance. This number can be truncated later during the assay validation phase.

Selection of the biological matrix to be analyzed in the assay may be influenced by PK–PD modeling information, for example, by an understanding of the target tissue and/or blood to tissue drug transition for the biotherapeutic. Biophysical characteristics of the biotherapeutic and expected assay interferences are also very important. Matrix additives, for example, protease inhibitors or anticoagulants, may be evaluated to assess potential impact on recovery or stability of the biotherapeutic candidate during sample collection or post collection processing. If any sample pretreatment is required as part of the assay procedure, all samples, including calibrators, QC samples, and study samples, are expected to be treated following the same procedure if possible.

Evaluation of the mathematical model used to fit the calibrator mean response values should be based on analysis of accuracy as a function of mean back-calculated values at different concentration levels. Such analysis will require repeat regression of the calibration curve data using various fitting algorithms (e.g., 4-parameter, 5-parameter) as well as evaluation of various weighting options.

Precision and accuracy are performance characteristics designed to depict the level of random and systematic error in the assay, respectively. Data need to be collected in multiple runs over several days. Intra-run and inter-run data analysis should be conducted. Although some target limits specifying allowable analytical error exist [23], assay-specific limits for precision and accuracy characteristics should be established during assay development, confirmed during assay validation, and used throughout assay life cycle.

Performance of the calibration standards should not be utilized to define the assay range of quantification. Instead, QC samples should be prepared at the intended lowest (LLOQ) and highest (ULOQ) level of quantification and tested for precision and accuracy performance.

Due to the involvement of multiple protein–protein interactions in a typical LBA method, the assay dynamic range may be generally narrow. As a result, an evaluation should be conducted during the assay development phase to establish whether highly concentrated samples can be diluted into the assay range with an acceptable accuracy or performance. Additionally, a prozone (or hook) effect is common particularly for some LBA assay formats. To address this concern, a dilutional linearity experiment should be performed in which a high-concentration sample is tested at dilutions intended to generate above ULOQ, within the range, and below LLOQ values.

Any changes to the assay conditions may impact overall assay performance, including assay precision and accuracy. Therefore during the assay development phase, assay parameters that may influence assay performance should be identified. Such evaluation will constitute assay robustness and ruggedness assessments and may include, but not be limited to, the effect of incubation temperature, incubation times, inter-analyst evaluation, manual versus automated pipetting, and use of multiple plate readers. Assay pre-study validation should be conducted within the assay parameter limits established during the assay development phase.

The specificity of the assay will be determined to a large degree by the assay format and the nature and specificity of the assay critical reagents, including capture and detector reagents. Generally assay specificity is evaluated by testing individual matrix samples that have been spiked with the drug material, commonly at LLOQ, ULOQ, and QC levels, and observing sample recovery in the assay. Substances that are expected to have an impact on the assay performance should be investigated. For example, effect of any concomitant medications, biological target for the drug, or other binding proteins should be investigated, if these are expected to be present in samples.

Any effect of specific components found in the disease matrix should be investigated. Individual disease matrix samples should be spiked with the drug as described above and recovery of the samples in the assay should be analyzed. In cases where high under- or over-recoveries are observed, an examination of assay critical reagents or a complete redesign of the assay may be required.

8.2.2 PK Assay Validation Considerations

PK assays that will be used in support of regulated nonclinical (GLP) or clinical studies are expected to be validated. Standard practices for validation of assays used for PK analysis are now well established in biotherapeutic laboratories following the recommendations provided in several industry-driven white papers, as well as guidances issued by regulatory agencies [22–27]. The white papers and guidances represent excellent resources that should be consulted by any laboratory developing these types of assays for use in regulated studies.

Typical parameters evaluated during pre-study validation of a PK assay are as follows:

- Specificity and selectivity—confirmed based on the assay conditions selected during assay development.
- Sample preparation and other assay conditions—confirmed based on the assay conditions selected during assay development.
- Confirm the range for the assay standard calibrator samples and selection of the model for the standard curve (e.g., 4PL vs. 5PL).
- Confirm assay range of quantification.
- Establish assay imprecision and inaccuracy parameters.
- Formally establish and in some cases initiate sample stability, for example, freeze/thaw and room temperature stability, $-80°C$ long-term stability.
- Confirm and establish assay dilutional linearity based on the confirmed assay quantification range.
- Establish assay robustness and ruggedness where possible.

Assay validation samples should be prepared in a matrix identical to study samples intended to be tested in the assay. General assay validation criteria are listed in Table 8.2.

Assay selectivity and specificity, often determined by the nature of the assay critical reagents selected during assay development phase, are confirmed during pre-study assay validation. Similarly, appropriateness of the assay format and any sample preparation procedure (e.g., MRD) are verified to ensure for a consistence performance.

Information on the performance of calibrator standards obtained during the assay development phase should be used when planning for the validation including number and concentrations of the calibrator standards to be used. For a typical 4PL or 5PL fit, at least six nonzero calibrator standards should be included within the assay calibration range. Because of the nonlinear nature of the concentration–signal response

TABLE 8.2. Validation parameters and target acceptance criteria for parameters evaluated during a typical LBA drug assay validation

Validation parameter	Target acceptance criteria
Minimum acceptable number of calibration standards within the assay range of quantification	6
Accuracy values for calibration standards within the range of quantification	≤20% (≤25% at LLOQ and ULOQ)
%CV of replicate raw assay signal values (e.g., ODs)	≤20%
Inter- and intra-assay precision (%CV) of validation samples (QCs)	≤20% (≤25% at LLOQ and ULOQ)
Inter- and intra-assay accuracy validation samples (QCs)	≤20% (≤25% at LLOQ and ULOQ)
Inter-run total error	≤30% (≤40% at LLOQ and ULOQ)
Selectivity	Unspiked matrix < LLOQ in >80% of tested lots; drug-spiked matrix ≤ 20% accuracy in >80% of tested lots
Sample stability	Similar to assay QC criteria

LLOQ, lower limit of quantification; ULOQ, upper limit of quantification; OD, optical density; CV, coefficient of variation; QC, quality control.

relationship of these types of assays, often additional standards positioned outside of the quantification range (referred to as "anchor points") are included to improve curve fitting.

Assay precision and accuracy parameters are tested by evaluating performance of several validation samples prepared at LLOQ, low QC, mid-QC, high QC, and ULOQ concentrations using an undiluted biological matrix. Precision and accuracy acceptance criteria are generally predetermined in the laboratory's PK assay validation Standard Operating Procedure (SOP) or assay validation plan. Criteria can be changed for a given procedure based on the assay performance and if a demonstrated and documented effort to improve the assay has been available. In addition to the precision and accuracy parameters, the use of a total error (a sum of precision, % coefficient of variation (CV), and absolute inaccuracy, % relative error (RE)) has been recommended [24, 25]. For a typical LBA assay, the total error value is expected to be within 30% with the exception of the LLOQ and ULOQ levels where the total error is expected to be within 40% [24]. It is suggested that the benefit of applying a total error parameter during assay pre-study validation is an improved rate of assay acceptance (reduced assay failure rate) during study execution phase. Overall, recommended QC acceptance criteria for LBA methods are based on results for a set of QCs (low, mid, and high level) commonly run in duplicates. The assay is considered to be acceptable when four out of six QCs are within 20% accuracy with at least one QC sample passing at each of the levels (a 4–6–20 rule) [25].

The assay quantification range can be established by using precision and accuracy information for the LLOQ and ULOQ samples prepared as described above. In some cases, several concentrations for the proposed LLOQ and ULOQ may be evaluated. Criteria established prior to initiation of the assay validation should be used to determine final assay LLOQ and ULOQ concentrations. Generally these criteria include precision, accuracy, and total error parameters (see Table 8.2). Based on the confirmed assay quantification range, linearity of dilution should be verified by utilizing a procedure similar to that described for the assay development phase.

Evaluation of sample stability should be formally conducted and/or initiated during the assay pre-study validation phase. For example, freeze/thaw and room temperature stability should be established. Long-term stability, for example, at −80°C, should be initiated at the time of pre-study assay validation is conducted. Long-term stability should be established before study sample analysis or, ideally, before in-life study initiation. Often, low and high QC samples are used in stability evaluation. It is also suggested to include a high-concentration sample mimicking the highest study sample concentration anticipated, should that information become available during the assay pre-study validation phase. In order for the stability samples to be acceptable, the same precision and accuracy acceptance criteria used for the assay QC samples should be used for the stability samples. Any possible trend should be evaluated, for example, if a consistently reduced recovery is observed with an increasing number of freeze/thaw cycles.

Based on the assay parameters established during pre-study validation phase, assay robustness and ruggedness should be evaluated to understand assay performance under various conditions.

Finally, assay acceptance criteria should be established based on the performance of the assay during the pre-study validation phase. It is important that all run information should be presented and included during assay performance evaluation. Runs should not be rejected during pre-study validation phase without a clearly assigned and documented cause; all data should be reported and analyzed for performance. In addition to the assay parameters that are evaluated during assay pre-study validation, incurred sample reanalysis and incurred sample dilutional linearity (parallelism) are performed when study samples become available [28, 29].

8.2.3 PK Assay Life Cycle

As the project matures through the drug development phases, the PK assay requirements may change. A fit-for-purpose, qualified assay is expected to suffice during support of an early exploratory study. Assay performance is expected to be understood better as the project is used for later phase support. Often, a generic assay, for example, one that detects any human immunoglobulin administered to a nonhuman species, is used in support of an early discovery study allowing for a shorter development time and no requirements for generation of highly specialized reagents. Validation of the analytical procedure will be necessary as the project is prepared for a submission with a regulatory agency, for example, IND submission with FDA. At this point, most of the projects will receive a specialized assay requiring access to highly specific and well-understood reagents. A transition from nonclinical to clinical phases necessitates development of a specialized PK assay purposely designed to measure the biotherapeutic in human matrix. Often this will require reevaluation or redesign of the assay including generation of new assay reagents. Similar to assays used for the support of regulated nonclinical investigations, clinical PK assays will require a validation in order to establish and confirm assay performance.

Often GLP or GMP grade reference material is not available at the time of assay development or even during assay validation. A change in the reference material lot or a batch would be expected prior to the initiation of the study support phase. In addition, changes in manufacturing process, the disease indication, patient population, or dosing regimen, that occur throughout the biotherapeutic candidate's life cycle could result in changes in the biophysical properties of the biotherapeutic or characteristics of the matrix, consequently affecting the original assay's performance or suitability for the intended analysis. In such cases, a careful cross-examination of the performance for the lots of the reference material, matrices, and dynamic range used during assay validation and currently available lots should be conducted to assure the assay is still suitable for its intended use. Ideally, the reference material lot used during the in-life phase of a clinical or nonclinical study should be applied to generate assay calibration standards and QCs.

A cross-qualification or cross-validation bioanalytical investigation would be required should a change in the batch of the drug reference material occur during any of the steps described above. Such cross-evaluation is needed to understand the relative performance of the new lot of the reference material in comparison with the material lot that was previously utilized during assay development, validation, or study support. Examples of the changes that may influence the performance of the assay reference material will include but are not limited to the characteristics of the reference material listed above (e.g., changes in aggregation, glycosylation, and co-purified material). Normally these are expected to be included as part of the certificate of analysis. As mentioned above, a timely transfer of the information from the manufacturing group to the bioanalytical lab is essential. Additionally, the bioanalytical site should be aware of any changes in the formulation buffer to help with the design and troubleshooting of the assay standard curve and QC preparation steps. Often later batches of the reference material are provided in the form of a drug product rather than drug substance material that may have been the source of reference material available at the earlier phase of development. Drug product could be formulated as lyophilized material requiring a reconstitution step. An effect of the change on the reference material performance in the assay should be carefully investigated. Usually, such an investigation necessitates a parallel analysis of the two material lots, for example, a lot used previously during assay development and a newly available drug product lot. Generally, precision and accuracy parameters for the LLOQ, ULOQ, and dilutional QC samples prepared using currently used and newly available lots of the drug material should be evaluated. QC and calibration standards prepared using current and newly available drug material lots are often analyzed against each other. For example, performance of QC samples prepared with the new lot is analyzed against the calibration curve prepared with the currently used drug material lot and vice versa. In some cases, previous drug material lot is not available or is in a limited supply. In that situation, assay performance for the newly prepared reference material lot can be analyzed to determine assay compatibility and compared to the assay acceptance criteria established during assay pre-study validation phase. Generally, such investigation will include an inter-day evaluation of precision and accuracy performance for the assay standards and QC samples.

The bioanalytical site should ensure availability of critical reagents and reference material during support of a given study. A change in either critical material and/or the drug material used to prepare QC and calibrator standards may result in a change in assay performance resulting in an impact on the reported study data. Such an impact on the bioanalytical data and study in general should be carefully investigated. Substantial changes in the assay performance may necessitate retesting of the study samples.

Assay performance should be carefully monitored during the study support phase, which for regulated nonclinical and clinical studies is also referred to as in-study validation phase. During this period, performance of the QC and calibrator standards should be examined for any altered trend in the QC recovery (inaccuracy) that may indicate a potential instability of QC samples and assay critical reagents or a change in other assay characteristic, for example, an impact of the plate lot change.

As mentioned above, LBA methodologies are highly sensitive to the matrix components. Effect of some components can be evaluated during assay development and validation phase when determining assay specificity and selectivity characteristics. Effect of other commonly encountered matrix components, such as binding or neutralizing anti-drug antibodies, can only be estimated qualitatively. For example, a neutralizing anti-drug antibody can effectively block interaction between the drug and specific capture reagent, for example, biological target, resulting in a reduced assay signal. As a result, drug concentration in the sample could be incorrectly underreported. To better understand an effect of anti-drug antibody development on the PK assays, data obtained in both the PK and anti-drug antibody testing should be analyzed in an integrated manner. Refer to Section 8.4 for the descriptions of the assays used to determine the presence and to characterize antibodies to biotherapeutics.

8.3 BIOMARKER ASSAYS

Following the trends in the wider pharmaceutical industry, biotherapeutic research and development has become increasingly more mechanistic with focus on acquiring deep knowledge of the mechanism of action of the biotherapeutic candidate complemented by developing an understanding of the interaction with the target and how the biological pathway is modulated. Biomarkers that are specific to target, pathway, as well as disease can provide quantitative evidence for target engagement, pathway modulation, and effect on disease progression, respectively. The quantitative measurement of biomarkers helps establish a link between drug exposure, target occupancy, pathway modulation, and effect on disease progression or efficacy, which can be described mathematically by constructing a PK/PD model. Ideally, a PK/PD model based on PK and quantitative biomarker data as well as information on key physiological and biological parameter intrinsic to the system that is studied ("biomeasures") is established preclinically and ultimately translated to the clinical situation where it may assist with the estimation of the dosing regimen and frequency [30]. In early clinical development, biomarkers are used to confirm the mechanism of action of the drug candidate and most biopharmaceutical companies also aspire to use biomarkers for personalized medicine, that is, for patient selection and individualized drug treatment.

In many ways, the measurement of biomarkers which are indicative for pathway modulation or disease progression is similar between biotherapeutic and small-molecule drug research and development. Those biomarkers (when they are molecular as opposed to physiological) can span a wide variety of types of endogenous molecules or events, which can be measured with a range of different analytical tools. Biomarkers that fall into this category include proteins, small molecules, lipids, nucleic acid-based molecules (DNA, single nucleotide polymorphisms, and mRNA expression profiles), or posttranslational modification such a phosphorylation events (e.g., indicative for a kinase activity in a signaling cascade), to name a few.

Most protein biotherapeutics are targeted against either soluble protein ligands across a broad spectrum of protein subclasses or against membrane-associated proteins such as receptors, cell adhesion molecules, enzymes, and others. The demonstration of target occupancy for biotherapeutics requires measuring the endogenous protein target in the presence of the biotherapeutic in the same sample. Quantitative assays for a target ligand measure either the free form (unbound), the portion that is bound to biotherapeutic, or the total ligand (sum of bound and free) [16]. A desirable measurement is oftentimes the free fraction of the target biomarker, that is, the portion of the target ligand that is not bound to the biotherapeutic, which can be directly correlated with PK, a downstream PD biomarker, or with an efficacy endpoint. While free target ligand data can be very informative, an accurate free ligand assay can be very challenging to develop because it is critical that the equilibrium between drug and target is not perturbed during the sample preparation procedure (sample dilutions and other manipulations). The required assay sensitivity needs to be considered when the target ligand at baseline is of very low abundance and when a significant reduction of free target ligand levels during therapy is expected. Furthermore, antibody reagents that bind to the same or nearby epitope on the target as the biotherapeutic are usually required to develop a free target assay, but will compete with the drug in a drug–target complex, potentially complicating a free target biomarker assay. In contrast, the measurement of total target ligand during treatment (sum of free and bound) follows the dynamics of the dose-dependent accumulation of the drug–target complex, possibly alleviating assay sensitivity challenges. As mentioned for assays used to monitor the PK of biotherapeutics, the availability of well-characterized assay reagents is critical to developing a reliable biomarker analytical protocol irrespective of the technology or format. Assay types for occupancy assessment by means of measuring the target biomarker include LBA, flow cytometry, and other cellular biology tools as well as liquid chromatography tandem mass spectrometry (LC-MS/MS).

LBA is often used for occupancy measurements of soluble protein targets, because of its good sensitivity and ease of use if suitable reagents are available. LBA technologies used for target biomarker assays are somewhat similar to biotherapeutic PK assays; however, due to a number of reasons, there are also some differences:

1. There could be an enormous concentration difference between drug and biomarker (potentially several orders of magnitude) requiring the use of the most sensitive immunoassays for low-abundance target biomarkers, particularly if the

free form is measured. Standard ELISAs may be insufficient at times. More sensitive assay formats, for example, using an electrochemiluminescence read-out, have become widespread (e.g., MSD). Some newer bead-based immunoassay formats can measure low-abundance proteins in the low and sub-picogram per milliliter range [31].

2. Oftentimes, reference standards for biomarkers are not available, or not well characterized and differ from the endogenous version at the protein-folding or posttranslational modification level, for example because they have been recombinantly expressed in a non-eukaryotic cell system. Thus, data must be interpreted carefully acknowledging this caveat and, as a consequence, results can potentially be only semi-quantitative in nature.

3. Analyte-free matrix may not be available because of the endogenous nature of the biomarker which can prompt the search for and use of an analyte-free surrogate matrix for assay calibration such as the same matrix from another species or buffer (requires investigation of parallelism). Complete or partial affinity removal techniques have also been successfully employed to generate analyte-free or analyte-reduced matrix for assay calibration.

LBAs are routinely employed for occupancy assessment for biotherapeutics in a preclinical setting as well as in the clinical studies. However, a target biomarker that is measured preclinically can be structurally different between the various preclinical species and ultimately human. This may require the use of different reagents for capture or detection in the different species. Reagents will have to be generated if they are not commercially available and the biomarker assay reagents (irrespective of the assay platform) will have to be planned for in a reagent generation and assay strategy.

Flow cytometry is a major platform for target biomarker assessment with growing importance in the development of biotherapeutics. This technology can analyze heterogeneous cellular systems and deliver multi-parameter information at the single cell level. Flow cytometry is primarily used for assessing the occupancy of membrane-bound targets if intact cells can be isolated for analysis. Blood is one of the most easily accessible tissues and can be extensively analyzed in flow cytometry-based biomarker assays. A simple flow cytometry-based occupancy assay uses a reagent antibody labeled with a fluorophor which binds to the same epitope on the target protein as the biotherapeutic. The lower the specific signal the higher the occupancy. More advanced receptor occupancy assays simultaneously assess relative target expression and require two reagent antibodies that bind the same target [32]. In one particular configuration, one reagent antibody is the therapeutic labeled with a fluorophor, while the second is directed against another, noncompeting epitope on the target and labeled with another fluorophor. In this assay format, the latter reagent will provide quantitation of the relative target expression compared to baseline, while the former will be indicative of target occupancy. The flow cytometry technique also offers the capability to perform intracellular target occupancy measurements if required. Flow cytometry is routinely employed preclinically and clinically; however, the implementation of standardized flow cytometry target biomarker assays in the clinic can be logistically challenging due to sample stability concerns and technical variations between instruments and laboratories [33]. Standardization of flow cytometry-based assays to be compliant with GLP guidelines is challenging but progress is currently being made.

Protein biomarker quantitation by LC-MS/MS has stimulated tremendous interest in the biopharmaceutical industry in recent years [34–36]. Assays typically require an immunoaffinity (IA) enrichment step for proteins in the low nanogram per milliliter to picogram per milliliter range followed by digestion of the biomarker proteins into tryptic peptides that can be monitored by LC-MS/MS with good selectivity and sensitivity. The IA enrichment can either happen at protein level (the equivalent to a protein capture step in an immunoassay) or at the tryptic peptide level in an approach termed *stable isotope standards and capture by anti-peptide antibodies* (SISCAPA) [37]. Which technique is employed depends on the availability of capture reagents, assay feasibility, and other bioanalytical goals. One key advantage of IA-LC-MS/MS over other bioanalytical tools is the high measurement specificity that can be achieved, which proves to be advantageous especially for biomarker measurements.

The quantitation of soluble protein target ligands in biotherapeutics research and development is a fairly new application area for LC-MS/MS, but offers some unique advantages over other assay formats from a reagent perspective. Unlike typical immunoassays, IA-LC-MS/MS-based occupancy assays require only a single capture antibody, which can result in a reduced time and resource commitment for assay development. Another key advantage of the IA-LC-MS/MS approach is that total assays can be developed for smaller protein and peptide target biomarkers, for which two reagent antibodies that bind to different epitopes from the therapeutic antibody and from each other do not exist or are difficult to generate. In addition, a total target assay format using LC-MS/MS offers the prospect of simultaneously assessing in a qualitative fashion what is bound to the target ligand, that is, binding partners such as shed receptors that can possibly modulate the normal function of the target or the ligand's characteristic posttranslational modifications.

Generally, target biomarkers are measured in a number of fluids and tissues. Disease-specific tissue measurements are also explored because they provide direct target assessment at the site of action. Depending on the analytical question at hand and the method employed, tissue measurements of target biomarkers can also be very challenging and technical components of the workflow need to be well controlled and understood, such as tissue disruption and homogenization as well target extraction efficiency. Furthermore, a main consideration for target engagement assays in tissue matrix is whether the biotherapeutic/target equilibrium can be maintained during the sample preparation process if free or complexed target is measured.

Independent of the platform employed, target biomarker assays are validated using a stage-dependant fit-for-purpose approach with increasing analytical rigor applied the further the biotherapeutic compound advances through the research and development stage gates [38,39]. The fit-for-purpose approach to biomarker assay validation considers the intended use of the data and the associated regulatory requirements.

In addition to target biomarker quantitation, the above assay platforms and their associated technological challenges and opportunities are also applicable to the determinations of PD biomarkers that measure pathway modulation or disease progression.

Other PD biomarker assay technologies range from immunohistochemistry, imaging techniques, fluorescent *in situ* hybridization (FISH), and real-time polymerase chain reaction (RT-PCR) genotyping tools. Emerging assay platforms that hold great promise for cellular PD biomarker measurements are Amnis and CyTOF, which exemplify the technological advances which are being made in this area. Amnis is an imaging flow cytometry technique which combines high-speed microscopy with flow cytometry for cellular biomarker analysis applications, such as internalization, co-localization, and cell–cell interactions [40]. The CyTOF technology is an adaptation of a mass cytometer combining flow cytometry with inductively coupled plasma mass spectrometry (ICP-MS) as an endpoint detection system [41]. The technology has high multiplexing capabilities facilitating quantitative insights into the functional complexity of biological systems at the single cell level with several orders of magnitude resolution without the need for compensation. These and other biomarker assay technologies will enable our ability to continue to acquire a more mechanistic understanding of the interaction of the biotherapeutic drug candidates with the target and how the biological pathway is modulated.

8.4 ASSAYS FOR DETECTION AND PREDICTION OF ANTI-DRUG ANTIBODIES

Most biotherapeutics have the potential to be immunogenic under some circumstances; therefore, development of anti-drug antibodies is typically monitored in nonclinical and clinical studies in order to assess the impact on PK, efficacy, and safety. Anti-drug antibody characterization (titers, neutralizing activity, antibody isotype or subclass, epitope specificity, and cross-reactivity) may be performed to better understand the potential impact on these parameters. Most human biotherapeutic candidates contain sequences or structures that are foreign in animals and therefore development of anti-drug antibodies may be expected to occur in nonclinical species and is not usually predictive of human anti-drug antibody development. However, assessment of anti-drug antibody development in animals is often needed to interpret nonclinical study results, in particular whether unusual PK, exposure, PD and efficacy, and safety can be attributed to anti-drug antibodies. In recent years in silico and *in vitro* assays designed to evaluate the effect of the biopharmaceutical on other elements of the immune response (e.g., binding propensity to the major histocompatability complex (MHC) molecules that are involved in antigen presentation or T-cell activation) have gained interest based on their potential to predict human immunogenicity risk as well as to understand the underlying cause of the anti-drug antibody response and guide immunogenicity mitigation approaches like humanization and immune tolerance induction.

Assay strategies to monitor for anti-drug antibody responses to biotherapeutics typically employ a tiered approach (Figure 8.1). That is, samples would be screened to determine if anti-drug antibodies are present, then additional analysis would be performed if necessary to confirm the specificity and determine the magnitude (typically reported as titer) of the response [42]. Samples that are confirmed positive for anti-drug antibodies in the screening and confirmatory assays may then be characterized in other assays, for example, for the presence of neutralizing activity, for isotype, subclass, or

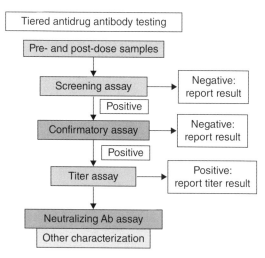

Figure 8.1. Tiered approach to anti-drug antibody testing. (See color insert)

epitope specificity, or for cross-reactivity with related proteins. In general, the type of characterization depends upon the need to interpret study findings or address theoretical concerns. For example, if hypersensitivity reactions are observed, IgE or IgG subclass characterization may be performed to understand the nature of the hypersensitivity. Alternatively, epitope specificity or cross-reactivity may be characterized for a biotherapeutic that has homology to an endogenous counterpart with a critical function.

One of the most important concepts to understand when considering the development of an anti-drug antibody assay is the unique nature of the "analyte" (anti-drug antibody) with respect to its changing characteristics. The initial phase of an antibody response is characterized by low affinity, high avidity, and restricted epitope IgM. With later maturation, other isotypes (IgG subclasses and IgA or IgE) with higher affinity and expanded epitope specificity may emerge. Consequently, the characteristics of the analyte are expected to differ from sample to sample with respect to its affinity and specificity for the biotherapeutic, its cross-reactivity with other molecules in the matrix, and its reactivity with detector and capture reagents. Thus it is not possible to develop the types of standards that are needed for relative quantitative assays. Although positive controls are used to optimize the assays and monitor performance over time, these controls are assumed not to be closely representative of most actual study samples in terms of the binding characteristics and reactivity with detection reagents. In addition, the results are usually reported as positive or negative based on whether the samples generate signal in the assay that equals or exceeds a designated cutpoint signal. When the assays are designed to compare magnitude of response between samples, the data are typically reported as titers based on dilutions at which the sample signal equals or exceeds the cutpoint. Thus the assays used to monitor anti-drug antibody development and to characterize the antibodies that develop (e.g., for neutralizing activity or isotype) are referred to as qualitative, semi-quantitative, or quasi-quantitative assays, designations used to describe assays for which no representative standard is available.

8.4.1 Anti-drug Antibody Screening Assays

8.4.1.1 Screening Assay Design Considerations. Recommendations for the development of anti-drug antibody screening assays were well described in the 2004 white paper authored by a group of scientists from multiple organizations experienced in using these assays [43]. The nature of the biotherapeutic, the species under investigation, and the intended use are important factors to consider in the selection of the assay platform for screening of anti-drug antibody formation. For example, for evaluation of human biotherapeutics in animals, it is possible to use direct binding formats in which the biotherapeutic is immobilized on a surface and a species-specific anti-immunoglobulin reagent is used to detect the anti-drug antibody. Ideally the biotherapeutic-specific positive control antibody reagent should be reactive with the species-specific detector. In addition, for most cloned proteins the direct binding assay format may be used with an antihuman immunoglobulin detector reagent to measure anti-drug antibodies in humans. Protein A and protein G can be used as nonspecific detector reagents in direct binding formats as well. However, most human immunoglobulin-binding detector reagents will cross-react with biotherapeutics containing human Fc regions and with immunoglobulin in human and nonhuman primate matrices; therefore, it is common to use an alternative format, the bridge assay format. In many cases the assays are intended for use in measuring biotherapeutics that have long half-lives, for which *in vivo* study samples often contain some level of drug that can interfere with the detection of anti-drug antibodies. Therefore, assay formats that decreased risk of drug interference have gained popularity. These platforms are often chemiluminescence based to provide greater sensitivity while incubation periods, coating densities, and use of acid dissociation are optimized to increase the chance of dissociation of the drug–antibody complex in samples and to favor rebinding of antibody to the assay drug capture reagent rather than to the drug in solution [44]. A more recently introduced platform is based on a miniaturized design that allows more rapid analysis with smaller volumes [45]. The common screening assay formats are listed in Table 8.3.

Because LBAs performed on biological samples are prone to nonspecific binding from irrelevant matrix components, specificity or confirmatory analysis is usually

TABLE 8.3. Common assay formats/platforms for anti-drug antibody screening assays

Format	Immobilized capture	Detector
Direct binding	Biotherapeutic	Anti-species immunoglobulinProtein A/G
Bridge assay	Biotherapeutic	Labeled biotherapeutic
Radioimmunoassay	Protein A/G	Radiolabeled biotherapeutic in solution
Electrochemilumiscence	Biotin-labeled biotherapeutic on avidin surface	Ruthenium-labeled biotherapeutic
Surface-plasmon resonance	Biotherapeutic on chip	None necessary

performed on samples that score positively in the initial screening assay and only samples that score in the confirmatory assay are designated positive. Several publications address optimization for anti-drug antibody screening assays [43–46]. Typically the steps for developing an assay are as follows:

- Identification of reagents:
  - Positive control source usually derived from immunized animals or from animals that develop antibodies during treatment with the biotherapeutic.
  - Identification of assay matrix, negative control, usually pool of serum from drug-naïve individuals from the species under study.
  - Generation of labeled reagents, develop standardized conjugation process to facilitate consistent quality for resupply.
- Optimization of assay conditions:
  - Explore buffers, coating concentration, conjugation conditions, conjugate concentrations, matrix starting dilution, incubation times, and wash conditions.
  - A factorial design of experiments (DOE) approach is recommended to identify and optimize interactions among these variables.
  - Goal: minimize irrelevant, nonspecific binding but allow detection of low affinity antibodies.
  - Define assay procedure for qualification and validation.
  - Select conditions for confirmatory assay.
- Assay qualification:
  - Establishes that assay meets minimum requirements for use in nonregulated studies or to proceed with validation for regulated studies. This could involve evaluating a subset of validation parameters in a reduced number of runs. For example, it may be appropriate to establish a preliminary assay cutpoint for distinguishing positive and negative samples (described in more detail below) and demonstrating that the assay has acceptable nonspecific background, precision, and sensitivity.

8.4.1.2   *Anti-drug Antibody Screening Assay Validation.*   Anti-drug antibody assays that will be used to analyze GLP or clinical samples are expected to be validated. Standard practices for validation of screening and confirmatory anti-drug antibody assays are now well established in biotherapeutic laboratories following the recommendations provided in several industry-driven white papers, as well as guidances issued by regulatory agencies [47–50]. The white papers and guidances represent excellent resources that should be consulted by any laboratory developing these types of assays for use in regulated studies.

Parameters typically evaluated include the following:

- Screening assay cutpoint—used for distinguishing positive and negative samples and calculating titers in the intended study population.

- Precision—usually intra-run, intra-day, and inter-day precision should be evaluated.
- Robustness—evaluates impact of small deliberate changes in assay conditions.
- Assay sensitivity—determined based on the concentration of a positive control that can be detected in the assay; because the sensitivity will be highly dependent on the affinity and other characteristics of the particular positive control chosen, the sensitivity for detection of an actual study sample is expected to differ.
- Potential sources of interference from matrix components.
- QC performance—acceptance criteria should be established for routine assay monitoring.
- Confirmatory assay cutpoint—established for assays that will incorporate a competitive binding confirmatory assay.

The assay cutpoint is defined as the signal in the assay that delineates positive and negative samples (Figure 8.2). A sample that generates a signal equal to or greater than the cutpoint would be considered positive for anti-drug antibody while a sample that generates a signal below the cutpoint would be considered negative for anti-drug antibody. In validated clinical assays, the cutpoint is usually established by testing a panel of samples from individuals who have not been exposed previously to the drug and determining the signal at which approximately 5% of the samples would score above the cutpoint to assure that all potential positive samples will be identified. The cutpoint based on a 5% false-positive rate can be established by determining the 95th percentile or 1.65 standard deviations [39, 43]. In order to distinguish true positive samples from false-positive samples, those samples that score positively in the screen assay may be further analyzed to confirm the positive result. The most common method for confirming a positive sample is a competitive inhibition assay in which an excess level of drug is

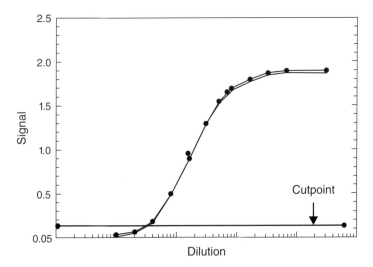

Figure 8.2. Anti-drug antibody titration curve and cutpoint.

TABLE 8.4. Common sources of matrix interference for anti-drug antibody assays

Natural antibodies
Anti-carbohydrate antibodies
Rheumatoid factor
Complement
Drug target

mixed with the sample prior to analysis. In this confirmatory assay format, a sample that is positive for anti-drug antibody should generate a reduced signal relative to the signal generated when no drug was added. The confirmatory assay cutpoint should also be determined statistically [47]. If the sample does not confirm positive it is reported as negative for the presence of anti-drug antibody. Additional characterization may be performed on a risk-based approach [42].

It is important to characterize normal healthy and disease state samples for potential sources of interferences that could cause false-positive or -negative results (Table 8.4). Common interferences include natural antibodies, irrelevant or disease-associated antibodies (e.g., rheumatoid factor), complement, or soluble target. Levels of some of these factors may be elevated in disease state serum and change after drug treatment. Such changes over time may have a significant impact on the ability to adequately monitor for anti-drug antibody development. In addition, if samples are collected when high concentrations of drug are still present, detection of anti-drug antibodies may be inhibited or prevented.

8.4.2 Neutralizing Antibody Assays

Biotechnology-derived therapeutics have the potential to induce anti-drug antibodies that can be generally classified as either non-neutralizing or neutralizing. Neutralizing antibodies (NAb) prevent target binding and block the biological activity of the therapeutic molecule, which may have a dramatic effect on the efficacy and may also impact safety of the drug product when the antibodies not only neutralize the administered drug product but cross-react with and neutralize the biological activity of the endogenous counterpart. For example, NAbs to megakaryocyte-derived growth factor (MDGF) and recombinant human erythropoietin (Epo) have been found to cause severe thrombocytopenia [51] and pure red cell aplasia (PRCA) [52,53], respectively. Decrease in efficacy has been observed in a wide variety of products including interferon beta [54,55], factor VIII [56], and alglucosidase alfa [57, 58]. In the case of enzyme replacement therapies to treat lysosomal storage disorders or recombinant factor VIII replacement therapy to treat hemophilia, patients will have their treatment deleteriously affected by the development of NAbs. As patient safety and therapeutic efficacy are of paramount importance, the development of assays to detect NAbs is critical for many product development programs. The type and extent of the immunogenicity monitoring program followed during product development should be based upon an immunogenicity risk assessment of the therapeutic and driven by the likelihood of the therapeutic protein eliciting an immune response and a consideration of the potential consequences due to that immune response [42, 59]. Differences in manufacturing or product attributes can have an impact on immunogenicity [60, 61]. Examples of increases in NAb development due to product differences or process-related manufacturing events have been observed with recombinant human interferon beta [62], and Epo [63], the latter example resulting in PRCA [53].

Testing for NAbs is implemented following the detection of screened and confirmed positive samples to determine if the detected antibodies have neutralizing potential. Regulatory guidances on immunogenicity state that assessing the neutralizing capacity of antibodies should preferably be done using functional bioassays (e.g. cell-based assays, CBA), which measure changes in the biologic activity of the therapeutic protein triggered by the presence of the antibody, while recognizing that competitive ligand-binding (CLB) assays may be substituted when neutralizing bioassays are inadequate or not feasible [64].

Based on the therapeutic protein drug's MOA and its immunogenicity risk, a bioassay may be deemed more appropriate for a given stage of the drug development program. However, in some instances a CLB approach may be found to be more suitable for NAb detection. This may be true if direct binding of the therapeutic to its target is sufficient to evoke all biological processes that define its MOA. However, if other functional aspects of the therapeutic may be affected by anti-drug antibody/NAb which do not directly interfere with target binding, then the CLB assay format may not be adequate [64]. Thus, a comprehensive understanding of the drug's MOA can be critical to adequately choose the most appropriate assay format.

Neutralizing bioassays are more complex than CLB assays and selection of a final assay format is dependent on the individual biopharmaceutical's MOA. Although CBAs are preferred for some complex MOA, they are neither the only option nor most informative for all MOA. The development of an *in vitro* CBA or a CLB assay will depend on a variety of factors. For instance, finding appropriate cell lines that yield adequate signal to noise responses and appropriate specificity may pose a challenge. In addition, CBAs tend to be more sensitive to matrix interference, less drug tolerant, less precise, less sensitive, and more time consuming than CLB assays. Factors to consider when choosing a particular NAb assay format include, but are not limited to, the risk assessment of the therapeutic, the stage of the program, the availability of reagents, and the ability of developing an appropriate fit-for-purpose assay. The overall CBA testing scheme requires a risk-based analysis of potential harm to patients should a NAb response occur. For high-risk products with an endogenous counterpart, it is recommended that a sensitive NAb assay that closely mimics the biological activity of the endogenous protein be implemented early in the program. In addition, it may be necessary to determine whether antibodies to the drug product cross-react with other family members of the endogenous protein. If so, then NAb assays for both the drug product and the endogenous protein would require implementation [65]. The ability to perform this assessment is dependent upon the availability of sufficient quantities of the purified native endogenous protein. Alternatively, a recombinant form of the endogenous protein may be used. For low-risk products such as those with no endogenous counterpart (e.g., MAb therapeutics, bacterially derived enzymes), a bioassay may not be necessary early on in the program.

8.4.2.1 Considerations for Cell-Based Neutralizing Antibody Assays.

The development of a CBA for the detection of NAbs poses unique challenges as compared to those encountered when developing immunoassays. Specific assay development considerations and qualification experiments have been extensively reviewed by Gupta et al. [65]. Factors to consider when embarking on the development of a CBA for NAb detection are depicted below, for example, appropriate selection of a drug-responsive cell line, assay endpoint and NAb assay format, impact of biological matrix, and positive control selection. In general, assay endpoints could utilize early or late biological responses that are triggered in the cells upon treatment with either the drug or by the ligand which inhibits the drug. A factorial design approach using a mathematical approach is very useful when developing and optimizing a NAb assay, as it enables the investigator to efficiently evaluate multiple parameters and interactions of important parameters in order to improve assay performance and assay robustness. The final outcome of the design will result in an optimized, robust assay. Statistical software such as Design Expert or JMP can assist with this type of factorial design.

NAB ASSAY DEVELOPMENT CONSIDERATIONS:
- Identification of drug-responsive cell line
- Identification of the assay format:
  ○ Direct versus indirect
- Selection of the assay endpoint
- Selection of drug product concentration
- Selection of positive control antibody
- Assay incubation conditions
- Adaptation of cells to biological matrix of interest
- Identification of assay matrix, sample preparation procedure, establishing MRD parameter
- Assay response range
- Qualitative versus quasi-quantitative NAb assay
- Matrix interference considerations

A successful CBA is one in which there are minimal matrix effects, minimal variability, a sufficiently high (at least fivefold) signal to background ratio in the presence of minimal amount of drug, and specific for the detection of NAbs against the drug product. Thus, one of the most critical parameters for the development of a CBA is the cell line selection. The selected cell line should respond with a good signal to background in the presence of biological matrix and with a low level of the therapeutic. Additionally, the ability of the cell line to produce a response with a measurable assay endpoint with selectivity and specificity to the therapeutic is critical. When applicable, it is important to address the specificity of signal by utilizing structurally related family members to the therapeutic [65]. A good starting point for identifying a drug-responsive cell line is to investigate the potency assay that is used to evaluate manufactured drug lots. The major advantage of this approach is that optimization of the cell line maintenance, culture conditions, and culture media have already been established. However, the potency assay may need to be adapted and optimized to function in the presence of the biological matrix of interest. This adaptation to matrix may result in loss of responsiveness to the drug and possibly loss of assay sensitivity. Cell subculturing and plating conditions play a vital role in determining how the cells respond to the drug. The increased variability sometimes observed with bioassays as compared with other bioanalytical techniques can be minimized by characterization and proper care and maintenance of the cell line. Most of the bioassays, particularly the cell-based *in vitro* assays, fail to detect NAbs when high level of the drug is present in the samples. The remaining drug, either a free molecule or in immune complex with anti-drug antibodies, can inhibit the neutralizing activity of the antibody and prevent detection. Many therapeutic proteins, particularly MAbs, have a long half-life resulting in high concentrations in the biologic system being tested. The high concentrations of drug remaining in the serum or plasma samples may interfere with the detection of anti-drug antibodies, in some cases resulting in false-negative results. Sample pretreatments, such as acid dissociation and affinity adsorption, can be used to remove drug from samples but these treatments may also have adverse effects on the NAbs present in the samples or may reduce or remove the NAbs. Assays that use less drug as part of the assay will be more sensitive to having false negatives when drug is present in samples.

Specific assay development considerations and appropriate qualification and validation experiments have been extensively reviewed in several industry-driven white papers, as well as guidances issued by regulatory agencies [47–50, 65, 66].

8.4.2.2 *Detection of Neutralizing Antibodies.*

The detection of NAbs is based on the principle that any sample containing NAbs would reduce or abolish the biological activity associated with the known concentration of drug product used in the NAb assay. Samples are deemed negative or positive based on a statistically derived assay cutpoint [47, 66], which is defined as an assay response value above (if the drug is an antagonist) or below (if the drug is an agonist) which a sample is defined to be positive for neutralizing activity toward the drug product. NAb assay formats can be direct or indirect and each has specific parameters that need to be evaluated and optimized. Biological therapeutics with agonistic properties (cytokines, growth factors, hormones, agonistic MAbs) exert their effect in a direct manner by directly binding to receptors on the target cell surface and inducing a measurable response. Biological therapeutics with antagonistic properties (antagonistic MAbs and soluble receptor) act by blocking a ligand that binds to the target receptor expressed on the surface of responsive cells thus influencing a cell's response in an indirect manner. Determining whether the therapeutic has agonistic or antagonist properties plays an important role in the decision of the NAb assay format that will be implemented and in the choice of critical assay components and reagents that will be utilized. Generally, NAb assays for biologics with agonist activity use the direct format while NAb assays for biologics with an antagonistic mechanism of action utilize the indirect assay format [65]. The critical components of a direct NAb assay include a drug-responsive cell line, the drug product, a positive control NAb, and test biological matrix. The critical components of an indirect NAb assay include a cell line that responds to the ligand that is blocked by the drug product, ligand, drug product, positive control Nab, and test biological matrix [65].

Both types of assay formats use a fixed concentration of cells, the drug product, and the biological matrix containing NAbs. Additionally, if the drug is an antagonist, ligand is also utilized. Relevant controls for indirect NAb assays include (a) cells alone; (b) cells with ligand; (c) cells, ligand, and drug; and (d) cells, ligand, drug, and positive control. For the direct NAb assay format, similar controls are used, however in the absence of ligand. The sensitivity of an indirect and direct NAb assay depends upon the concentration of ligand (which will determine the concentration of drug) and drug that will be used in the assay, respectively. As a general rule, the amount of ligand (indirect assay) or drug (direct assay) used should be the lowest possible concentration that elicits $\geq 50\%$ response [49, 65].

The positive control antibody for a NAb assay is used to monitor the performance of the assay over time and to demonstrate that the assay can reliably detect neutralization of the drug used in the assay. Positive controls are generally antisera obtained from hyper-immunized animals; however, monoclonal NAbs used alone or as a "cocktail" may be employed as well. The positive control antibodies are spiked into a biological matrix at a specific concentration for use as positive controls.

8.4.2.3 Qualification of Neutralizing Cell-Based Assays.
Once the NAb assay is developed and optimized, qualification allows for a better understanding of how the assay performs, whether it will withstand assay validation, and for identification of the target acceptance criteria that will be included in the validation protocol [65]. Depicted below are some typical experiments performed during qualification. NAb assays tend to be more susceptible to drug presence than anti-drug antibody screening assays and several factors will contribute to the magnitude of interference (drug concentration, the characteristics of the positive control antibody, and the design of the assay), thus an establishment of a single "drug tolerance level" is not possible. However, the extent of drug interference should be addressed regarding susceptibility of the assay to drug interference [65].

NAb assay qualification considerations:

- Characteristics of the drug product standard curve
- Characteristics of the positive control antibody curve
- Assay cutpoints (drug-naïve individual healthy donor or target untreated diseased state sera)
- Assay cutpoint for the matrix interference assay (drug-naïve individuals)
- Assay sensitivity
- Assay precision
- Matrix interference
- Drug product interference

*8.4.2.4 **NAb Assay Validation Considerations.*** NAb assays that will be used in support of clinical studies are expected to be validated. At the present time, only EMA has issued regulatory guidance governing the validation and performance acceptance criteria for immunogenicity assays [48, 50]. However, a draft guidance by the FDA [46] on immunogenicity testing have also been issued [49]. In an effort to harmonize approaches that may be employed during development and validation of NAb assays to support immunogenicity assessments, drug industry researchers have published white papers outlining current industry practices [65, 66]. The white papers and guidances represent excellent resources that should be consulted by any laboratory developing these types of assays [47–50, 65, 66].

Typical parameters evaluated during validation of a NAb assay:

- Cutpoint:
  - The assay cutpoint should be derived using a statistically adequate number of individual sera (typically 30 or more).
- Sensitivity:
  - Using a surrogate positive control antibody to ensure that the assay performs as validated over time.
- Assay precision:
  - Qualitative NAb assay (intra-assay and inter-assay precision).
  - Quasi-quantitative NAb assay (low positive control and one or two concentrations of the high positive control).
- Robustness.
- Ruggedness.
- Assay specificity.
- Selectivity/drug product interference.
- Stability.
- System suitability.

In conclusion, selection of an appropriate NAb assay is a critical component of an immunogenicity evaluation. In a clinical immunogenicity program, the primary goal is to provide a reliable data set that ensures the ability to evaluate the impact of antibodies on safety and efficacy; a validated assay will serve this purpose.

8.4.3 Assays for Immunogenicity Prediction

A variety of product-, patient-, and treatment-related factors are thought to influence the development of anti-drug antibodies [67, 68]. The extent to which foreign amino acid sequences or biophysical features, for example, unusual glycosylation, degradants, and impurities, are present is among those risk factors cited. As a widening variety of biotherapeutics enter development and clinical use, there is a growing interest in developing methods that may be used to predict the impact of these risk factors in order to reduce various risks that are currently not easy to predict, for example, the risk that a novel biotherapeutic candidate would induce significant clinical immunogenicity, or the risk that changes in manufacturing process of an established product would result in changes in the biophysical properties that consequently alter the immunogenicity profile, or the risk that changing from intravenous to subcutaneous route of administration would result in a higher rate of clinically significant anti-drug antibody induction. Methods developed and used for vaccine and autoimmunity research are now being evaluated for applicability in predicting these types of immunogenicity risk factors for biotherapeutics. Currently the methods that are most often cited as having applicability are in silico tools that predict whether the biotherapeutic has epitopes that bind to the MHC human leukocyte antigens (HLA) that are required to present antigen to T cells and *in vitro* assays using human peripheral blood to measure activation of peripheral blood lymphocytes, particularly T cells, after exposure to the biotherapeutic [69–71]. In general, T cells are considered to be necessary for the generation of robust and sustained anti-drug antibody responses and therefore there is a high level of interest in identifying and removing the amino acid sequences in the biotherapeutic that may be potential T-cell epitopes.

The in silico methods are based on databases such as the publically accessible Immune Epitope Database [72], founded in 2004 by the US National Institute for Allergy and Infectious Diseases to curate all experimentally defined immune epitopes available from the literature and from independent submissions. The in silico predictions most widely used are those that predict HLA-binding and T-cell epitopes. Due to the complex nature of the epitopes that antibodies can recognize, the antibody or B-cell epitope predictors are considered less reliable. The clear advantages to utilizing in silico methods are the low cost and high throughput, with no need to provide either whole protein or peptide test articles for analysis, and therefore these are often applied in early candidate selection. However, these methods are thought to over-predict in that not all epitopes that have been identified were shown to actually bind HLA or to result in T-cell activation. Therefore, it is common to combine in silico prediction with *in vitro* HLA-binding assays as well as cellular activation assays (proliferation, cytokine production) to improve the chance for identification of relevant epitopes [71].

Assays used to detect peripheral blood lymphocyte or T-cell responses to the drug typically measure proliferation (measured as ^{3}H-thymidine incorporation or by flow cytometry) or release of cytokines (usually IL-2 or IFNγ) which can be measured by immunoassay or ELISpot assays [71, 73]. The flow cytometry-based methods allow measurements of other activation markers as well. Whole protein or peptides can be used in these assays.

Although incidence and magnitude of immunogenicity observed during typical animal studies conducted during development of human biotherapeutics are not considered predictive of human immunogenicity [74], some animal models, for example, HLA transgenic models, have been used to assess whether a peptide can be presented and stimulate responses in the context of HLA *in vivo* [75]. In addition, cell culture systems that mimic some aspects of the complex architecture of human lymph nodes have been considered for use in the identification of less immunogenic biotherapeutics [76].

There are few reports available [77] demonstrating that the HLA types and T-cell epitopes predicted by these methods are representative of HLA types and activated

T-cell specificity of patients dosed with the biotherapeutic. Therefore, overall there is a need for more extensive clinical validation of these methods to demonstrate they do appropriately predict the risk factor they are intended to predict.

8.5 NEW TRENDS: BIOSIMILARS, BIOBETTERS, ANTIBODY–DRUG CONJUGATES

Early decades of the modern biotherapeutics industry were initially characterized by the development of clone versions of normal human proteins followed later by the introduction of humanized MAbs and Fc fusion proteins. As the industry enters its fourth decade, a range of first-generation biotherapeutics are now undergoing improvements in their manufacturing processes by original manufacturers or are being redeveloped as biosimilars by competitors. At the same time, an unprecedented diversity of novel biotherapeutic protein scaffolds, delivery, and treatment approaches are entering development [1–5]. Alternative approaches being considered for delivery of biotherapeutics encompass both novel formulations that may extend the duration of delivery and alternative routes of administration to provide greater convenience of use. Many of these novel approaches are directed toward established targets and fall into the category of "biobetters." All of these approaches introduce an equally diverse set of development challenges, including those related to bioanalytical, PD, and immunogenicity assays. The bioanalytical, PD, and immunogenicity assays will be pivotal to the success of these next-generation biotherapeutic treatments and therefore some of the key considerations are briefly highlighted in this section.

Demonstration of comparability, biosimilarity, or "biobetterness" usually requires evaluations that are comparative in nature. The bioanalytical strategy must accommodate the need to demonstrate that a new manufacturing process or biosimilar is comparable or highly similar to the original or that the biobetter has the improved PK, PD, and/or safety characteristics intended by its design.

For both PK and anti-drug antibody assays, best practices have evolved considerably since the early generation of biotherapeutics. Rather than applying bioanalytical and immunogenicity assays that resemble those used in the development of the original or reference product, developers of biosimilars are expected to apply state-of-the-art assays [78]. However, any information about the assays used for the reference product and the results will be highly useful in designing the assays for the biosimilar program. Because of the requirement for biophysical and biochemical similarity, it is reasonable to expect that the reference and biosimilar products will perform similarly in a PK assay using either product as the calibrator; however, this will need to be demonstrated in the validation if a single assay is used to measure both reference and biosimilar concentration. In contrast, the immunogenicity evaluation is intended to identify marked differences in the response to the product and therefore it is necessary to utilize separate assays for the reference and biosimilar in order to fully describe the immune response to each, although once antibodies are identified, evaluation for cross-reactivity against the other product may provide additional useful information about any differences in immune response against the two products.

The regulatory pathways for "biobetters" may not require direct comparison studies with the original products; however, a strategy to demonstrate differentiation from the first-generation product will be needed. The PK, biomarker/biomeasure, and immunogenicity analysis should consider the aspect of the product that is to be improved relative to the first-generation product to assure that the differentiation can be adequately demonstrated. For all products, state-of–the-art assays and analysis strategies are expected to be used, and the implications of these are especially important to consider for immunogenicity testing. Although product labels contain specific text cautioning that results of anti-drug antibody assays are highly dependent on the sensitivity of the assay, as well as sample collection times and other factors, and thus a comparison of immunogenicity across products is expected to be misleading, nonetheless such comparisons may seem inevitable as treaters and patients will be seeking to understand the relative safety of these products. Such comparisons will be particularly challenging when the reported rates of anti-drug antibody development are being compared to those of the original products that were developed with the use of less sensitive assays or fewer/different sample collection timepoints.

Protein–antibody–drug conjugates are typically composed of a MAb or other protein against a specific target cell conjugated to a therapeutically active or cytotoxic component (referred to as the "payload") via a linker system designed to release the payload at the intended site of action, that is, after cellular internalization. For many of these conjugates, unintended *in vivo* instability of the conjugate has been an issue and therefore some free antibody or payload may be present after *in vivo* administration. Evaluation of PK usually requires the utilization of multiple assays; separate assays are typically developed to measure the free and total payload component, the total protein or antibody component, and the intact conjugate [5, 79, 80]. The stage at which these assays are applied would depend on the novelty of the construct design. During candidate selection for new linkers/payloads, it is important to demonstrate that the intact conjugate is stable *in vivo* and to determine the extent to which the PK or biodistribution of the conjugate follows that of the antibody component.

In summary, emerging trends in the biotherapeutics industry highlight the need to develop stage-appropriate, risk-based strategies to assure that the PK, PD, and immunogenicity profiles can be adequately described.

8.6 CONCLUSIONS

Appropriate characterization of PK, PD, and immunogenicity profiles of any biotherapeutic candidate in nonclinical and clinical studies is highly dependent on the nature and quality of the assays. Common practices are now widely implemented for the development, application, and interpretation of these assays. The complexity and heterogeneity of the biotherapeutic candidate need to be considered in designing and maintaining the assays. With a growing number and diversity of candidates across a wider range of modalities entering development, it is essential that biotherapeutic development teams integrate knowledge of the biophysical and biochemical characteristics with their PK, PD, and immunogenicity data in the design and selection of candidates, the execution of nonclinical and clinical studies, and the design and control of their manufacturing processes.

REFERENCES

1. Nelson, A.L., Dhimolea, E., and Reichert, J.M. (2010) Development trends for human monoclonal antibody therapeutics. *Nat. Rev. Drug Discov.*, **9**, 767–774.

2. Skerra, A. (2007) Alternative non-antibody scaffolds for molecular recognition. *Curr. Opin. Biotechnol.*, **18**, 295–304.

3. Hosse, R.J., Rothe, A., and Power, B.E. (2006) A new generation of protein display scaffolds for molecular recognition. *Protein Sci.*, **15**, 14–27.

4. Schaefer, W., et al. (2011) Immunoglobulin domain crossover as a generic approach for the production of bispecific IgG antibodies. *Proc. Natl. Acad. Sci. U.S.A.*, **108**, 11187–11192.

5. Hughes, B. (2010) Antibody–drug conjugates for cancer: poised to deliver? *Nat. Rev. Drug Discov.*, **9**, 665–667.

6. Tremblay, G.A. and Oldfield, P.R. (2009) Bioanalysis of siRNA and oligonucleotide therapeutics in biological fluids and tissues. *Bioanalysis*, **1**, 595–609.

7. Tremblay, G.A., et al. (2011) Dual ligation hybridization assay for the specific determination of oligonucleotide therapeutics. *Bioanalysis*, **3**, 499–508.

8. van Dongen, W.D. and Niessen, W.M.A. (2011) Bioanalytical LC-MS of therapeutic oligonucleotides. *Bioanalysis*, **3**, 541–564.

9. Wang, L. (2011) Oligonucleotide bioanalysis: sensitivity versus specificity. *Bioanalysis*, **3**, 1299–1303.

10. Daugherty, A.L. and Mrsny, R.J. (2006) Formulation and delivery issues for monoclonal antibody therapeutics. *Adv. Drug Deliv. Rev.*, **58**, 686–706.

11. Ponce, R., et al. (2009) Immunogenicity of biologically-derived therapeutics: assessment and interpretation of nonclinical safety studies. *Reg. Toxicol. Pharmacol.*, **54**, 164–182.

12. ICH Topic S6. *International Conference on Harmonisation of Technical Requirements for Registration of Pharmaceuticals for Human Use. ICH Harmonised Tripartite Guideline. Note for Guidance on Preclinical Safety Evaluation of Biotechnology-Derived Pharmaceuticals (CHMP/ICH/302/95).*

13. ICH Topic S6 (R1). *International Conference on Harmonisation of Technical Requirements for Registration of Pharmaceuticals for Human Use. ICH Harmonised Tripartite Guideline. Addendum to ICH S6: Preclinical Safety Evaluation of Biotechnology-Derived Pharmaceuticals.*

14. Wang, W., Wang, E.Q., and Balthasar, J.P. (2008) Monoclonal antibody pharmacokinetics and pharmacodynamics. *Clin. Pharmacol. Ther.*, **84**, 548–558.

15. Roskos, L.K., et al. (2011) PK–PD modeling of protein drugs: implications in assay development. *Bioanalysis*, **3**, 659–675.

16. Lee, J.W., et al. (2011) Bioanalytical approaches to quantify "total" and "free" therapeutic antibodies and their targets: technical challenges and PK/PD applications over the course of drug development. *Am. Assoc. Pharm. Sci. J.*, **13**, 99–110.

17. Mesmin, C., Fenaille, F., Ezan, E., and Becher, F. (2001) MS based approaches for studying the pharmacokinetics of protein drugs. *Bioanalysis*, **3**, 477–480.

18. Heudi, O., et al. (2008) Towards absolute quantification of therapeutic monoclonal antibody in serum by LC–MS/MS using isotope-labeled antibody standard and protein cleavage isotope dilution mass spectrometry. *Anal. Chem.*, **80**, 4200–4207.

19. O'Hara, D.M., et al. (2012) Ligand binding assays in the 21st century: I. Recommendations for characterization and supply of critical reagents. *Am. Assoc. Pharm. Sci. J.*, **14**, 316–328.

20. Rup, B. and O'Hara, D. (2007) Critical ligand binding reagent preparation/selection: when specificity depends on reagents. *Am. Assoc. Pharm. Sci. J.*, **9**, E148–E155.

21. ICH Topic Q 6 B. *Specifications: Test Procedures and Acceptance Criteria for Biotechnological/Biological Products.* September 1999. CPMP/ICH/365/96

22. DeSilva, B., Smith, W., and Weiner, R. (2003) Recommendations for the bioanalytical method validation of ligand-binding assays to support pharmacokinetic assessments of macromolecules. *Pharm. Res.*, **20**, 1885–1900.

23. Nowatzke, W. and Woolf, E. (2007) Best practices during bioanalytical method validation for the characterization of assay reagents and the evaluation of analyte stability in assay standards, quality controls, and study samples. *Am. Assoc. Pharm. Sci. J.*, **9**, E117–E122.

24. Findlay, J.W.A. and Dillard, R.F. (2007) Appropriate calibration curve fitting in ligand binding assays. *Am. Assoc. Pharm. Sci. J.*, **9**, E260–E267.

25. Viswanathan, C.T., et al. (2007) Workshop/Conference report—quantitative bioanalytical methods validation and implementation: best practices for chromatographic and ligand binding assays. *Am. Assoc. Pharm. Sci. J.*, **9**, E30–E42.

26. US FDA. (2001) *Guidance for Industry: Bioanalytical Method Validation* [Docket No. 98D–1195]. Federal Register 66, 28526–28527.

27. European Medicines Agency, Committee for Medicinal Products for Human Use (CHMP). *Guideline on Bioanalytical Method Validation.* July 21, 2011. EMEA/CHMP/EWP/192217/2009.

28. Rocci, M.L., Devanarayan, V., Haughey, D.B., and Jardieu, P. (2007) Confirmatory reanalysis of incurred bioanalytical samples. *Am. Assoc. Pharm. Sci. J.*, **9**, E336–E343.

29. Kelley, M. and DeSilva, B. (2007) Key elements of bioanalytical method validation for macromolecules. *Am. Assoc. Pharm. Sci. J.*, **9**, E156–E163

30. Lowe, P.J., et al. (2010) On setting the first dose in man: quantitating biotherapeutic drug-target binding through pharmacokinetic and pharmacodynamic models. *Basic Clin. Pharmacol. Toxicol.*, **106**, 195–209.

31. Todd, J., et al. (2009) Ultrasensitive flow-based immunoassays using single-molecule counting. *Clin. Chem.*, **53**, 1990–1995.

32. Metz, D.P., et al. (2009) Defining dose-response relationships in the therapeutic blockade of B7RP-1-dependent immune responses. *Eur. J. Pharmacol.*, **610**, 110–118.

33. Wu, D.Y., Patti-Diaz, L., and Hill, C.G. (2010) Development and validation of flow cytometry methods for pharmacodynamic clinical biomarkers. *Bioanalysis*, **9**, 1617–1626.

34. McIntosh, M. and Fitzgibbon, M. (2009) Biomarker validation by targeted mass spectrometry. *Nat. Biotechnol.*, **27**, 622–623.

35. Neubert, H., Gale, J., and Muirhead, D. (2010) Online high-flow peptide immunoaffinity enrichment and nanoflow LC-MS/MS: assay development for total salivary pepsin/pepsinogen. *Clin. Chem.*, **56**, 1413–1423.

36. Ocaña, M.F. and Neubert, H. (2010) An immunoaffinity liquid chromatography-tandem mass spectrometry assay for the quantitation of matrix metalloproteinase 9 in mouse serum. *Anal. Biochem.*, **399**, 202–210.

37. Anderson, N.L., et al. (2004) Mass spectrometric quantitation of peptides and proteins using stable isotope standards and capture by anti-peptide antibodies (SISCAPA). *J. Proteome Res.*, **3**, 235–44.

38. Lee, J.W., et al. (2006) Fit-for-purpose method development and validation for successful biomarker measurement. *Pharm. Res.*, **23**, 312–328.

39. Wagner, J.A. (2008) Strategic approach to fit-for-purpose biomarkers in drug development. *Ann. Rev. Pharm. Toxicol.*, **48**, 631–651.

40. Elliott, G.S. (2009) Moving pictures: imaging flow cytometry for drug development. *Comb. Chem. High Throughput Screen*, **12**, 849–59.

41. Bendall, S.C., et al. (2011) Single-cell mass cytometry of differential immune and drug responses across a human hematopoietic continuum. *Science*, **332**, 687–696

42. Koren, E., et al. (2008) Recommendations on risk-based strategies for detection and characterization of antibodies against biotechnology products. *J. Immunol. Methods*, **333**, 1–9.

43. Mire-Sluis, A.R., et al. (2004) Recommendations for the design and optimization of immunoassays used in the detection of host antibodies against biotechnology products. *J. Immunol. Methods*, **289**, 1–16.

44. Lofgren, J.A., et al. (2007) Comparing ELISA and surface plasmon resonance for assessing clinical immunogenicity of panitumumab. *J. Immunol.*, **178**, 7467–7472.

45. Mora, J.R., Obenauer-Kutner, L., and Patel, V. (2010) Application of the Gyrolab™ platform to ligand-binding assays: a user's perspective. *Bioanalysis*, **2**, 1711–1715.

46. Dodge, R., Daus, C., and Yaskanin, D. (2009) Challenges in developing antidrug antibody screening assays. *Bioanalysis*, **1**, 699–704.

47. Shankar, G., et al. (2008) Recommendations for the validation of immunoassays used for detection of host antibodies against biotechnology products. *J. Pharm. Biomed. Anal.*, **48**, 1267–1281.

48. European Medicines Agency Committee for Medicinal Products for Human Use (CHMP). (2008) *Guideline on Immunogenicity Assessment of Biotechnology-Derived Therapeutic Proteins.* EMEA/CHMP/BMWP/ 14327/2006.

49. US Food and Drug Administration. (2009) *Draft Guidance for Industry: Assay Development for Immunogenicity Testing of Therapeutic Proteins.*

50. European Medicines Agency Committee for Medicinal Products for Human Use (CHMP). (2010) *Draft Guideline on Immunogenicity Assessment of Monoclonal Antibodies Intended for In Vivo Clinical Use.* EMA/CHMP/BMWP/86289/2010.

51. Nicolino, M., et al. (2009) Clinical outcomes after long-term treatment with alglucosidase alfa in infants and children with advanced Pompe disease. *Genet. Med.*, **11**, 210–219.

52. Li, J., et al. (2001) Thrombocytopenia caused by the development of antibodies to thrombopoietin. *Blood*, **98**, 3241–3248.

53. Casadevall, N., et al. (2002) Pure red-cell aplasia and antierythropoietin antibodies in patients treated with recombinant erythropoietin. *N. Engl. J. Med.*, **346**, 469–475.

54. Sharma, B., et al. (2004) Technical investigations into the cause of the increased incidence of antibody-mediated pure red cell aplasia associated with Eprex®. *Eur. J. Hosp. Pharm.*, **5**, 86–91.

55. Hartung, H.P., et al. (2007) Neutralising antibodies to interferon beta in multiple sclerosis: expert panel report. *J. Neurol.*, **254**, 827–837.

56. Hochuli, E. (1997) Interferon immunogenicity: technical evaluation of interferon-alpha 2a. *J. Interferon Cytokine Res.*, **17**(Suppl 1), S15–S21.

57. Bray, G.L., et al. (1994) A multicenter study of recombinant factor VIII(recombinate): safety, efficacy, and inhibitor risk in previously untreated patients with hemophilia A. The Recombinate Study Group. *Blood*, **83**, 2428–2435.

58. van der Ploeg, A.T., et al. (2010) A randomized study of alglucosidase alfa in late-onset Pompe's disease. *N. Engl. J. Med.*, **362**, 1396–1406.

59. Shankar, G., Pendley, C., and Stein, K.E. (2007) A risk-based bioanalytical strategy for the assessment of antibody immune responses against biological drugs. *Nat. Biotechnol.*, **25**, 555–561.

60. Sharma, B. (2007) Immunogenicity of therapeutic proteins. Part 3: impact of manufacturing changes. *Biotechnol. Adv.*, **25**, 325–331.

61. Singh, S.K. (2011) Impact of product-related factors on immunogenicity of biotherapeutics. *J. Pharm. Sci.*, **100**, 354–387.

62. Sorensen, P.S., et al.; Danish Multiple Sclerosis Study Group. (2003) Clinical importance of neutralizing antibodies against interferon-beta in patients with relapsing-remitting multiple sclerosis. *Lancet*, **362**, 1184–1191.

63. Boven, K., et al. (2005) The increased incidence of pure red cell aplasia with an Eprex formulation in uncoated rubber stopper syringes. *Kidney Int.*, **67**, 2346–2353.

64. Finco, D., et al. (2011) Comparison of competitive ligand-binding assay and bioassay formats for the measurement of neutralizing antibodies to protein therapeutics. *J. Pharm. Biomed. Anal.*, **54**, 351–358.

65. Gupta, S., et al. (2007) Recommendations for the design, optimization, and qualification of cell-based assays used for the detection of neutralizing antibody responses elicited to biological therapeutics. *J. Immunol. Methods*, **321**, 1–18.

66. Gupta, S., et al. (2011) Recommendations for the validation of cell-based assays used for the detection of neutralizing antibody immune responses elicited against biological therapeutics. *J. Pharm. Biomed. Anal.*, **15**, 878–888.

67. Rosenberg, A.S. and Worobec, A.S. (2004) A risk-based approach to immunogenicity concerns of therapeutic protein products. Part I. Considering consequences of the immune response to a protein. *BioPharm Int.*, **17**, 22–26.

68. Rosenberg, A.S. and Worobec, A.S. (2004) A risk-based approach to immunogenicity concerns of therapeutic protein products. Part II. Considering host-specific and product-specific factors impacting immunogenicity. *BioPharm Int.*, **17**, 34–42.

69. DeGroot, A.S. and Scott, D.W. (2007) Immunogenicity of protein therapeutics. *Trends Immunol.*, **28**, 482–290.

70. Bryson, C.J., Jones, T.D., and Baker, M.P. (2010) Prediction of immunogenicity of therapeutic proteins: validity of computational tools. *Biodrugs*, **24**, 1–8.

71. Van Walle, I., et al. (2007) Immunogenicity screening in protein drug development. *Expert Opin. Biol. Ther.*, **7**, 405–418.

72. Vita, R., et al. (2010) The Immune Epitope Database 2.0. *Nucl. Acids Res.*, **38**, D854–D862.

73. Jaber, A. and Baker, M. (2007) Assessment of immunogenicity of different interferon beta-1a formulations using ex vivo assays. *J. Pharm. Biomed. Anal.*, **43**, 1256–1261.

74. Bugelski, P.J. and Treacy, G. (2004) Predictive power of preclinical studies in animals for the immunogenicity of recombinant therapeutic proteins in humans. *Curr. Opin. Mol. Ther.*, **6**, 10–16.

75. Depil, S., et al. (2006) Peptide-binding assays and HLA II transgenic Aβ° are consistent and complementary tools for identifying HLA II-restricted peptides. *Vaccine*, **24**, 2225–2229.

76. Giese, C., et al. (2009) Immunological substance testing on human lymphatic micro-organoids in vitro. *J. Biotechnol.*, **148**, 38–45.

77. Koren, E., et al. (2007) Clinical validation of the "in silico" prediction of immunogenicity of a human recombinant therapeutic protein. *Clin. Immunol.*, **124**, 26–32.

78. Committee for Medicinal Products for Human Use (CHMP). *Guideline on Similar Biological Medicinal Products Containing Biotechnology-Derived Proteins as Active Substance: Non-clinical and Clinical Issues.* EMEA/CHMP/BMWP/42832/2005.

79. Roberts, S.A. et al. (2013) Considerations for the nonclinical safety evaluation of antibody drug conjugates for oncology. *Reg. Tox. and Pharm.*, **67**, 382–391.

80. Kaur, S. (2013) Bioanalysis special focus issue on antibody–drug conjugates. *Bioanalysis*, **5**, 981–983.

9

STRUCTURES AND DYNAMICS OF PROTEINS PROBED BY UV RESONANCE RAMAN SPECTROSCOPY

Brian S. Leigh, Diana E. Schlamadinger, and Judy E. Kim

Department of Chemistry and Biochemistry, University of California at San Diego, La Jolla, CA, USA

Biophysical Methods for Biotherapeutics: Discovery and Development Applications, First Edition. Edited by Tapan K. Das.
© 2014 John Wiley & Sons, Inc. Published 2014 by John Wiley & Sons, Inc.

9.1   INTRODUCTION

9.1.1   Background and Historical Perspective

The Raman process is an inelastic scattering effect that was first observed in 1928 by Indian physicist C.V. Raman, who was awarded the Nobel prize in Physics two years after the experimental discovery [1, 2]. The early observation of the Raman effect is remarkable given that we now understand that a monochromatic light source is essential for Raman spectroscopy; such instrumentation was not available at the time of these initial studies and in fact, C.V. Raman utilized focused sunlight and a combination of blue and yellow filters to observe the inelastically scattered light by eye. In-depth experimental studies of the Raman effect became more facile with the use of mercury arc lamp emission lines and photographic detection. With the advent of lasers and sensitive electronic detectors in the last several decades, Raman spectroscopy has exploded to become a major analytical tool in the physical and biological sciences. Here, we focus on the application of Raman spectroscopy to probe structures and environments of complex biomolecules. We first discuss the fundamental processes of Raman and resonance Raman (RR) scattering, followed by a presentation of the experimental considerations for this technique. We present examples of applications of RR spectroscopy and conclude with a discussion of the challenges and benefits of this method.

A brief discussion on the principles of Raman spectroscopy is presented here, and we direct the reader to comprehensive and in-depth treatments of this topic [3, 4]. In the Raman process, an incident photon interacts with and is inelastically scattered by a molecule; the energy of this scattered photon (E_{scatt}) may be greater or less than that of the incident excitation photon (E_{exc}). It is important to note that the inelastic scattering process is far less efficient than elastic scattering, called Rayleigh scattering, in which E_{scatt} and E_{exc} are equivalent. Nonetheless, molecular insight from Raman scattering is significant. The difference in energy between incident and scattered radiation is called the Raman shift, and this Raman shift is equivalent to the vibrational energy of the molecule. A qualitative description of this shift is that the scattered photon has deposited energy into or gained energy from the molecule, and this energy is derived from vibrations of the molecule. The condition where $E_{scatt} < E_{exc}$ is called Raman Stokes scattering whereas the reverse situation in which $E_{scatt} > E_{exc}$ is called Raman anti-Stokes scattering. Since anti-Stokes scattering requires population in a vibrationally excited state of the molecule, anti-Stokes intensities are typically much smaller than Stokes intensities and depend strongly on molecular temperature. For this reason, the vast majority of Raman experiments probe the Stokes scattering process. Figure 9.1 shows a schematic of the Raman process for a single vibration. Biomolecules are of course quite complex. Their Raman spectra consist of multiple peaks, each of which corresponds to one of the

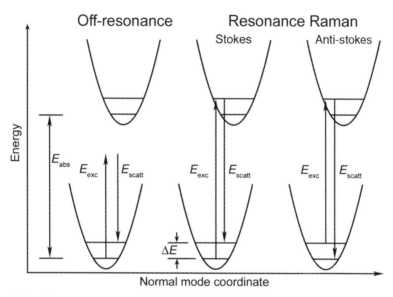

Figure 9.1. Schematic of Raman process for a single, harmonic normal mode. Left: Off-resonance Stokes Raman spectroscopy in which E_{exc} does not overlap with an absorption band. The difference in energy between E_{exc} and E_{scatt}, ΔE, is the Raman shift and reflects the vibrational energy of a specific normal mode. When the excitation wavelength overlaps with an allowed absorption band E_{abs}, Stokes (middle) or anti-Stokes (right) resonance Raman scattering is observed. The efficiency of RR scattering is several orders of magnitude greater than off-resonance Raman scattering, enabling RR spectroscopy to be a selective analytical tool.

Raman-active vibrational normal modes of the molecule, which in turn is a subset of the $3N$-6 possible normal modes (N is the number of atoms in the protein).

9.1.2 Resonance Raman Scattering

The energy of the incident photon is an important consideration because the Raman scattering intensity in photons per second, P_{R}, depends on the energies of the excitation and scattered photons to the fourth power: $P_{\text{R}} \propto E_{\text{exc}}(E_{\text{exc}} \pm E_{\text{scatt}})^3$. One can immediately see that excitation with blue photons is preferred over excitation with infrared photons from the viewpoint of Raman intensity. Additional enhancement of the Raman signal is obtained in the resonance condition in which the energy of the incident photon falls within an absorption band of a molecule (see Figure 9.1). This process, called RR scattering, increases the Raman signal by several orders of magnitude, and the quantitative basis for this enhancement has been reviewed extensively [5]. It is straightforward to see that in addition to enhancing the sensitivity of an experiment, RR spectroscopy enables selectivity; by tuning the excitation wavelength to lie within an absorption band of a molecule of interest, an RR spectrum from the absorbing species is easily observed without obfuscation from the other molecules in solution, such as solvent and buffer. This selectivity enables RR spectroscopy to be a powerful tool for studies of complex molecules, such as large biological systems, that are comprised of different chromophores.

As a vibrational technique, RR spectroscopy is a sensitive reporter of structure and environment of molecules of interest. The inherent sensitivity of Raman spectroscopy to molecular bonds differentiates it from other optical techniques based on electronic transitions, such as absorption and fluorescence, which typically report on environment but not structure. All proteins and nucleic acids absorb UV light and as such, UV resonance Raman (UVRR) spectroscopy is a valuable tool for the study of biomolecules. However, in practice, UVRR spectroscopy is used to probe a subset of moieties that exhibit relatively large absorption cross-sections in UV regions that are experimentally accessible: amide backbone, aromatic amino acids, proline, and, to a lesser extent, sulfur-containing residues [6,7]. Given the increasing absorbance of the carbonyl $\pi{\rightarrow}\pi^*$ transition at wavelengths less than ~210 nm, excitation near 200 nm preferentially probes secondary structure while lower energy excitation (~230 nm) probes side chain structure. An example of this selectivity is shown in UVRR spectra of the peptide melittin, which has the primary sequence GIGAVLKVLTTGLPALISWIKRKRQQ. Figure 9.2 illustrates that 210 nm excitation of melittin gives rise to a UVRR spectrum of the backbone that is drastically different from that of 230 nm spectrum of the single tryptophan residue at position 19. A summary of vibrational normal modes and their correlation to structure/environment in UVRR spectra of proteins [6–9] is presented in Table 9.1. It should be noted that the subset of proteins with chromophores that absorb visible light, such as hemes and retinals, can also be studied with visible RR spectroscopy to interrogate the prosthetic group.

9.1.3 Secondary Structure

The UVRR spectrum of amide backbone reveals secondary structure because of the sensitivity of vibrational frequencies to hydrogen-bonding environment. Strong UVRR intensities appear for amide I, II, and III modes, which are vibrations that involve coupled displacements of the carbonyl and amide functional groups; the reader is encouraged to consult the classic biophysics reference by Cantor and Schimmel for further discussion and diagrams [10]. In addition to these well-known amide modes, the amide IIp mode of proline and amide S mode are also utilized in UVRR spectroscopy [7, 11, 12]. Analysis of the amide III (~1250 to ~1350 cm^{-1}), amide II (~1520–1560 cm^{-1}), amide I (~1630–1680 cm^{-1}), and amide S (~1390 cm^{-1}) regions via decomposition of the UVRR bands using a least-squares fitting technique reveals the relative content of random coil, α-helix, and β-sheet secondary structures in proteins and peptides [11, 13, 14].

It is important to note that the information gained from UVRR spectra of the backbone may be compared to those from the widely utilized technique of circular dichroism (CD). Because normal mode frequencies largely reflect variations in backbone dihedral angles, UVRR can report on a single turn of an α-helix [15]. In contrast, exciton coupling dominates the features in a typical CD spectrum and as a result, CD reports on the global average of the secondary structure of the peptide [16]. A consequence of this difference is that UVRR intensities scale linearly with the fraction of α-helix

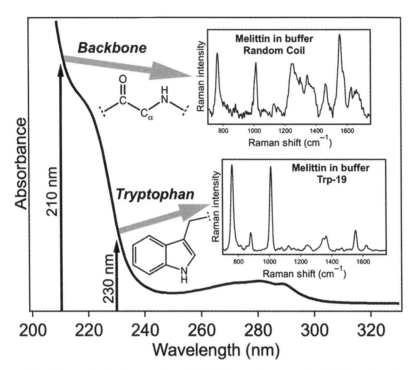

Figure 9.2. Absorption (main graph) and UVRR (insets) spectra of ~40 μM melittin in phosphate buffer where the peptide is a random coil. UVRR spectra with 210 nm excitation primarily reflects backbone vibrations; peaks in the region ~1200–1700 cm^{-1} can be assigned to amide I, II, III, and S modes while peaks at ~750–1020 cm^{-1} primarily reflect tryptophan peaks. UVRR spectra with 230 nm enhances signal from the single tryptophan residue in tryptophan; all strong peaks in this spectrum can be assigned to tryptophan normal modes. Spectra were acquired using apparatus shown in Figure 9.3. Sample was flowed in 100 μm (230 nm) or 250 μm (210 nm) quartz capillaries with 5 mW (230 nm) or 3 mW (210 nm) of UV power at the sample.

whereas in CD spectroscopy, the molar ellipticity per residue decreases as α-helical content decreases, and a threshold length of α-helix must persist in order to give reliable results in a CD spectrum [17]. Exciton coupling is especially important in CD spectra of short peptides; short α-helical peptides result in CD spectra with variations in peak positions and relative intensities [18]. An additional challenge inherent to CD spectra of membrane-bound peptides is that differential light scattering and absorption flattening from membrane vesicles may alter the observed shapes and relative intensities in CD spectra [19]. In UVRR spectroscopy, enhanced scattering from vesicles may impact the signal-to-noise ratio, but peak positions and relative intensities are unaffected under the appropriate experimental conditions. Thus, while UVRR experiments are more challenging than CD studies, secondary structure assignment based on UVRR analysis is reliable regardless of peptide length.

TABLE 9.1. Summary of UVRR bands of protein backbone and residues from references 6 to 9

	UVRR band assignments			
Chemical moiety	Designation	Description	Raman shift (cm^{-1})	Properties
Amide backbone	Amide I	C=O stretch	1630–1680	Secondary structure
Amide backbone	Amide II	C–N stretch + N–H bend	1520–1560	Secondary structure
Amide backbone	Amide III	C–N stretch + N–H bend	1250–1350	Secondary structure
Amide backbone	Amide S	Cα–H bend	1390	Secondary structure
Proline	Amide IIp	C–N stretch of X-Pro	1460	H-bonding, *cis/trans*
Tryptophan	W1	Benzene ring stretch	1620	Environment polarity
Tryptophan	W3	C–C pyrrole stretch	1552	Structure
Tryptophan	W4	C–H bend + N–H bend	1494	H-bonding
Tryptophan	W6	N–C–C stretch + N–H bend	1435	H-bonding
Tryptophan	W7 doublet	Fermi doublet	1360 and 1340	Environment polarity, π-interactions
Tryptophan	W8	C–C stretch + N–H bend	1307	H-bonding
Tryptophan	W9, W10	C–H bend + N–H bend	1230–1250	H-bonding
Tryptophan	W16	Benzene ring breathing	1004	Environment polarity
Tryptophan	W17	N–H bend	879	H-bonding
Tryptophan	HOOP	Hydrogen out of plane	770–820	π-interactions
Tryptophan	W18	Indole breathing	760	Environment polarity, π interactions
Tyrosine	Y8a	Ring stretch	1617	H-bonding, protonation state
Tyrosine	Y8b	Ring stretch	1601	H-bonding, protonation state
Tyrosine	Y7a′	C–O stretch	1263	Proton-accepting strength
Tyrosine	Y7a	C–C stretch	1210	Proton-donating strength
Tyrosine	Y9a	C–H bend + O–H bend	1180	Structure

(Continueud)

TABLE 9.1. (Continued)

		UVRR band assignments		
Chemical moiety	Designation	Description	Raman shift (cm^{-1})	Properties
Tyrosine	Y1 + 2Y16a	Fermi doublet	850 and 830	H-bonding
Histidine	N/A	N/A	1577	Neutral histidine
Histidine	N/A	Ring deformation	1323	Neutral histidine
Histidine	N/A	N–C–N stretch + N–H bend	1242	Neutral histidine
Histidine	N/A	Ring deformation + N–H bend	1166	Neutral histidine
Histidine-H+	N/A	N–C–N stretch + N–H bend	1202	Protonated histidine
Cysteine–cysteine[a]	S–S	S–S stretch	510–540	C–S–S–C torsion angles
Cystine,[a] cysteine,[a] methionine[a]	C–S	C–S stretch	630–760	Conformation

Molecular and environment properties that are correlated to the band positions and intensities are also indicated. Designations and normal mode descriptions that are not available are indicated N/A.
[a]Sulfur-containing residues are not strongly enhanced with excitation wavelengths above 200 nm.

9.1.4 Aromatic Amino Acids

The majority of UVRR studies on side chains focus on tryptophan, tyrosine, and histidine because these aromatic residues have numerous and strong Raman peaks that report on structure, environment, and protonation state [8, 20–27]. Off-resonance Raman studies of aromatic residues have also played a critical role in interpretation of UVRR spectra [28–31]. These and other Raman studies have established important empirical relationships between Raman frequencies/intensities and molecular detail of aromatic side chains, including hydrogen-bonding strength, microenvironment, static structure, cation–π interactions, and protonation state. Tryptophan is by far the most extensively studied amino acid, and systematic UVRR studies reveal spectral signatures that report upon local hydrophobicity, strength of indole N–H hydrogen bond, and torsional angle, χ2,1 about the C-2-C-3-Cβ-Cα linkage. The wealth of vibrational information makes tryptophan an ideal probe in many different types of biophysical studies, such as protein folding, dynamics, and protein–protein interactions.

The ability of UVRR to report on molecular details of tryptophan is advantageous because tryptophan is one of the most important residues in terms of protein structure, function, and dynamics. It is the least abundant residue in soluble proteins, accounting for only 1.1% of the amino acids expressed in cytoplasmic proteins [32], but is more prevalent in membrane proteins, with an abundance of 2.9% in transmembrane α-helical domains [33]. This aromatic residue typically plays key functional roles in proteins because of its unique properties among the 20 natural amino acids: tryptophan exhibits the largest accessible nonpolar surface area that is highly polarizable, possesses an

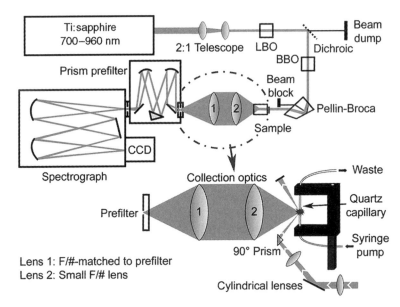

Figure 9.3. Schematic of a UVRR apparatus used in the Department of Chemistry and Biochemistry at UC San Diego. The output of a 1 kHz, nanosecond Ti:sapphire laser is telescoped and passed through a lithium triborate (LBO) crystal for second-harmonic generation. The doubled blue light is separated from the fundamental infrared beam by a dichroic mirror. The blue light is then doubled in a β-barium borate (BBO) crystal to generate UV light; these beams are separated by a fused silica Pellin-Broca prism. The UV excitation beam is passed through a pair of cylindrical lenses to a spot size of ∼300 μm (height) × 75 μm (width) and focused on a fused silica capillary through which sample is pumped vertically via a syringe pump. The Raman scattered light is collected and focused onto the entrance slit of a prism-based prefilter, dispersed in a spectrograph, and imaged onto a CCD detector.

indole N–H moiety that is capable of hydrogen bond donation, and displays the greatest electrostatic potential for cation–π interactions [34, 35].

The important physical properties render tryptophan an ideal amphiphilic residue with the greatest propensity to reside in the interfacial region of a membrane protein compared to any other naturally occurring amino acid [36]. For example, tryptophan has been found to stabilize membrane-spanning proteins and peptides by acting as anchors along the interface of the bilayer [37, 38]. Replacement of tryptophan residues with phenylalanine in the 325-residue integral membrane protein, outer membrane protein A, destabilizes the protein relative to wild type when folded into lipid bilayers [39, 40]. Tryptophan residues in membrane-associated antimicrobial peptides (AMPs) also play important functional roles in hemolytic and/or bactericidal activity [41, 42]. In the antibiotic channel peptide gramicidin A, substitution of tryptophan for phenylalanine residues results in reduction of antibacterial activity [43, 44]. These and other examples illustrate that the presence, location, and environment of tryptophan residues are critical in the study of folding and insertion of membrane proteins and membrane-associated peptides.

9.1.5   Considerations for UVRR

The high sensitivity and selectivity of UVRR offer advantages over other techniques in studies of complex biomolecules. Most importantly, protein samples do not need to be labeled with extrinsic chromophores or isotopically substituted moieties, as is often necessary for fluorescence, NMR, or FTIR measurements. Additionally, small sample volumes on the order of 0.08–2 mL of ~10 μM sample is required for a typical UVRR experiment. UVRR data may be acquired at any temperature, including cryogenic temperatures. A key advantage of UVRR is that time-resolved studies can be pursued relatively easily. These benefits have encouraged the widespread use of UVRR to probe a variety of topics in biophysics, including folding of soluble proteins [45–48], allosteric transitions [49–51], activation of G-protein-coupled receptors [52, 53], and ultrafast protein dynamics [53–55].

As with other biophysical techniques, UVRR has limitations. The Achilles' heel of UVRR (or visible RR) is fluorescence. In Stokes scattering where $E_{scatt} < E_{exc}$, the Raman-shifted photon may overlap with the fluorescence spectrum. This overlap in RR and fluorescence emission may significantly reduce the signal-to-noise ratio of the UVRR spectra because the process of fluorescence is typically orders of magnitude more efficient than the RR effect. One way to address this problem is by taking advantage of the fact that RR scattering tracks with the excitation wavelength whereas fluorescence emission is typically static. In tryptophan, fluorescence photons that are emitted with a quantum yield of ~0.15 are absent from the Stokes UVRR window by utilizing Raman excitation wavelengths near 230 nm; this excitation wavelength is within the strong absorption B_b band of indole so the resonance condition is preserved. With 230.0 nm excitation, Raman-shifted photons for high-frequency vibrations near 1800 cm^{-1} appear at 239.9 nm, which is shifted from the fluorescence maximum of ~320–350 nm for tryptophan. In contrast, excitation of tryptophan with 280.0 nm, at the peak of the tryptophan L_{ab} absorption band, results in Raman-shifted photons at 294.9 nm that would be far outnumbered by fluorescence photons at this wavelength. An additional limitation to UVRR is that a small subset of side chains may be probed. As mentioned above, in practice, the majority of UVRR measurements on side chains are limited to tryptophan, tyrosine, histidine, and proline. If the protein of interest contains multiple tryptophan residues, UVRR is unable to unambiguously isolate signal from any single residue since each tryptophan residue gives rise to similar UVRR spectra. In this scenario, it is beneficial to generate protein mutants with a single residue of choice or to incorporate isotopically substituted residues. Despite these limitations, UVRR remains a powerful optical tool that offers molecular insight into structures and dynamics of biomolecules (see Table 9.1).

9.2   EXPERIMENTAL

An ideal UVRR setup has the following components: (1) narrow bandwidth and tunable excitation source from ~195 to 240 nm; (2) low-volume sample cell that permits constant replenishment of fresh sample; and (3) detection system that is capable of sharp rejection of Rayleigh scattering. To our knowledge, there is no commercial source of a UVRR system that includes each of these components. Instead, part or all of a UVRR apparatus is designed and built by individual research groups based on their needs. The level of customization is varied, and may be as minimal as assembling a sample flow cell that is coupled to a commercial UVRR microscope or as extensive as purchasing each component separately. Numerous reviews on UVRR technology have been published elsewhere [56–63]. Here, we summarize some of the options and briefly discuss the advantages and disadvantages.

9.2.1 Excitation Source

A variety of laser excitation sources are utilized for UVRR. Intracavity-doubled continuous-wave lasers are excellent for UVRR experiments because they offer inherently narrow spectral bandwidths that give rise to high-resolution spectra. On the other hand, continuous-wave lasers are not tunable, with wavelengths limited to harmonics of the lasing medium; useful wavelengths for UVRR studies of proteins are 228.9 and 206.5 nm from doubled argon and krypton ion lasers, respectively.

Greater tunability is achieved with nanosecond Nd:YAG lasers that pump other media. Nd:YAG lasers have been used to pump dye lasers whose visible output is frequency doubled or mixed with the Nd:YAG fundamental to access the deep-UV region. This method offers tunable UV wavelengths, but has the disadvantage that several dyes may be required to cover the UV range of interest. Additionally, the dyes themselves are relatively easily degraded. A more robust and commonly used excitation source is Raman-shifted lines from low repetition rate ($\sim$20 Hz) Nd:YAG lasers. Here, the harmonics of an Nd:YAG laser are focused into a commercially available or custom-built Raman shifter that is filled with an appropriate gas, such as H_2 or D_2. The pulse energies of the Nd:YAG input are sufficiently high to stimulate Raman scattering via a four-wave mixing process, resulting in numerous UV lines that differ in energy by the number of quanta of H_2 or D_2 vibrations. The advantage of this Raman-shifting method is that it is economical and straightforward; however, the disadvantage is that the high-pulse energies required for efficient Raman shifting may result in unwanted, nonlinear optical phenomena associated with the protein of interest during data collection. Additionally, the low repetition rate of Nd:YAG lasers requires somewhat long acquisition times.

UVRR systems assembled in the past several years utilize kilohertz nanosecond Ti:sapphire lasers combined with harmonic generation. In this setup, the laser is tunable from $\sim$700–960 nm and the third and fourth harmonics provide wavelengths in the UV region of approximately 195–240 nm. The advantages of this system are numerous: the high repetition rate minimizes the probability for nonlinear phenomena in the sample, data collection time is reasonable, the nanosecond pulses ensure narrow bandwidth that are suitable for UVRR, and the system is entirely solid state and therefore relatively robust and straightforward to operate. We operate such a laser in our lab as part of our UVRR system (Figure 9.3).

9.2.2 Sample Cell

Important considerations in constructing a sample cell are volume and flow. UV excitation results in irreversible photodamage to the protein and causes the population of photolyzed protein to increase in the solution with extended UV exposure; this buildup may give rise to artifacts in the spectrum. The extent of photodamage may be minimized by using low laser powers; however, if the power is too low, data collection time becomes unreasonable and other potential artifacts may impact the spectra, such as drift in laser power and alignment. Additionally, because proteins typically have high absorbances in the deep-UV region, a backscattering geometry is preferred over a right-angle orientation.

Ideally, one would like to flow small volumes of sample in a single-pass configuration such that photolyzed protein is discarded after exposure to a single laser pulse. The advantage of this configuration is that the laser power may be high within the linear regime of the power dependence (see 9.2.4 Other Considerations), thereby reducing data collection time. This setup requires optimization of the sample flow system and focusing geometry. An example of such a single-pass flow system is in our lab; in this setup, less than 2 mL of sample may be pumped through a vertically mounted, 100 μm inner diameter (160 μm outer diameter) quartz capillary at a rate such that each laser pulse interrogates fresh sample (see Figure 9.3) [64]. A laser power of ~3 mW at the sample is used to obtain high signal-to-noise UVRR data in 10 minutes. Analysis of protein sample after it has passed through the UV beam indicates that less than 20% of the bulk protein is photodamaged. Typically, the sample flows at a rate such that two interrogated volumes are passed through the area of the focused beam per laser pulse, so the amount of protein damaged per pulse is ~40%. This rate of photodamage would likely result in rapid buildup of damaged protein in a recirculating system, but in this single-pass configuration, the damaged protein does not contribute to the spectra.

If volumes of less than 1 mL are needed, one may utilize recirculating systems or low-temperature cells. An example of a recirculating setup that requires less than 1 mL of volume has been described [65]. In addition to requiring small volumes, this clever setup allows the sample to flow through an open jet, thereby eliminating contribution from capillary walls in the UVRR spectra. Sub-milliliter volumes may be interrogated in spinning cells. In one example, 500 μL is placed in a Suprasil quartz NMR tube connected to a centric spinner and spun at approximately 10 Hz [49]. In another example, sample volume as low as 80 μL is placed in a rotating cell connected to a small electric motor that spins at an adjustable rate of 1–30 Hz [66]. Low-temperature Raman spectra may also be acquired to reduce heat-induced protein damage from the laser [67, 68].

9.2.3 Detection System

The most challenging aspect of a UVRR experiment is elimination of stray light from Rayleigh scattering and visible fluorescence. This stray light may be reduced with UV filters, modified asymmetric double and triple spectrographs, and prism prefilters [69]. The Littrow prism prefilter shown in Figure 9.3 is ideal because it has relatively high throughput of 60% and allows acquisition of UVRR spectra that are within 200 cm^{-1} of the laser line [69]. After passing through the prefilter, the light is directed to a spectrograph where it is further dispersed via a high-dispersion grating, such as a holographic UV grating with 3600 grooves/mm or a ruled visible grating with

1200 grooves/mm operated in second order. The light is then imaged onto a UV-enhanced charge-coupled device (CCD) detector.

9.2.4 Other Considerations

It is critical that the conditions for UVRR experiments be carefully considered in order to obtain reliable and reproducible UVRR spectra. Laser power and focusing geometry should be optimized to reduce the probability for nonlinear phenomenon. For backscattering light collection, a cylindrically focused beam is ideal. Additionally, power-dependence experiments should be performed to determine the range in which Raman photons respond linearly to laser power; experiments should be performed well within this range. Sample concentration should be optimized such that self-absorption is minimized. For example, a 1000 µM solution of tryptophan has an optical density of 0.2 per 140 µm pathlength at the excitation wavelength of 228 nm. The Beer–Lambert law and knowledge of the absorption spectrum allow one to calculate the percentage of excitation and Raman-shifted photons to be absorbed in this sample. In this example, 37% of the excitation beam is absorbed, which is a relatively high value. Therefore, it is ideal to reduce the absorbance, and hence tryptophan concentration, to make the solution more transparent. Reduction of the concentration to 100 µM decreases the absorbance to 0.02, or 4.5% absorption. Finally, because many molecules absorb light in the UV region that is relevant for UVRR, all efforts must be taken to avoid fluorescent impurities in the sample.

9.3 APPLICATIONS OF UVRR SPECTROSCOPY TO MEMBRANE-ASSOCIATED PEPTIDES

Soluble proteins and peptides have been investigated extensively with UVRR spectroscopy. Far fewer UVRR studies of membrane-associated biomolecules are reported, but this number is increasing in parallel with the general expansion of biophysical and structural investigations of membrane proteins. Here, we illustrate the type of information that may be learned from UVRR studies of peptides in the context of folding and interactions with membranes. The choice of biological systems that are discussed here is intended to be representative of the array of biomolecules that may be probed as well as the types of scientific questions that may be investigated.

9.3.1 Model Peptides for Soluble and Membrane Protein Folding

As discussed above, UVRR spectroscopy is a sensitive technique capable of probing the local environment and noncovalent interactions of aromatic amino acids as well as identifying secondary structure elements in proteins and peptides. Because of the high sensitivity of UVRR experiments and wealth of molecular information that may be gained, small peptides provide an excellent opportunity to study soluble and membrane protein folding. The advantage of peptides is that they often form secondary structures that are identical to those in large proteins, yet peptides are less complex and typically easier to handle than proteins. Model peptides are especially important for the study of membrane protein folding because very little is known about structures and dynamics of membrane proteins relative to soluble proteins (only 180 unique membrane protein structures exist in the PDB as of 2009 [70]). Additionally, the expression and purification of large quantities of hydrophobic proteins is a challenging task whereas membrane-associated peptides can be synthesized in sufficient quantities. UVRR spectroscopy is well suited for the study of membrane-associated biomolecules because the presence of lipid vesicles and detergents required to solubilize and fold membrane peptides and proteins does not complicate the spectra [64].

Many groups have measured UVRR spectra of small peptides to study the structures and kinetics of soluble protein folding. Asher and coworkers, for example, have studied the kinetics and thermal stability of many α-helix-like soluble peptides (see Reference 71 and references therein). However, fewer model membrane peptides have been studied. The hydrophobic model membrane peptide *N*-acetyl-tryptophan-pentaleucine (AcWL5) was found to have β-sheet-type structure in lipid vesicles based on results obtained from UVRR spectra [72]; this finding is in accordance with the β-sheet oligomeric structure previously reported for AcWL5 [73]. The role of the important tryptophan residue in the structure of AcWL5 was also revealed with UVRR studies. The tryptophan residue of AcWL5 was found to be located in a hydrophobic, hydrogen-bonded position within the lipid bilayer. This finding may have implications for elucidating the mechanisms of peptide insertion and oligomeric assembly since tryptophan residues are known to act as anchors in membrane protein folding [37]. The hydrophobic α-helical model membrane peptide, ME1, has also been investigated using UVRR spectroscopy. ME1 is a 32-residue transmembrane helix segment of the natural membrane protein glycophorin A that has five additional mutations. Comparison of the UVRR spectrum of this peptide to that of the soluble α-helical protein myoglobin led to the conclusion that the intensity of UVRR amide bands serves as a reliable marker to identify lipid-solubilized and solvent-exposed helical structure in proteins [74].

9.3.2 Antimicrobial Peptides (AMPs)

While the membrane peptides described above serve as models for studying membrane protein folding, some small membrane-active peptides fold and insert into lipid bilayers and function as membrane disruptors. One such class of peptides is AMPs. AMPs are an ancient component of innate immunity based on the observation that they are present in both plant and animal kingdoms. These peptides are selectively toxic to bacteria, causing disruption of the membrane that eventually leads to cell death [75]. AMPs are crucial for the survival of many organisms that do not possess lymphocytes or antibodies important for immunity, such as insects [76]. AMPs typically have less than 50 residues and generally have segregated regions of cationic and hydrophobic amino acid residues that give rise to an ideal amphiphilic structure for binding to membranes [76]. They exist as linear peptides that form α-helices or rigid β-sheet structures with disulfide bridges and often exhibit primary sequences with an unusually high abundance of select amino acids [75, 77]. AMPs are water soluble yet spontaneously insert into membranes and, therefore, have properties of both soluble and membrane proteins. The observation that bacteria do not develop resistance to AMPs suggests that AMPs may serve as a novel class of antibiotics to combat antibiotic-resistant bacteria. This potentially transformative use of AMPs motivates UVRR and other studies of the mechanisms of membrane disruption.

The mechanisms of AMP insertion and folding into membranes are not well understood. Electrostatic interaction between cationic AMPs and anionic bacterial membranes may be responsible for initial peptide–membrane binding. However, the formation of salt bridges does not fully explain the selectivity of AMPs since nonselective peptide toxins may also be cationic. Several models of AMP disruption mechanisms have been proposed. In one model, AMPs carpet the membrane causing an increase in permeability of the membrane. AMPs may form pores that cause ions and molecules to leak out through the membrane. It has also been proposed that AMPs simply act to dissolve the membrane in a manner similar to a detergent [75]. Many questions about AMP behavior continue to cloud our understanding of the folding, insertion, and disruption mechanisms.

UVRR spectroscopy is a useful technique to probe AMP folding, insertion, and mechanisms of cell killing. As discussed, this technique probes the vibrational structure of the amide backbone and offers benefits over CD spectroscopy. Differences in amide band frequencies and intensities are indicative of protein secondary structural types [13, 14]. For many AMPs, the formation of secondary structure occurs during peptide insertion into the bilayer, and this insertion process is likely responsible for disruption of the membrane. In addition to backbone structure, UVRR spectroscopy reports the vibrational structure of aromatic amino acid residues. Band frequencies and relative intensity ratios are sensitive to local environment and noncovalent interactions of the aromatic residue [6, 8, 9, 78]. Mutation or deletion of tryptophan residues, in particular, has been found to be detrimental to the activity of many AMPs [41–44, 79, 80].

The AMPs lactoferricin B (LfB) and pEM-2 have been investigated using UVRR spectroscopy [81]. LfB is an 11-residue derivative of an α-helical 25-residue peptide composed of a segment of bovine lactoferrin, and pEM-2 is a modified 13-residue peptide derived from myotoxin isolated from *Bothrops asper* snake venom. Both peptides have tryptophan residues that are important for AMP activity and broad-spectrum activity against Gram-negative and -positive bacteria [82, 83]. Based on results obtained with UVRR experiments, the hydrophobicity of tryptophan environments increased when both peptides were introduced into the helix inducing solvent 2,2,2-trifluoroethanol. Additionally, the hydrophobicity near these tryptophan residues of LfB did not increase in the presence of purely zwitterionic lipid vesicles (eukaryotic membrane mimic). However, the hydrophobicity was enhanced in the presence of anionic lipid vesicles (bacterial membrane mimic), indicating that LfB preferentially folds and inserts into lipid bilayers that mimic bacterial membranes.

Similar results were found for the anoplin peptide, an α-helical peptide isolated from the venom of the wasp *Anoplius samariensis* [84]. Modifications in the primary structure of the peptide, including full substitutions of L-amino acids for D-amino acids and C-terminus deamidation, did not change the overall secondary structure of the peptide based on the results obtained with UVRR spectroscopy. Furthermore, it was found that all modified forms of the peptide adopted α-helical secondary structure in the presence of anionic lipid vesicles. Very little peptide structure was observed in the presence of zwitterionic lipid vesicles, further confirming the preferential binding of AMPs for anionic synthetic lipid bilayers.

The structure of the human cathelicidin AMP LL-37 has also been investigated using UVRR spectroscopy [85]. In this study, phenylalanine residues at positions 6 and 17 were replaced with tryptophan residues and yielded the two single tryptophan mutant peptides F6W and F17W. These mutations did not alter the antimicrobial activity nor the overall secondary structure compared to wild-type peptide LL-37. Based on results obtained from fluorescence quenching experiments and UVRR spectroscopy, the positions of tryptophan residues for both mutants were found to be deeply buried in the lipid bilayer (~12 Å from bilayer center) of mixed anionic/zwitterionic lipid vesicles (bacterial membrane mimics). A decrease in hydrogen bonding of both tryptophan residues was also observed upon folding and insertion of the peptide into the lipid bilayer, and this finding is consistent with deep insertion of the aromatic residues into the bilayer.

9.3.3 Toxins

Peptide toxins differ from AMPs because they show no selectivity of cell types; both eukaryotic and prokaryotic membranes, oftentimes including red blood cells, are efficiently damaged by toxins. They are similar to AMPs in that toxins act on the lipid bilayers by forming pores within or disrupting the membrane, causing cell death. Melittin, a 26 amino acid peptide that is the main component in European honeybee venom, is one of the most well-studied hemolytic, toxic peptides [86]. Melittin is cationic (+6 net charge) and exhibits amphipathic α-helical secondary structure upon spontaneous insertion into lipid membranes. The peptide is highly soluble in water despite the fact that residues 1–20 are predominantly hydrophobic [87]. Under some aqueous conditions, melittin self-associates to form α-helical tetramers. In this tetrameric conformation, the charged and polar residues are located on the outside of the structure, and a hydrophobic pocket remains in the center. A tryptophan residue crucial for the activity of this peptide (trp-19) is located within this hydrophobic region [41, 79, 88].

UVRR spectra of self-associated tetrameric melittin, as well as the membrane-bound form, have been measured. In one study, UVRR bands originating from trp-19 in both forms of melittin were found to be two to three times more intense than bands originating from a fully solvated tryptophan model compound, indicating a general increase in band intensity for tryptophan residues in hydrophobic residues [23]. This increase in UVRR intensity is consistent with the typical red-shift of the tryptophan absorption spectrum in a more hydrophobic environment; this red-shift enhances the Raman scattering cross-sections at the wavelength of study. In another report, trp-19 of melittin was investigated in the unfolded state (melittin in buffer only), in the tetrameric state, and in the presence of two different types of lipid vesicles: purely zwitterionic lipid vesicles and a mixture of anionic and zwitterionic lipid vesicles [8]. As expected, the UVRR spectra of trp-19 of melittin unfolded in buffer indicated a solvent-exposed local environment. In the presence of anionic membranes or as a soluble tetrameric form, the trp-19 residue of melittin was observed to be in a hydrophobic environment; this finding is consistent with deep burial of the peptide in the bilayer and hence efficient disruption of membranes. A surprising finding was that trp-19 was in a relatively hydrophilic environment and hydrogen bonded in the presence of purely zwitterionic membranes. This finding suggests that despite the similarity in potency for disruption of zwitterionic and anionic membranes, the structures of melittin may vary in these different membrane compositions.

A UVRR analysis of the amide backbone of melittin in folded, α-helical structures in the tetrameric and anionic membrane-bound species was also performed [89]. Furthermore, spectra of melittin in buffer suggested an unfolded random coil structure confirming the observations that trp-19 of melittin in buffer is solvent exposed. Figure 9.4 shows 210 nm UVRR spectra of melittin. The relative intensities and positions of the amide peaks indicate that melittin is >95% folded with α-helical secondary structure in the mixed anionic/zwitterionic lipid bilayer and random coil in buffer. The amide IIp peak is also evident and indicates weakened hydrogen bond in the folded state.

9.3.4 Engineered AMPs for Enhanced Efficacy

Several reports have indicated that mutagenesis can improve the potency of selectively toxic AMPs. The addition of tryptophan residues, for example, improves the antimicrobial activity of several AMPs and creates an opportunity to measure the structure and dynamics of the tryptophan residue during a peptide folding and insertion event using UVRR spectroscopy. Additionally, hybrid peptides that incorporate a segment of an AMP and a segment of peptide toxin may retain the strong antimicrobial properties of toxins while exhibiting the selectivity and nonhemolytic behavior of AMPs [90]. We recently published a UVRR comparison of the peptide–membrane interactions of a toxin, AMP, and an engineered hybrid [91]. While these UVRR experiments are challenging, such studies are likely to contribute to our understanding of AMP action and may help design new, more effective AMPs which may address the growing problem of bacterial resistance to conventional antibiotics [76].

9.3.5 Fibril-Forming Peptides

Amyloid-β (Aβ) is an intrinsically unstructured peptide comprised of 29–43 amino acids and is derived from proteolysis of the larger Aβ precursor protein. For reasons that are not yet understood, Aβ forms well-ordered insoluble aggregates, or fibrils, that collect as plaques in the brains of Alzheimer's patients. Because of the unique properties of this peptide, techniques typically utilized to elucidate protein structure, such as X-ray crystallography and solution NMR, are not well suited for studying the structures of the biologically relevant, unfolded forms of this peptide [92]. UVRR spectroscopy, however, does not require well-defined static structures and therefore can be used to measure structures of disordered peptides and proteins, including fibrils and heterogeneous aggregates. An additional benefit is that UVRR spectroscopy specifically reports on hydrogen-bonding networks that give rise to peptide secondary structure, local hydrophobicity, and tertiary contacts that are known to be associated with aggregate formation [93–95]. In the context of these advantages, UVRR spectroscopy, in particular

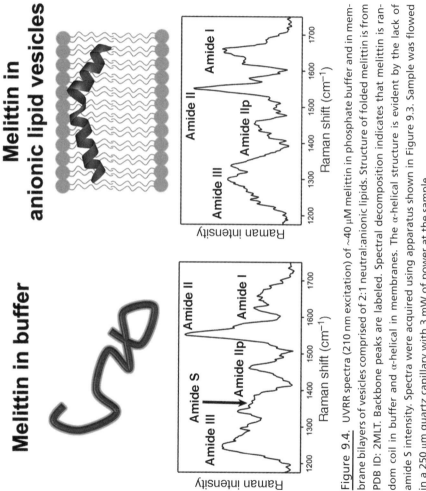

Figure 9.4. UVRR spectra (210 nm excitation) of ~40 μM melittin in phosphate buffer and in membrane bilayers of vesicles comprised of 2:1 neutral:anionic lipids. Structure of folded melittin is from PDB ID: 2MLT. Backbone peaks are labeled. Spectral decomposition indicates that melittin is random coil in buffer and α-helical in membranes. The α-helical structure is evident by the lack of amide S intensity. Spectra were acquired using apparatus shown in Figure 9.3. Sample was flowed in a 250 μm quartz capillary with 3 mW of power at the sample.

deep-UV (<200 nm excitation), is ideally suited for the study of Aβ and other peptides that may be linked to neurodegenerative disorders.

A relatively small number of UVRR studies of Aβ peptides have been published. The deep UVRR spectra of Aβ peptides that form parallel and antiparallel β-sheets have been characterized, and it was reported that these different β-sheet motifs of amyloid fibrils could be distinguished via UVRR [96]. This study further indicated that the parallel β-sheet conformation in fibrils is different from the analogous β-sheets in soluble proteins. In contrast, antiparallel β-sheet conformations were described to be similar for both fibrils and soluble proteins. In another study, myricetin, a flavonoid known to interact with Aβ, was observed to inhibit amyloid formation of Aβ via a thioflavin T assay. Additionally, results from CD and UVRR experiments revealed that myricetin altered the conformation of the hydrophobic segment of the peptide and may interact with the aromatic amino acids present in Aβ. This finding indicates that UVRR spectroscopy may be sensitive to changes in peptide structure when anti-amyloidogenic small molecules are introduced into the sample [97]. These discoveries on the structure of Aβ in amyloid fibrils underscore the flexibility of UVRR spectroscopy and the level of molecular detail that can be attained by this technique.

9.4 PROTEIN CONFORMATIONAL CHANGES

The sensitivity of UVRR signal to structure and local environment makes this tool a good reporter of conformational changes; therefore, UVRR may be a valuable complementary technique in the field of biotherapeutics. Alterations in vibrational frequencies and/or peak intensities are straightforward to detect and provide insights into molecular details of the system. Many examples of UVRR-detected conformational changes are reported in the literature. Local conformational changes in the form of enhanced RR cross-sections of small peptides can be observed as a function of temperature [98]. Temperature-induced global changes may also be detected by monitoring specific hydrogen-bonding signals, as well as backbone markers, such as evolution of α-helix to β-sheet [99, 100].

Other types of conformational alterations may also be detected. Protein reactions and signaling events may involve modification of the charge of a metal center. Redox changes, such as reduction of a 2Fe-2S cluster in the oxidative stress sensor protein [101] or oxidation of the heme in the direct oxygen-sensor protein [102], cause metal–ligand bonds to contract or stretch. These subtle structural changes have been detected by shifts in Raman peaks, and RR is especially useful because both UVRR and visible RR may be utilized to probe residues and metal centers, respectively.

Binding events that cause global and/or local perturbations may also be interrogated with RR spectroscopy. UVRR exhibits a high degree of sensitivity and has been reported to distinguish subtle enzyme conformational changes associated with binding of different diatomic molecules, such as CO, NO, and O_2 [103]. Specific molecular interactions associated with binding of flavonoids to peptides has also been described [97]. Other binding events have also been successfully probed with UVRR, including drug–protein interactions [104] and nucleotide substrates [105].

Detection of conformational changes is not limited to proteins. UVRR is also capable of monitoring changes in DNA structure. Global changes in overall super-helicity [106] as well as switching between right- and left-handed Z-DNA can be monitored [107].

Because UVRR is vibrational technique, it is one of the few analytical tools capable of reporting on changes in solution-phase structures at the molecular level. Even subtle modifications, such as strengthening of a hydrogen bond associated with a single residue or rotation of a side chain about its backbone, lead to changes in normal mode force constants that may be detected. Given the complexity of biomolecules, it is advantageous, but not always possible, to synthesize model compounds or generate truncated proteins that isolate the region of interest. The combination of model compounds and development of library of spectral markers makes UVRR a valuable method to probe conformational changes in biomolecules.

9.5 CHALLENGES AND BENEFITS OF UVRR SPECTROSCOPY

The primary difficulty in UVRR is that there is a small subset of side chains that may be probed. As discussed above, tryptophan, tyrosine, and histidine are the main focus of UVRR experiments on residues. Ligands may also be probed, but the absorption profiles would need to overlap with current UV lasers and the vibrational spectra should exhibit unique markers that distinguish it from side chains. Signal is not easily isolated among multiple residues/ligands; for example, in the common scenario where there are several tryptophan residues in a protein, the signal that is observed will reflect the structures of all the tryptophan residues. Confirmation of the specific residue that undergoes structural changes can only be achieved with mutagenesis or isotopic labeling of the individual residues. If signal from the residue of interest can be isolated, the insights gained on these residues can be significant. However, despite the gain in knowledge of these few residues, it is not possible to utilize UVRR to explore other noncovalent interactions, such as salt bridges or alterations in structure near a hydrophobic patch of, say, leucine residues.

The use of UV light is also challenging from an experimental viewpoint. UV-transparent optics, which may be quite costly, are required. Additionally, rejection of Rayleigh light is difficult in the UV region where the energy density per nanometer is high. Commercial filters in the UV region are available, such as from Semrock, but the cutoffs are not sharp and the available wavelengths are limited. For this reason, many UVRR spectroscopists utilize custom-designed prefilters to reject Rayleigh scattering. However, the idea of custom design may not be ideal in an industry setting.

The Raman scattering process is weak, and even under resonance conditions where signals are enhanced by at least 10^5 [3, 108], Rayleigh scattering will dominate the signal. For very dilute solutions, or molecules that exhibit low Raman cross-sections, acquisition of a decent spectrum may take on the order of an hour. In this case, the buildup of irreversibly damaged sample may become an issue if a recirculating system is utilized; the damage caused by UV photons can be significant and has been documented for many decades [109, 110]. The practical solution is to utilize a single-pass system, but this apparatus requires fresh sample for the duration of the experiment.

The problem of scattering is especially significant in the presence of large particles. In solutions that contain macroscopic oligomeric complexes, aggregated proteins, or vesicles, the Rayleigh signal will be significant and extra effort must be devoted to elimination of this elastically scattered light. Because the efficiency for Rayleigh scattering also scales as λ^{-4}, the problem of scattering is more significant in the UV than in the visible region. A consequence of the presence of strong Rayleigh signal is that low-frequency UVRR peaks are difficult to detect, and a larger background signal may be present.

There are several benefits of UVRR spectroscopy. One advantage of this technique is avoidance of fluorescence. In contrast to visible RR where the Raman-shifted photons often overlap the fluorescence spectrum, UVRR analysis with less than 230 nm excitation allows spectral separation of the Raman and fluorescence signals. The fact that Raman scattering scales as λ^{-4} indicates that UV excitation is much more efficient that NIR or visible excitation. The Raman process with 228 nm excitation will result in 25-fold greater number of scattered photons relative to excitation with 514 nm. This enhanced efficiency offers numerous advantages, including reduction in data collection time from 1 hour (514 nm) to 2.5 minutes (228 nm). Alternatively, one may prepare a 25-fold dilute sample for UVRR analysis or decrease laser power significantly relative to the visible/NIR case. A final advantage of UVRR over visible/NIR experiments is the absence of stray ambient light. For example, one may acquire a standard 5 minute UVRR spectrum in a lit room.

9.6 CONCLUSION

UVRR is a useful tool that reports on structure and local environment. An inherent advantage of this vibrational technique is that it provides molecular information of biomolecules and may reveal site-specific information about backbone and aromatic residues, including noncovalent interactions. When combined with other biophysical tools, such as fluorescence, calorimetry, and mass spectrometry, a rich data set that characterizes physical properties of complex biomolecules, such as stability, structure, and thermodynamics, may be developed.

ACKNOWLEDGMENTS

We are grateful to the National Science Foundation for supporting our biophysical studies of proteins using RR and fluorescence spectroscopy.

REFERENCES

1. Raman, C.V. and Krishnan, K.S. (1928) A new type of secondary radiation. *Nature*, **121**, 501–502.

2. Singh, R. (2002) C.V. Raman and the discovery of the Raman effect. *Phys. Perspect.*, **4**, 399–420.

3. McCreery, R.L. (2000) *Raman Spectroscopy for Chemical Analysis*. Wiley Interscience, New York.

4. Long, D.A. (1977) *Raman Spectroscopy*. McGraw-Hill, Inc., New York.

5. Myers, A.B. and Mathies, R.A. (1984) Resonance Raman intensities: a probe of excited-state structure and dynamics. In: Spiro, T.G. editor. *Biological Applications of Raman Spectroscopy*. John Wiley & Sons, Inc., pp. 1–58.

6. Harada, I. and Takeuchi, H. (1986) Raman and ultraviolet resonance Raman spectra of proteins and related compounds. In: Clark, R.J.H. and Hester, R.E. editors. *Spectroscopy of Biological Systems*. John Wiley & Sons, Ltd., Chichester.

7. Austin, J.C., Jordan, T., and Spiro, T.G. (1993) Ultraviolet resonance Raman studies of proteins and related model compounds. In: Clark, R.J.H. and Hester, R.E. editors. *Biomolecular Spectroscopy, Part A*. John Wiley & Sons, New York, pp. 55–127.

8. Schlamadinger, D.E., Gable, J.E., and Kim, J.E. (2009) Hydrogen-bonding and solvent polarity markers in the UV resonance Raman spectrum of tryptophan: application to membrane proteins. *J. Phys. Chem. B*, **113**, 14769–14778.

9. Schlamadinger, D.E., Leigh, B.S., and Kim, J.E. (2012) UV resonance Raman study of TrpZip2 and related peptides: π–π interactions of tryptophan. *J. Raman Spectrosc.*, **43**, 1459–1464.

10. Cantor, C.R. and Schimmel, P.R. (1980) *Biophysical Chemistry*. W. H. Freeman and Company, New York.

11. Wang, Y., Purrello, R., Jordan, T., and Spiro, T.G. (1991) UVRR spectroscopy of the peptide bond. 1. Amide S, a nonhelical structure marker, is a CαH bending mode, *J. Am. Chem. Soc.*, **113**, 6359–6368.

12. Jordan, T., Mukerji, I., Wang, Y., and Spiro, T.G. (1996) UV resonance Raman spectroscopy and hydrogen bonding of the proline peptide bond, *J. Mol. Struct.*, **379**, 51–64.

13. Huang, C.Y., Balakrishnan, G., and Spiro, T.G. (2006) Protein secondary structure from deep-UV resonance Raman spectroscopy. *J. Raman Spectrosc.*, **37**, 277–282.

14. Chi, Z., Chen, X.G., Holtz, J.S.W., and Asher, S.A. (1998) UV resonance Raman-selective amide vibrational enhancement: quantitative methodology for determining protein secondary structure. *Biochemistry*, **36**, 2854–2864.

15. Ozdemir, A., Lednev, I.K., and Asher, S.A. (2002) Comparison between UV Raman and circular dichroism detection of short α-helices in bombolitin III. *Biochemistry*, **41**, 1893–1896.

16. Sreerama, N. and Woody, R.W. (2000) *Circular Dichroism of Peptides and Proteins*. Wiley & Sons, Hoboken, NJ.

17. Woody, R.W. (1996) *Theory of Circular Dichroism of Proteins*. Plenum Press, New York.

18. Chin, D.H., Woody, R.W., Rohl, C.A., and Baldwin, R.L. (2002) Circular dichroism spectra of short, fixed-nucleus alanine helices. *Proc. Natl. Acad. Sci. U.S.A.*, **99**, 15416–15421.

19. Wallace, B.A., et al. (2003) Analyses of circular dichroism spectra of membrane proteins. *Protein Sci.*, **12**, 875–884.

20. Fodor, S.P.A., Copeland, R.A., Grygon, C.A., and Spiro, T.G. (1989) Deep-ultraviolet Raman excitation profiles and vibronic scattering mechanisms of phenylalanine, tyrosine, and tryptophan. *J. Am. Chem. Soc.*, **111**, 5509–5518.

21. Sweeney, J.A. and Asher, S.A. (1990) Tryptophan UV resonance Raman excitation profiles. *J. Phys. Chem.*, **94**, 4784–4791.

22. Chi, Z. and Asher, S.A. (1998) UV Raman determination of the environment and solvent exposure of tyr and trp residues. *J. Phys. Chem. B*, **102**, 9595–9602.

23. Efremov, R.G., Feofanov, A.V., and Nabiev, I.R. (1992) Effect of hydrophobic environment on the resonance Raman spectra of tryptophan residues in proteins. *J. Raman Spectrosc.*, **23**, 69–73.

24. Hildebrandt, P.G., et al. (1988) Tyrosine hydrogen-bonding and environmental effects in proteins probed by ultraviolet resonance Raman spectroscopy. *Biochemistry*, **27**, 5426–5433.

25. Ludwig, B. and Asher, S.A. (1988) Ultraviolet resonance Raman excitation profiles of tyrosine: dependence of Raman cross sections on excited-state intermediates. *J. Am. Chem. Soc.*, **110**, 1005–1011.

26. Asher, S.A. and Murtaugh, J.L. (1988) UV Raman excitation profiles of imidazole, imidazolium, and water, *Appl. Spectrosc.*, **42**, 83–90.

27. Takeuchi, H. (2003) Raman structural markers of tryptophan and histidine side chains in proteins. *Biopolymers*, **72**, 305–317.

28. Miura, T., Takeuchi, H., and Harada, I. (1988) Characterization of individual tryptophan side chains in proteins using Raman spectroscopy and hydrogen-deuterium exchange kinetics. *Biochemistry*, **27**, 88–94.

29. Miura, T., Takeuchi, H., and Harada, I. (1989) Tryptophan Raman bands sensitive to hydrogen bonding and side-chain conformation. *J. Raman Spectrosc.*, **20**, 667–671.

30. Harada, I., Miura, T., and Takeuchi, H. (1986) Origin of the doublet at 1360 and 1340 cm-1 in the Raman spectra of tryptophan and related compounds. *Spectrochim. Acta,* **42A**, 307–312.

31. Takeuchi, H., Watanabe, Y., Satoh, Y., and Harada, I. (1989) Effects of hydrogen bonding on the tyrosine Raman bands in the 1300–1150 cm-1 region. *J. Raman Spectrosc.*, **20**, 233–237.

32. The UniProt Consortium. (2008) The Universal Protein Resource (UniProt). *Nucleic Acids Res.*, **36**, D190–D195.

33. White, S.H., Ladokhin, A.S., Jayasinghe, S., and Hristova, K. (2001) How membranes shape protein structure. *J. Biol. Chem.*, **276**, 32395–32398.

34. Gallivan, J.P., and Dougherty, D.A. (1999) Cation-π interactions in structural biology. *Proc. Natl. Acad. Sci. U.S.A.*, **96**, 9459–9464.

35. Millefiori, S., Alparone, A., Millefiori, A., and Vanella, A. (2008) Electronic and vibrational polarizabilities of the twenty naturally occurring amino acids. *Biophys. Chem.* **132**, 139–147.

36. Wimley, W.C. and White, S.H. (1996) Experimentally determined hydrophobicity scale for proteins at membrane interfaces. *Nat. Struct. Biol.*, **3**, 842–848.

37. Killian, J.A. and von Heijne, G. (2000) How proteins adapt to a membrane-water interface. *Trends Biochem. Sci.*, **25**, 429–434.

38. White, S.H. and Wimley, W.C. (1999) Membrane protein folding and stability: physical principles. *Annu. Rev. Biophys. Biomol. Struct.*, **28**, 319–365.

39. Sanchez, K.M., Gable, J.E., Schlamadinger, D.E., and Kim, J.E. (2008) Effects of tryptophan microenvironment, soluble domain, and vesicle size on the thermodynamics of membrane protein folding: lessons from the transmembrane protein OmpA. *Biochemistry*, **47**, 12844–12852.

40. Hong, H., et al. (2007) Role of aromatic side chains in the folding and thermodynamic stability of integral membrane proteins. *J. Am. Chem. Soc.*, **129**, 8320–8327.

41. Blondelle, S.E. and Houghten, R.A. (1991) Probing the relationships between the structure and hemolytic activity of melittin with a complete set of leucine substitution analogs. *Peptide Res.*, **4**, 12–18.

42. Andreu, D., Merrifeld, R.B., Steiner, H., and Boman, H.G. (1985) N-terminal analogues of cecropin A: synthesis, antibacterial activity, and conformational properties. *Biochemistry*, **24**, 1683–1688.

43. Salom, D., Perez-Paya, E., Pascal, J., and Abad, C. (1998) Environment- and sequence-dependent modulation of the double-stranded to single-stranded conformational transition of gramicidin A. *Biochemistry*, **37**, 14279–14291.

44. Becker, M.D., Greathouse, D.V., Koeppe, R.E., and Andersen, O.S. (1991) Amino acid sequence modulation of gramicidin channel function: effects of tryptophan-to-phenylalanine substitutions on the single-channel conductance and duration. *Biochemistry*, **30**, 8830–8839.

45. Ahmed, A., Beta, I.A., Mikhonin, A.V., and Asher, S.A. (2005) UV-resonance Raman thermal unfolding study of Trp-cage shows that it is not a simple two-state miniprotein. *J. Am. Chem. Soc.*, **127**, 10943–10950.

46. Chi, Z. and Asher, S.A. (1999) Ultraviolet resonance Raman examination of horse apomyoglobin acid unfolding intermediates. *Biochemistry*, **38**, 8196–8203.

47. Rodriguez-Mendieta, I.R., et al. (2005) Ultraviolet resonance Raman studies reveal the environment of tryptophan and tyrosine residues in the native and partially unfolded states of the E colicin-binding immunity protein Im7. *Biochemistry*, **44**, 3306–3315.

48. Huang, C.Y., Balakrishnan, G., and Spiro, T.G. (2005) Early events in apomyoglobin unfolding probed by laser T-jump/UV resonance Raman spectroscopy. *Biochemistry*, **44**, 15734–15742.

49. Rodgers, K.R., Su, C., Subramaniam, S., and Spiro, T.G. (1992) Hemoglobin R to T structural dynamics from simultaneous monitoring of tyrosine and tryptophan time-resolved UV resonance Raman signals. *J. Am. Chem. Soc.*, **114**, 3697–3709.

50. Jayaraman, V., Rodgers, K.R., Mukerji, I., and Spiro, T.G. (1995) Hemoglobin allostery: resonance Raman spectroscopy of kinetic intermediates. *Science*, **269**, 1843–1848.

51. Cho, N., Song, S., and Asher, S.A. (1994) UV resonance Raman and excited-state relaxation rate studies of hemoglobin. *Biochemistry*, **33**, 5932–5941.

52. Kochendoerfer, G.G., Kaminaka, S., and Mathies, R.A. (1997) Ultraviolet resonance Raman examination of the light-induced protein structural changes in rhodopsin activation. *Biochemistry*, **36**, 13153–13159.

53. Kim, J., Pan, D., and Mathies, R.A. (2003) Picosecond dynamics of G-protein coupled receptor activation in rhodopsin from time-resolved UV resonance Raman spectroscopy. *Biochemistry*, **42**, 5169–5175.

54. Mizuno, M., Hamada, N., Tokunaga, F., and Mizutani, Y. (2007) Picosecond protein response to the chromophore isomerization of photoactive yellow protein: selective observation of tyrosine and tryptophan residues by time-resolved ultraviolet resonance Raman spectroscopy. *J. Phys. Chem. B*, **111**, 6293–6296.

55. Sato, A., Gao, Y., Kitagawa, T., and Mizutani, Y. (2007) Primary protein response after ligand photodissociation in carbonmonoxy myoglobin. *Proc. Natl. Acad. Sci. U.S.A.*, **104**, 9627–9632.

56. Balakrishnan, G., Hu, Y., Nielsen, S.N., and Spiro, T.G. (2005) Tunable kHz deep ultraviolet (193–210 nm) laser for Raman applications. *Appl. Spectrosc.*, **59**, 776–781.

57. Zhao, X.J., et al. (1999) Time-resolved Raman spectroscopy with a tuneable ultraviolet kilohertz nanosecond laser. *J. Raman Spectrosc.*, **30**, 773–776.

58. Austin, J.C., Rodgers, K.R., and Spiro, T.G. (1993) Protein structure from ultraviolet resonance Raman spectroscopy. *Methods Enzymol.*, **226**, 374–396.

59. Bykov, S., et al. (2005) Steady-state and transient ultraviolet resonance Raman spectrometer for the 193–270 nm spectral region. *Appl. Spectrosc.*, **59**, 1541–1552.

60. Asher, S.A. (1988) UV resonance Raman studies of molecular structure and dynamics: applications in physical and biophysical chemistry. *Ann. Rev. Phys. Chem.*, **39**, 537–588.

61. Efremov, R.G., Ariese, F., and Gooijer, C. (2008) Achievements in resonance Raman spectroscopy. review of a technique with a distinct analytical chemistry potential. *Anal. Chim. Acta*, **606**, 119–134.

62. Hudson, B.S. and Mayne, L. (1986) Ultraviolet resonance Raman spectroscopy of biopolymers. *Methods Enzymol.*, **130**, 331–350.

63. Efremov, R.G., Feofanov, A.V., and Nabiev, I.R. (1991) Quantitative treatment of UV resonance Raman spectra of biological molecules: application to the study of membrane-bound proteins. *Appl. Spectrosc.*, **45**, 272–278.

64. Sanchez, K.M., Neary, T.J., and Kim, J.E. (2008) UV resonance Raman spectroscopy of folded and unfolded states of an integral membrane protein. *J. Phys. Chem. B*, **112**, 9507–9511.

65. Chen, J. and Barry, B.A. (2008) Ultraviolet resonance Raman microprobe spectroscopy of photosystem II. *Photochem. Photobiol.*, **84**, 815–818.

66. Kim, M., Mathies, R.A., Hoff, W.D., and Hellingwerf, K.J. (1995) Resonance Raman evidence that the thioester-linked 4-hydroxycinnamyl chromophore of photoactive yellow protein is deprotonated. *Biochemistry*, **34**, 12669–12672.

67. Czernuszewicz, R.S. and Johnson, M.K. (1983) A simple low-temperature cryostat for resonance Raman studies of frozen protein solutions. *Appl. Spectrosc.*, **37**, 297–298.

68. Miller, F.A. and Harney, B.M. (1969) Variable temperature sample holder for Raman spectroscopy. *Appl. Spectrosc.*, **24**, 291–292.

69. Kaminaka, S. and Mathies, R.A. (1998) High-throughput large-aperture prism prefilter for ultraviolet resonance Raman spectroscopy. *Appl. Spectrosc.*, **52**, 469–473.

70. White, S.H. (2009) Biophysical dissection of membrane proteins. *Nature*, **459**, 344–346.

71. Oladepo, S.A., Xiong, K., Hong, Z., and Asher, S.A. (2011) Elucidating peptide and protein structure and dynamics: UV resonance Raman spectroscopy. *J. Phys. Chem. Lett.*, **2**, 334–344.

72. Shafaat, H.S., Sanchez, K.M., Neary, T.J., and Kim, J.E. (2009) Ultraviolet resonance Raman spectroscopy of a β-sheet peptide: a model for membrane protein folding. *J. Raman Spectrosc.*, **40**, 1060–1064.

73. Wimley, W.C., et al. (1998) Folding of β-sheet membrane proteins: a hydrophobic hexapeptide model. *J. Mol. Biol.*, **277**, 1091–1110.

74. Halsey, C.M., et al. (2011) Simultaneous observation of peptide backbone lipid solvation and α-helical structure by deep-UV resonance Raman spectroscopy. *ChemBioChem.*, **12**, 2125–2128.

75. Brogden, K.A. (2005) Antimicrobial peptides: pore formers or metabolic inhibitors in bacteria? *Nat. Rev. Microbiol.*, **3**, 238–250.

76. Zasloff, M. (2002) Antimicrobial peptides of multicellular organisms. *Nature*, **415**, 389–395.

77. Epand, R.M. and Vogel, H.J. (1999) Diversity of antimicrobial peptides and their mechanisms of action. *Biochim. Biophys. Acta*, **1462**, 11–28.

78. Schlamadinger, D.E., Daschbach, M.M., Gokel, G.W., and Kim, J.E. (2011) UV resonance Raman study of cation-π interactions in an indole crown ether. *J. Raman Spectrosc.*, **42**, 633–638.

79. Blondelle, S.E., and Houghten, R.A. (1991) Hemolytic and antimicrobial activities of the twenty-four individual omission analogs of melittin. *Biochemistry*, **30**, 4671–4678.

80. Wei, G., et al. (2010) Trp-26 imparts functional versatility to human α-defensin HNP1. *J. Biol. Chem.*, **285**, 16275–16285.

81. Quan, B. and Ianoul, A. (2009) UV resonance Raman spectroscopy probes the localization of tryptophan-containing antimicrobial peptides in lipid vesicles. *J. Raman Spectrosc.*, **40**, 260–263.

82. Strøm, M.B., Rekdal, Ø., and Svendsen, J.S. (2000) Antibacterial activity of 15-residue lactoferricin derivatives. *J. Peptide Res.*, **56**, 265–274.

83. Santamaría, C., et al. (2005) Bactericidal and antiendotoxic properties of short cationic peptides derived from a snake venom Lys49 phospholipase A_2. *Antimicrob. Agents Chemother.*, **49**, 1340–1345.

84. Pripotnev, S., Won, A., and Ianoul, A. (2009) The effects of L-to D-isomerization and C-terminus deamidation on the secondary structure of antimicrobial peptide anoplin in aqueous and membrane mimicking environment. *J. Raman Spectrosc.*, **41**, 1645–1649.

85. Gable, J.E., et al. (2009) Fluorescence and UV resonance Raman study of peptide-vesicle interactions of human cathelicidin LL-37 and its F6 W and F17 W mutants. *Biochemistry*, **48**, 11264–11272.

86. Habermann, E. (1972) Bee and wasp venoms. *Science*, **177**, 314–322.

87. Raghuraman, H. and Chattopadhyay, A. (2007) Melittin: a membrane-active peptide with diverse functions. *Biosci. Rep.*, **27**, 189–223.

88. Terwilliger, T.C. and Eisenberg, D. (1982) The structure of melittin. II. Interpretation of the structure. *J. Biol. Chem.* **257**, 6016–6022.

89. Schlamadinger, D.E., Gable, J.E., and Kim, J.E. (2009) Toxins and antimicrobial peptides: interactions with membranes. *Proc. SPIE*, **7397**, 73970 J-73971–73970 J-73913.

90. Rivas, L. and Andreu, D. (2003) Cecropin-melittin hybrid peptides as versatile templates in the development of membrane-active antibiotic agents. In: Menestrina, G., Serra, M.D., and Lazarovici, P. editors. *Pore-Forming Peptides and Protein Toxins.* Taylor & Francis, Inc., New York, NY.

91. Schlamadinger, D.E., Wang, Y., McCammon, J.A., and Kim, J.E. (2012) Spectroscopic and computational study of melittin, cecropin A, and the hybrid peptide CM15. *J. Phys. Chem. B* **116**, 10600–10608.

92. Lührs, T., et al. (2005) 3D structure of Alzheimer's amyloid-β(1–42) fibrils. *Proc. Natl. Acad. Sci. U.S.A.*, **102**, 17342–17347.

93. Sikirzhytski, V., et al. (2008) Genetic engineering combined with deep UV resonance Raman spectroscopy for structural characterization of amyloid-like fibrils. *J. Am. Chem. Soc.*, **130**, 5853–5853.

94. Pawar, A.P., et al. (2005) Prediction of "aggregation-prone" and "aggregation-susceptible" regions in proteins associated with neurodegenerative diseases. *J. Mol. Biol.*, **350**, 379–392.

95. DuBay, K.F., et al. (2004) Prediction of the absolute aggregation rates of amyloidogenic polypeptide chains. *J. Mol. Biol.*, **341**, 1317–1326.

96. Popova, L.A., Kodali, R., Wetzel, R., and Lednev, I.K. (2010) Structural variations in the cross-β core of amyloid β fibrils revealed by deep UV resonance Raman spectroscopy. *J. Am. Chem. Soc.*, **132**, 6324–6328.

97. Wang, M. and Jiji, R.D. (2011) Resolution of localized small molecule-Aβ interactions by deep-ultraviolet resonance Raman spectroscopy. *Biophys. Chem.*, **158**, 96–103.

98. Pieridou, G., et al. (2010) UV resonance Raman study of TTR(105–115) structural evolution as a function of temperature. *J. Phys. Chem. B*, **115**, 4088–4098.

99. Ahmed, Z., Scaffidi, J.P., and Asher, S.A. (2008) Circular dichroism and UV-resonance Raman investigation of the temperature dependence of the conformations of linear and cyclic elastin. *Biopolymers*, **91**, 52–60.

100. Jiji, R., Balakrishnan, G., Hu, Y., and Spiro, T.G. (2006) Intermediacy of poly(L-proline) II and beta-strand conformations in poly(L-lysine) beta-sheet formation probed by temperature-jump/UV resonance Raman spectroscopy. *Biochemistry*, **45**, 34–41.

101. Kobayashi, K., Mizuno, M., Fujikawa, M., and Mizutani, Y. (2011) Protein conformational changes of the oxidative stress sensor, SoxR, upon redox changes of the [2Fe-2S] cluster probed with ultraviolet resonance Raman spectroscopy. *Biochemistry*, **50**, 9468–9474.

102. El-Mashtoly, S.F., et al. (2008) Resonance Raman investigation of redox-induced structural changes of protein and heme in the sensor domain of Ec DOS protein. *J. Raman Spectrosc.*, **39**, 1614–1626.

103. El-Mashtoly, S.F., et al. (2008) Protein conformation changes of HemAT-Bs upon ligand binding probed by ultraviolet resonance Raman spectroscopy. *J. Biol. Chem.*, **283**, 6942–6949.

104. Couling, V.W., Fischer, P., Klenerman, D., and Huber, W. (1998) Ultraviolet resonance Raman study of drug binding in dihydrofolate reductase, gyrase, and catechol o-methyltransferase. *Biophys. J.*, **75**, 1097–1106.

105. Wojtuszewski, K. and Mukerji, I. (2004) The HU-DNA binding interaction probed with UV resonance Raman spectroscopy: structural elements of specificity. *Protein Sci.*, **13**, 2416–2428.

106. Neugebauer, U., et al. (2007) DNA tertiary structure and changes in DNA supercoiling upon interaction with ethidium bromide and gyrase monitored by UV resonance Raman spectroscopy. *J. Raman Spectrosc.* **38**, 1246–1258.

107. Klump, H.H., Schmid, E., and Wosgein, M. (1993) Energetics of Z-DNA formation in poly d(A-T), poly d(G-C), and poly d(A-C) poly d(G-T). *Nucl. Acids Res.* **21**, 2343–2348.

108. Myers, A.B. and Mathies, R. (1987) Resonance Raman intensities: a probe of excited-state structure and dynamics. In: Spiro, T.G. editor. *Biological Applications of Raman Spectroscopy*. John Wiley & Sons, New York City, pp. 3–58.

109. Augenstine, L.G. and Ghiron, C.A. (1961) The inactivation of trypsin by ultraviolet light. I. The correlation of inactivation with the disruption of constituent cystine. *Proc. Natl. Acad. Sci. U. S. A.*, **47**, 1530–1547.

110. McLaren, A.D. and Luse, R.A. (1961) Mechanism of inactivation of enzyme proteins by ultra-violet light. *Science (Washington, DC, U.S.A)*, **134**, 836–837.

10

FREEZING- AND DRYING-INDUCED MICRO- AND NANO-HETEROGENEITY IN BIOLOGICAL SOLUTIONS

Alptekin Aksan[1], Vishard Ragoonanan[2], and Carol Hirschmugl[3,4]

[1]*Biostabilization Laboratory, Mechanical Engineering Department and BioTechnology Institute, University of Minnesota, Minneapolis, MN, USA*
[2]*Department of Pharmaceutics, University of Minnesota, Minneapolis, MN, USA*
[3]*Department of Physics, University of Wisconsin, Milwaukee, WI, USA*
[4]*Synchrotron Resource Center, University of Wisconsin, Stoughton, WI, USA*

Biophysical Methods for Biotherapeutics: Discovery and Development Applications, First Edition. Edited by Tapan K. Das.
© 2014 John Wiley & Sons, Inc. Published 2014 by John Wiley & Sons, Inc.

10.1   INTRODUCTION

Freezing and drying are surface-induced processes (i.e., heat and mass fluxes are maximum at the surface of a system) that produce gradients of temperature and concentration within the system [1,2]. Heat transfer (both by conduction and advection) and molecular mobility (diffusion down the concentration gradient and by random Brownian motion) work to smooth out these gradients. However, with decreasing temperature and increasing solute concentration, molecular mobility decreases very significantly and without its smoothing effects, the concentration gradients are then permanently etched into the fabric of the frozen and the desiccated systems. In general terms, spatial variations in the concentrations of solutes, the solvent, and the thermodynamic state/phase within a system is called heterogeneity [2,3]. One can then envision that spatial heterogeneity increases with the specific surface area of the system (surface area/system volume) and the relative magnitude of the heat/mass flux at the surface of the system with respect to the thermal inertia of the system and the diffusivity of water within the system.

Persistent gradients are not the only sources of spatial heterogeneity in a system undergoing freezing or drying. Presence of liquid–liquid phase partitioning at the macroscale- [4] and molecular-level demixing [5, 6] contributes very significantly to freezing- and drying-induced heterogeneity. Note that the freezing process, by itself, fuels heterogeneity through ice nucleation and solute rejection. In addition, preferential interaction of the different solutes within the solution with each other and with interfaces (e.g., phase interfaces as well as solid and gas interfaces) and the converse, preferential exclusion of solutes, as well as temperature- and concentration-induced changes in solubility cause further heterogeneity in a system undergoing cooling, freezing, or drying [7].

Freezing is a very complex process initiating and driving many mechanisms that induce heterogeneity. The ice nucleation temperature (i.e., the degree of supercooling, which affects the pre-freeze molecular mobility, as well as the free energy of phase change and therefore the critical ice nuclei size) and the post-freeze cooling rate (which determines ice crystal growth rate and deviation from thermodynamic equilibrium) are strong factors that determine the final state and heterogeneity of the frozen product. Other phenomena of significance that are driven by the freezing process are the change in the pH [8] of the freeze-concentrated liquid (FCL), formation of gas bubbles (due to reduced solubility of the gases in the FCL and in the ice phase), and the formation of interfaces (solid, liquid, or gas) with high free energy.

In the drying process, solvent flows within the system toward the surface (called the primary flow). Spatial variations in the surface mass flux are commonly seen in sessile droplet and film drying either due to the nonuniformities in the temperature and/or relative humidity of the surrounding atmosphere or its flow pattern, the curvature of the liquid–gas interface (a factor that is very dominant, for example, in sessile droplet drying [9, 10]), or the geometry of the container that encloses the product (e.g., a geometry that allows limited circulation of the surrounding atmosphere or exposes the product to interfaces). Spatial variation of the mass flux at the surface causes spatial variations in the surface temperature (due to latent heat of evaporation), which creates surface tension gradients that fuel secondary flows (the Marangoni instability [11]). The combined effects of the primary and the secondary flows generate complex flow patterns within the drying product, even causing separation of the solutes (of different size, charge, etc.) [12].

Freezing and drying are not simple processes. They generate heterogeneity within the system at different length scales (ranging from molecular level to macroscale) and also amplify the defects that may exist within the system (undissolved solute crystals, aggregates, container surface defects, etc.). The outcome of heterogeneity is that the frozen/dried product may not be uniform at the same thermodynamic state, which means that there are driving forces (e.g., chemical potential gradients) present within the system and therefore the state of the product will continue to evolve over time during storage. Besides, the product in different regions within the system will be exposed to different microenvironmental conditions and will therefore show variations in storage stability, quality, and so on. Moreover, even small changes in processing parameters may result in significant changes in the homogeneity of the product, generating a nonlinear outcome that is hard to predict. And probably, one of the most dramatic effects is that the product to be stabilized may not even "see" the lyo-/cryoprotectant agent, which is there to protect it [13].

So far, no work has been done to predict the losses in product yield and quality associated with processing heterogeneity and very little work has focused on minimizing the effects of heterogeneity on industrial processes. One of the main reasons has been the unavailability of the detection methods, especially at smaller length scales. However, this is changing. We are confident that once a basic level of understanding of heterogeneity is reached and the technologies that enable detection find widespread use, the product losses associated with heterogeneity will easily be quantified fueling research efforts for minimizing product heterogeneity to decrease industrial-scale processing and storage losses. In this chapter, we focus on experimental detection and quantification of heterogeneity in systems of pharmaceutical interest.

10.2   FREEZING-INDUCED HETEROGENEITY

The simplest system of interest is an aqueous buffer (added to the solution to presumably protect—to some extent—against fluctuations in pH) that contains a macromolecule (an enzyme, a protein, etc.) or a slightly more complex system (e.g., liposome-encapsulated drug) and a cryo-/lyoprotectant agent. In this context, the macromolecule is the "product" that one would want to stabilize by freezing, drying, or freeze-drying and store until use. The cryo-/lyoprotectant agent could be a sugar (a disaccharide such as trehalose or sucrose and/or a polysaccharide such as dextran that could also serve as a caking agent), a polyol (such as glycerol that is known to be effective against low-temperature damage and serves as a space-filler in vitrification [14]), an organic solvent (such as dimethyl sulfoxide, DMSO) added, for example, to increase solubility of the protein or to preserve small peptides, or even an amino acid (or another protein).

In a typical freezing process, at a temperature below the melting temperature of the solution, ice crystals form (assuming that water is the dominant solvent) (Figure 10.1). The degree of supercooling (the difference between the thermodynamic freezing

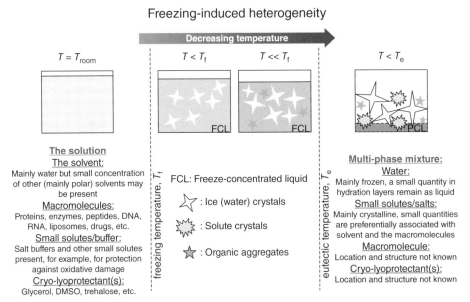

Figure 10.1. Transition of a "simple" solution to a multi-phase mixture during freezing. Note that the actual freezing-induced heterogeneity is much more complicated than what is shown here. In many cases, there are no clear boundaries among different phases but there are heterogeneous regions composed of different constituents (e.g., proteins entrapped in ice; see Reference 3). The amount of remaining water at temperatures below the eutectic temperature is a complex function of the thermal history of the specimen as well as its specific constituents. (See color insert)

temperature as given by the state diagram and the actual freezing temperature) depends on the chemistry of the solution and the pre-freeze cooling rate. The freezing process creates distinct ice crystals, which reject solutes and generate an FCL. An FCL region surrounding the ice crystals can easily be visualized under a transmitted light microscope. Depending on the temperature, the FCL could be a liquid or may have vitrified (turned into an amorphous, high-viscosity liquid, called a glass). The shapes and the size distribution of the ice crystals are dictated by the thermal history of the specimen as well as the chemical and physical properties of the FCL. In the absence of vitrification, the solute concentration in the FCL depends on temperature and increases with decreasing temperature with more of the liquid joining the ice phase (the solute concentration in the FCL can be as high as 50 times that of the initial solution [13]). During continued cooling some of the solutes start to precipitate and aggregate due to change in solubility (Figure 10.1). Below the eutectic temperature, no liquid phase remains and all of the solutes crystallize or aggregate. Note that residual amount of solvent may remain in liquid form in the closest hydration layers of the solutes and the macromolecules. The outcome of this process is that the macromolecule may end up completely separated from the lyo-/cryoprotectant agent, lose the majority of the water molecules in its hydration layers (or even desiccate completely), and aggregate or even crystallize. Another interesting factor in heterogeneity (especially in protein solutions) is the variation in the ice surface area (which is influenced by the temperature of ice crystal formation and ice crystal size) that would alter ice–protein interactions and cause denaturation or aggregation of the proteins in solution either through ice surface accumulation [1, 3, 15] of the proteins or inclusion within the growing ice crystals [3, 15–17]. Heterogeneous ice formation can also create processing issues during freeze-drying where the size and formation of the ice crystals will affect process time and product quality [18].

Vitrification is a nonequilibrium process and therefore does not really follow any path described by a conventional state diagram. During vitrification, the viscosity of the liquid increases very significantly, decreasing molecular mobility and therefore eliminating crystallization. Of course, this is true as long as the cooling (or drying) rate is sufficiently fast to bypass crystallization. Therefore, with sufficiently fast cooling, the eutectic transition might be eliminated and the FCL vitrifies (transitions into a glass) at the glass transition temperature of Tg'. Note that a glass is defined as the metastable state where the translational motions of the main constituent molecule (the α-relaxations) have slowed down to a stop practically for all experimental timescales [19–23]. However, even though the molecular motions are slowed down, a glass is still at a metastable state and therefore will continue to evolve toward thermodynamic equilibrium (to the crystalline state). The bottom line is that given sufficient time every glass will crystallize; the heterogeneity present in the FCL will be locked into the glass and will only increase with its transition into a crystal.

Since a glass is a noncrystalline substance without order, its degree of inherent heterogeneity depends on its water content [24]. Tg of a binary, fragile glass (e.g., trehalose, sucrose) decreases very rapidly with increasing water content (as opposed to a strong glass whose Tg does not change very dramatically with water content [25]). On the other hand, since vitrification is a kinetic process, it is possible to vitrify a binary solution of any composition if the cooling rate is sufficiently fast (proven experimentally and theoretically). Therefore, it is indeed possible to produce glasses of varying water content at the same temperature. Naturally, mobility of solutes and the solvent will be different in each glass due to the difference in the ultrastructure of the glass formed. However, all these concentrated solutions, for all practical purposes, will be called a glass, even though they are different from one another (another side effect of the "vitrified state" not being a thermodynamic state) [24].

Some of the systems may even phase separate (sometimes at the molecular level, in a process called demixing) at room temperature (even before the onset of freezing or drying) [4, 26]. This would aggravate the heterogeneity of the system during freezing. Another factor is the shift in the pH in buffer solutions, which have been shown to be very significant during freezing and drying [8]. The shift in pH is due to the precipitation of the lesser soluble buffer components during supersaturation that occurs with freezing and drying. Freezing-induced heterogeneity is also responsible for the increase in electric charge in the freeze concentrate due to the selective incorporation of ions into the ice crystal lattice [27, 28]. This selective incorporation may also cause shifts in pH.

The presence of a second solvent at high concentrations within the solution may create very different freezing behaviors (Figure 10.2), changing the specific surface area

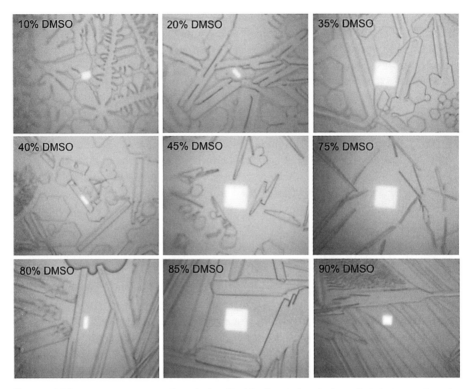

Figure 10.2. Frozen aqueous dimethyl sulfoxide (DMSO) solutions (w/w). Note that the eutectic concentration of aqueous DMSO is ~52% w/w. In solutions less than 50% DMSO w/w, ice crystals form whereas at concentrations higher than 70% DMSO w/w, DMSO crystals form. Large bright rectangles are 200 μm × 200 μm.

(and even the type) of the ice crystals and the characteristics of the FCL. This may generate different interactions of the product with the solution and the frozen crystals further damaging it.

10.3   DRYING-INDUCED HETEROGENEITY

For drying protein solutions, available methods are freeze-drying (lyophilization) [29, 30], foam vitrification, spray-drying [31, 32], and isothermal diffusive/convective drying [33, 34]. Depending on the specific application, all of these methods have major economical and process advantages that make them useful/efficient for a specific application. For example, one advantage of isothermal drying, with respect to lyophilization (especially at commercial production scales), is that the process does not require specialized cryogenic and low-pressure equipment or process, thus reducing processing costs. Spray-drying benefits from low processing cost, ease and speed of processing while it suffers from high surface area of the droplets that can cause denaturation or surface aggregation of certain proteins, and high temperatures that also can contribute to protein denaturation

Sessile droplet drying is of special interest due to its widespread use in deposition of thin films in surface patterning [35], bioreactive and biosensing coatings [36], ink-jet printing [37], protein microarray technologies [38–40], and, more recently, early diagnosis of disease in bodily fluids using proteinaceous biomarkers [41]. Desiccation of a solution in the form of a sessile droplet drying in contact with a surface is mainly affected by the environmental conditions (such as temperature and relative humidity), the chemical composition of the solution, and the interactions of the solution with the surface it is dried on [12]. Environmental conditions act on the surface of the solution, mainly affecting the evaporation rate of the solvent. Chemical composition of the solution dictates its physical, chemical, and rheological properties, the diffusivity of the solvent (and the co-solutes) in the solution, the solubility and crystallization rate of the co-solutes, and, ultimately, the final thermodynamic state after drying.

Hydrophilic surfaces induce contact line pinning for sessile droplets, causing ring-like deposits to form in the periphery of the droplet after drying [10, 12, 42] (Figure 10.3). One of the main factors contributing to spatial heterogeneity in a drying sessile droplet is the spatial variation of solvent evaporation flux on the droplet surface. The evaporation flux is lowest at the center of the droplet but increases toward the pinned periphery. This is associated with the higher probability of escape for the solvent molecules located at the periphery of the droplet [10]. Higher evaporation flux at the droplet periphery induces a radially outward flow within the desiccating droplet (called the peripheral flow) [10]. The peripheral flow continues as long as the driving force (i.e., the evaporation flux at the surface) is higher than the opposing force (the viscous drag on the solute particles). A secondary flow, called the Marangoni flow, is fueled by the temperature and solute concentration dependence of the surface tension [43] and is superimposed on the primary flow [11, 44]. Due to nonuniform evaporation flux on the sessile droplet surface, the temperature at the surface is lowest at the periphery and highest at the center. Due to peripheral flow, on the other hand, the solute concentration is higher at the periphery. These two factors interact to create Marangoni instability. In some cases, recirculation zones appear within the drying droplet due to the synergistic effects of the Marangoni flow with the peripheral flow [44, 45]. In other cases, the Marangoni flow may weaken the effects of the peripheral flow by carrying the solutes back to the center of the droplet.

These complex flow patterns combined with the specific attraction of certain proteins to interfaces and the resulting outcome (such as the interface-induced denaturation behavior of lysozyme [46], induced by the increased adsorption kinetics of these proteins at the air–liquid interface [47]) create very significant levels of heterogeneity where the protein is separated from the other constituents [12] in the solution (Figure 10.3). For example, as shown in Figure 10.3, there is a very significantly distinct difference between the distributions of trehalose and lysozyme within a desiccated sessile droplet; while the lysozyme distribution reaches a peak at the very edge of the droplet, the trehalose peak reaches its maximum earlier. This shows that at the very edge, where there is a considerable population of lysozyme, there is a very low trehalose to lysozyme ratio.

By manipulating either the magnitude of the evaporation flux (e.g., by controlling the environmental relative humidity) and/or its distribution on the droplet surface, the

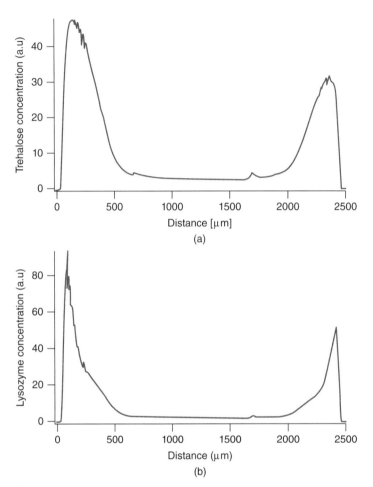

Figure 10.3. Distribution of (a) trehalose and (b) lysozyme along the centerline of a desiccated sessile droplet. Data collected at the InfraRed ENvironmental Imaging (IRENI) beamline in Synchrotron Resource Center, Stoughton WI. Pixel size: 0.54 μm.

resulting flow patterns within the droplet and hence the solute distribution in the dried product can be significantly altered. By preventing the peripheral and Marangoni flows, a more uniform product can be obtained. The heterogeneous distribution of the lyoprotectant with respect to the protein may also be reduced by using additives, which may increase the effectiveness of the lyoprotectant. For example, during drying of lysozyme, LiCl acts synergistically with trehalose to preserve the native structure of lysozyme in the dried droplet [12]. For a detailed analysis of drying-induced sessile droplet heterogeneity, and chemical/physical methods of minimizing this for increased process outcome and efficiency, see Ragoonanan and Aksan [12].

Desiccation kinetics of a glass-forming solution depend on the diffusivity of the solvent within the solution and in the surrounding environment [48]. In order to have a fairly uniform concentration gradient in the solution, the diffusion rate of the solvent within the solution should be comparable to that in the surrounding air [2]. With desiccation (and therefore increasing solute concentration) however, the diffusivity of the solvent in the solution decreases. For high glass transition temperature solutions, at the gas–liquid interface where evaporation takes place, a glassy skin forms slowing down (if not completely halting) the solvent mobility even further [2, 48]. The skin is of very small thickness and has been thought not to present a major barrier against desiccation. This may be true for binary solutions where the solvent molecular size is significantly smaller than the molecular free volume of the vitrified matrix (e.g., in high molecular weight aqueous dextran solutions). However, for smaller glass-forming solute molecules, it may represent a problem. The effects of the skin formation could be lessened by using small amounts (0.05–0.1 g/L) of plasticizers (such as glycerol) in the formulation [14, 49].

In formulation development for the desiccated/vitrified state preservation of biomolecules, the tendency of the biopreservation media to form a glassy skin at the interface causing water entrapment and nonuniform concentration gradients should be taken into consideration. *The solutions with high glass transition temperatures, inevitably, will yield to nonuniform drying resulting in very steep concentration gradients within the desiccated/vitrified product* [2].

10.4 METHODS OF DETECTION

Phase separation during desiccation, freezing [50], and lyophilization [51] processes has been previously observed and quantified in bulk solutions, mainly by thermal, kinetic, and calorimetric analyses of biopreservation solutions [52–55] and solutions containing proteins [13, 56–58]. The ultrastructure of the processed products was visualized by scanning electron microscopy (SEM) [26, 59] and analyzed by IR imaging [57] and spectroscopy of the bulk product [60]. However, to date, irrespective of the stabilization process applied, very little effort has been spent to determine how the protein populations are distributed among different thermodynamic regions within the same product and what role(s) the population distribution plays in determining the stabilization efficacy of the biopreservation solution or the process and the post-thaw activity of the bulk product.

Analysis of heterogeneity in frozen and dried media has been hindered due to the lack of technology to detect it at high spatial resolution and the lack of scientific evidence that linked the success/failure of a specific stabilization technique or protocol to the homogeneity of the product ultrastructure and the mechanistic information that can be used to engineer the ultrastructure [13]. Even though we are far from clearly exploring and understanding heterogeneity, recent advances in microscopy and spectroscopy started to open the door for analysis. In general, the main experimental methods that have been applied to detect heterogeneity and phase separation are conventional microscopy (transmitted light and fluorescence microscopy), electron microscopy (scanning/transmission electron microscopy (SEM/TEM)), thermal analysis (differential scanning calorimetry (DSC)), and spectroscopy (mainly, Fourier transform infrared

(FTIR) and Raman spectroscopy). Numerical methods are not within the scope of this review.

10.4.1 Conventional Microscopy

Liquid–liquid phase separation can be examined by traditional transmitted light microscopy analysis if the separation of the liquids generates sufficient optical contrast [61]. Similarly, fluorescent probes that respond to the changes in local viscosity or pH can be used to explore the effects of phase separation, crystallization, and skin formation [2]. For a detailed analysis of skin formation and desiccation-induced heterogeneity in glass-forming carbohydrate solutions, please see Aksan et al. [2].

10.4.2 Electron Microscopy

Uchida et al. have perfected TEM imaging of frozen disaccharide solutions by resolving features smaller than 10 nm in the FCL region [62]. Combining TEM observations with powder X-ray diffraction measurements, they identified trehalose dihydrate crystal precipitates dispersed within the trehalose glass in the FCL region [63]. Using atomic force microscopy, the presence of the crystal precipitates in the amorphous region was confirmed by Wu et al. [64]. Experimental evidence shows the nonuniform distribution of sugars and water within an aqueous solution (at the molecular level) to be present even at room temperature. This may be one of the factors that contribute to the onset of crystallization of trehalose in the FCL [5, 65, 66]. SEM photographs of lyophilized polyethylene glycol (PEG)/dextran solutions have also shown phase separation [26]. Note that however, the SEM/TEM-based characterization techniques only provide qualitative visual clues and require interpretation and may sometimes be prone to sample preparation artifacts.

10.4.3 Thermal Analysis

DSC-based techniques are in principle used to measure heat absorbed/rejected by a small specimen (1–20 mg) while it experiences a defined thermal history. The most commonly used protocols involve a combination of cooling and heating of the specimen from room temperature at a constant rate of 1–10 K/min. Heating analysis is usually conducted to determine (1) the melting temperatures of crystals or frozen solutions, (2) the denaturation temperatures and denaturation enthalpies (using the area under the curve of the endotherm) of proteins in different solutions (to determine their stabilization effect), (3) the water content of the sample (by heating the sample above 100°C, converting the liquid water in the sample to vapor, which escapes from the specimen pan through a small hole drilled at the top of the pan), and (4) recrystallization and melting of polymers and sugars.

Cooling analysis is generally used (1) to determine the freezing temperature of the solution (peak temperature of the exotherm) and to measure the latent heat release (using the area under the exotherm), (2) to determine the glass transition temperature (Tg) of the solution (from a baseline shift), and (3) for recrystallization of the solution.

The advantages of DSC are that it requires very small amount of sample (approximately milligrams), minimum sample preparation, and ease of operation over a large temperature range (generally, -180 to $200°C$) and that highly accurate thermal information can be obtained with minimum data processing. The disadvantage is the need to evaluate the signal and interpret the different reasons for the specific thermal profile obtained. Moreover, the method does not afford visual access and collects data from the bulk sample without any spatial information. Generally, the presence of multiple freezing exotherms or baseline shifts during cooling has been attributed to phase separation and heterogeneity. Similarly, multiple reverse shifts in baseline, multiple endotherms or even broadening of the endotherms have been attributed to the existent heterogeneity within the sample.

DSC has been used to detect freezing/thawing and vitrification of bulk solutions. The presence of multiple peaks has been first attributed to the sequential transitions of different regions within the same specimen by Her et al. [67]. They developed a methodology based on varying the constituents in a solution (such as salts and polymers) to determine the change in the Tg' of sugar solutions, which was then attributed to the changes in the water content of the FCL and liquid–liquid phase separation [67, 68]. This methodology has later been used to determine the presence of phase separation in freeze-concentrated protein–polymer and polymer–polymer systems [50, 69, 70].

10.4.4 Spectroscopy

By FTIR examination of the lyophilized bulk samples of a mixture of hemoglobin, PEG, and dextran, Heller et al. [60] concluded that polymer phase separation was responsible for the deterioration of hemoglobin secondary structure during processing. They also showed that PEG–dextran phase separation could be eliminated by changing the salt species in the formulation from NaCl to KCl. FTIR spectroscopy using focal plane array detectors (that enable detection of spatial variation at a resolution of 10–40 μm) was utilized to examine the heterogeneity of the protein–sugar solutions after lyophilization [57]. The information that can be easily obtained from FTIR spectroscopy is as follows: Water content can be determined from the $(\nu 2 + \nu 3)$ bands of water in the NIR range of ~ 4500–5500 cm^{-1} [5]; the amide II band can be used to determine the protein content and examine the changes in the secondary structures of proteins in the range 1400–1550 cm^{-1}; and the sugar content can be determined from the characteristic peaks in the fingerprint region in the range 400–1200 cm^{-1}. Some groups also use amide I band (1550–1700 cm^{-1}) to determine the changes in protein secondary structure. However, this needs to be done very carefully since the water bending band (δ-OH) contributes very significantly to the amide I band and in formulations with significantly high water content can easily overlap with the protein signal.

Especially in dry formulations, the ease of use of spectroscopic techniques with minimum sample preparation has resulted in its rapid adoption in drug processing and food sciences [71, 72]. However, examination of liquid and frozen formulations remains a challenge. This has limited the research very significantly to a handful of papers [3]. Our group was the first to apply high spatial resolution confocal Raman spectroscopy to detect heterogeneity in frozen protein formulations [3] and developed spectral analysis tools to quantify the extent of protein accumulation in the ice phase (with respect to FCL) and identify the changes in the secondary structures of frozen proteins. Using these techniques, in similar studies (unpublished data), we have detected that in specific freezing conditions accumulation of albumin on ice surfaces can reach as high as 18–30% of the concentration of albumin in the solution. The structure of albumin is known to be very sensitive to ice interactions, and this would cause significant damage to the albumin accumulated on the ice surface. More recently, Padilla et al. have started to utilize Raman spectroscopy to detect phase separation in freeze-dried model polymer systems composed of polyvinylpyrrolidone (PVD), dextran, and different salts [73–75].

Using confocal Raman spectroscopy, our group has also characterized the heterogeneity in frozen mammalian cells [76] and reported for the first time, that against the predictions of the current paradigm, ice forms in frozen cells irrespective of the cooling rate induced. However, we have detected very significant differences in the distribution of the ice phase within and around the frozen cells and varying degree of heterogeneity depending on the cooling rates applied. This suggests that heterogeneity induced during freezing also has very significant effects on cryopreservation of mammalian cells. The other advantage of the spectroscopic techniques is that Raman and IR spectroscopy can also be used to determine the extent of hydrogen bonding between the lyoprotectant and other components in the system [77,78], providing valuable information at the molecular level.

10.5 SUMMARY

Freezing and drying are processes that inherently induce heterogeneity in the product. However, the impact of microscopic heterogeneity on product quality loss, storage lifetime, and so on has not been quantified yet especially due to unavailability of the techniques to detect heterogeneity in frozen and desiccated formulations at high spatial resolution. We outline the causes of heterogeneity and summarize the techniques that are available to determine the extent of heterogeneity in the product. Once the degree of freezing- and desiccation-induced heterogeneity in products of industrial and commercial value is established, we believe, the demand for the development of techniques to minimize heterogeneity (and they do exist) will increase exponentially.

ACKNOWLEDGMENTS

This research was supported by NSF (CBET-0644784) and McKnight Land-Grant Professorship to AA.

REFERENCES

1. Butler, M.F. (2002) Freeze concentration of solutes at the ice/solution interface studied by optical interferometry. *Cryst. Growth Des.*, **2**(6), 541–548.
2. Aksan, A., et al. (2006) Desiccation kinetics of biopreservation solutions in microchannels. *J. Appl. Phys.*, **99**, 064703.

3. Dong, J., et al. (2009) Freezing-induced phase separation and spatial microheterogeneity in protein solutions. *J. Phys. Chem. B*, **113**(30), 10081–10087.

4. Zaslavsky, B.Y. (1995) *Aqueous Two-Phase Partitioning: Physical Chemistry and Bioanalytical Applications*. Marcel Dekker, New York.

5. Malsam, J. and Aksan, A. (2009) Hydrogen bonding kinetics of water in high concentration trehalose solutions at cryogenic temperatures. *J. Phys. Chem. B*, **113**(19), 6792–6799.

6. Malsam, J. and Aksan, A. (2010) Hydrogen bonding and compartmentalization of water in supercooled and frozen aqueous acetone solutions. *J. Phys. Chem. B*, **114**(12), 4238–4245.

7. Webb, S.D., et al. (2002) Surface adsorption of recombinant human interferon-gamma in lyophilized and spray-lyophilized formulations. *J. Pharm. Sci.*, **91**(6), 1474–1487.

8. Sundaramurthi, P., Shalaev, E., and Suryanarayanan, R. (2010) Calorimetric and diffractometric evidence for the sequential crystallization of buffer components and the consequential pH swing in frozen solutions. *J. Phys. Chem. B*, **114**(14), 4915–4923.

9. Deegan, R.D., et al. (1997) Capillary flow as the cause of ring stains from dried liquid drops. *Nature*, **389**, 827–839.

10. Deegan, R.D., et al. (2000) Contact line deposits in an evaporating drop. *Phys. Rev. E*, **62**(1), 756–765.

11. Ha, V.M. and Lai, C.L. (2002) Onset of Marangoni instability of a two-component evaporating droplet. *Int. J. Heat Mass Transf.*, **45**(26), 5143–5158.

12. Ragoonanan, V. and Aksan, A. (2008) Heterogeneity in desiccated solutions: implications for biostabilization. *Biophys. J.*, **94**(6), 2212–2227.

13. Randolph, T.W. (1997) Phase separation of excipients during lyophilization: effects on protein stability. *J. Pharm. Sci.*, **86**(11), 1198–1203.

14. Cicerone, M.T., et al. (2003) Substantially improved stability of biological agents in dried form. *Bioprocess Int.*, **1**(1), 36–47.

15. Schwegman, J.J., Carpenter, J.F., and Nail, S.L. (2009) Evidence of partial unfolding of proteins at the ice/freeze-concentrate interface by infrared microscopy. *J. Pharm. Sci.*, **98**(9), 3239–3246.

16. Strambini, G.B. and Gonnelli, M. (2007) Protein stability in ice. *Biophys. J.*, **92**, 2131–2138.

17. Goff, H.D., Verespej, E., and Jermann, D. (2003) Glass transitions in frozen sucrose solutions are influenced by solute inclusions within ice crystals. *Thermochim. Acta*, **399**(1–2), 43–55.

18. Searles, J.A., Carpenter, J.F., and Randolph, T.W. (2001) Annealing to optimize the primary drying rate, reduce freezing-induced drying rate heterogeneity, and determine Tg′ in pharmaceutical lyophilization. *J. Pharm. Sci.*, **90**(7), 872–887.

19. Green, J.L. and Angell, C.A. (1989) Phase relations and vitrification in saccharide-water solutions and the trehalose anomaly. *J. Phys. Chem.*, **93**, 2880–2882.

20. Aksan, A. and Toner, M. (2004) Isothermal desiccation and vitrification kinetics of trehalose-dextran solutions. *Langmuir*, **20**, 5521–5529.

21. Aksan, A. and Toner, M. (2006) Role of thermodynamic state and molecular mobility in biopreservation. In: *Tissue Engineering and Artificial Organs*. Bronzino, J.D., editor. Taylor & Francis, Boca Raton, FL. pp. 41.1–41.20.

22. Gotze, W. (1999) Recent tests of the mode coupling theory for glassy dynamics. *J. Phys. Condens. Matter*, **11**, A1–A45.

23. Conrad, P.B. and de Pablo, J.J. (1999) Computer simulation of cryoprotectant disaccharide a,a-trehalose in aqueous solution. *J. Phys. Chem. A*, **103**, 4049–4055.

24. Le, L. and Molinero, V. (2010) Nanophase segregation in supercooled aqueous solutions and their glasses driven by the polyamorphism of water. *J. Phys. Chem. A*, **115**(23), 5900–5907.

25. Martinez, L.M. and Angell, C.A. (2001) A thermodynamic connection to the fragility of glass-forming liquids. *Nature*, **410**(6829), 663–667.

26. Heller, M.C., Carpenter, J.F., and Randolph, T.W. (1997) Manipulation of lyophilization-induced phase separation: implications for pharmaceutical proteins. *Biotechnol. Progr.*, **13**(5), 590–596.

27. Cheng, J., et al. (2009) Confocal fluorescence microscopy of the morphology and composition of interstitial fluids in freezing electrolyte solutions. *J. Phys. Chem. Lett.*, **1**(1), 374–378.

28. Wilson, P.W. and Haymet, A.D.J. (2008) Workman–Reynolds freezing potential measurements between ice and dilute salt solutions for single ice crystal faces. *J. Phys. Chem. B*, **112**(37), 11750–11755.

29. Carpenter, J.F. and Chang, B.S. (1996) Lyophilization of protein pharmaceuticals. In: *Biotechnology and Biopharmaceutical Manufacturing, Processing and Preservation*. Avis, K. and Wu, V., editors. Intep0harm Press, Buffalo Grove, IN.

30. Rey, L. and May, J.C., editors (1999) Freeze-drying/lyophilization of pharmaceutical and biological products. In: *Drugs and the Pharmaceutical Sciences*. Swarbrick, J. editor. Marcel Dekker, New York.

31. Webb, S.J. (1965) *Bound Water in Biological Activity*. Charles C. Thomas, Springfield, IL.

32. Suihko, E.J., Forbes, R.T., and Apperley, D.C. (2005) A solid-state NMR study of molecular mobility and phase separation in co-spray-dried protein-sugar particles. *Eur. J. Pharm. Sci.*, **25**, 105–112.

33. Norris, M.M., et al. (2006) 3-O-Methyl-D-glucose improves desiccation tolerance of keratinocytes. *Tissue Eng.*, **12**(7), 1–7.

34. Bhowmick, S., et al. (2003) Desiccation tolerance of spermatozoa at ambient temperature: Production of fetal mice. *Biol. Reprod.*, **68**(5), 1779–1786.

35. de Gans, B.-J. and Schubert, U.S. (2004) Inkjet printing of well-defined polymer dots and arrays. *Langmuir*, **20**, 7789–7793.

36. Flickinger, M.C., et al. (2007) Painting and printing living bacteria: engineering nanoporous biocatalytic coatings to preserve microbial viability and intensify reactivity. *Biotechnol. Progr.*, **23**(1), 2–17.

37. Park, J. and Moon, J. (2006) Control of colloidal particle deposition patterns within picoliter droplets ejected by ink-jet printing. *Langmuir*, **22**, 3506–3513.

38. Dugas, V., Broutin, J., and Souteyrand, E. (2005) Droplet evaporation study applied to DNA chip manufacturing. *Langmuir*, **21**, 9130–9136.

39. Blossey, R. and Bosio, A. (2002) Contact line deposits on cDNA microarrays: a "twin spot effect". *Langmuir*, **18**, 2952–2954.

40. Templin, M.F., et al. (2002) Protein microarray technology. *Trends Biotechnol.*, **20**(4), 160–166.

41. Tarasevich, Y.Y. and Pravoslavnova, D.M. (2007) Segregation in desiccated sessile drops of biological fluids. *Eur. Phys. J. E*, **22**, 311–314.

42. Caddock, B.D. and Hull, D. (2002) Influence of humidity on the cracking patterns formed during the drying of sol-gel drops. *J. Mater. Sci.*, **37**, 825–834.

43. Kaminskii, V.A. and Dil'man, V.V. (2003) Marangoni instability in evaporation of binary mixtures. *Theor. Found. Chem. Eng.*, **37**(6), 533–538.

44. Hu, H. and Larson, R.G. (2006) Marangoni effect reverses coffee-ring depositions. *J. Phys. Chem. B*, **110**, 7090–7094.

45. Hu, H. and Larson, R.G. (2005) Analysis of the microfluidic flow in an evaporating sessile droplet. *Langmuir*, **21**, 3963–3971.

46. Lu, J.R., et al. (1998) Structural conformation of lysozyme layers at the air/water interface studied by neutron reflection. *J. Chem. Soc. Faraday Trans.*, **94**, 3279–3287.

47. Guzey, D., McClements, D.J., and Weiss, J. (2003) Adsorption kinetics of BSA at air-sugar solution interfaces as affected by sugar type an concentration. *Food Res. Int.*, **36**, 649–660.

48. Aksan, A., Morris, S.C., and Toner, M. (2005) Analysis of desiccation and vitrification characteristics of carbohydrate films by shear wave resonators. *Langmuir*, **21**(7), 2847–2854.

49. Cicerone, M.T. and Soles, C.L. (2004) Fast dynamics and stabilization of proteins: binary glasses of trehalose and glycerol. *Biophys. J.*, **86**(6), 3836–3845.

50. Izutsu, K. and Kojima, S. (2000) Freeze-concentration separates proteins and polymer excipients into different amorphous phases. *Pharm. Res.*, **17**(10), 1316–1322.

51. Heller, M.C., Carpenter, J.F., and Randolph, T.W. (1999) Protein formulation and lyophilization cycle design: prevention of damage due to freeze-concentration induced phase separation. *Biotechnol. Bioeng.*, **63**(2), 166–174.

52. Suzuki, T. and Franks, F. (1993) Solid-liquid phase transitions and amorphous states in ternary sucrose-glycine-water systems. *J. Chem. Soc. Faraday Trans.*, **89**(17), 3283–3288.

53. Shalaev, E.Y. and Kanev, A.N. (1994) Study of the solid-liquid state diagram of the water-glycine-sucrose system. *Cryobiology*, **31**(4), 374–382.

54. Shalaev, E.Y. and Franks, F. (1996) Changes in the physical state of model mixtures during freezing and drying: impact on product quality. *Cryobiology*, **33**, 11–26.

55. Murase, N. and Franks, F. (1989) Salt precipitation during the freeze-concentration of phosphate buffer solutions. *Biophys. Chem.*, **34**, 293–300.

56. Heller, M.C., Carpenter, J.F., and Randolph, T.W. (1999) Application of a thermodynamic model to the prediction of phase separations in freeze-concentrated formulations for protein lyophilization. *Arch. Biochem. Biophys.*, **363**(2), 191–201.

57. Jovanovic, N., et al. (2006) Near-infrared imaging for studying homogeneity of protein-sugar mixtures. *Pharm. Res.*, **23**(9), 2002–2013.

58. Kolhe, P. and Badkar, A. (2011) Protein and solute distribution in drug substance containers during frozen storage and post-thawing: a tool to understand and define freezing–thawing parameters in biotechnology process development. *Biotechnol. Progr.*, **27**(2), 494–504.

59. Murase, N., Echlin, P. and Franks, F. (1991) The structural states of freeze-concentrated and freeze-dried phosphates studied by scanning electron microscopy and differential scanning calorimetry. *Cryobiology*, **28**, 364–375.

60. Heller, M.C., Carpenter, J.F., and Randolph, T.W. (1996) Effects of phase separating systems on lyophilized hemoglobin. *J. Pharm. Sci.*, **85**(12), 1358–1362.

61. Choi, J.H. and Bischof, J.C. (2008) A quantitative analysis on the thermal properties of phosphate buffered saline with glycerol at subzero temperatures. *Int. J. Heat Mass Transf.*, **51**(3–4), 640–649.

62. Uchida, T., et al. (2007) Morphological investigations of disaccharide molecules for growth inhibition of ice crystals. *J. Cryst. Growth*, **299**(1), 125–135.

63. Uchida, T. and Takeya, S. (2010) Powder X-ray diffraction observations of ice crystals formed from disaccharide solutions. *Phys. Chem. Chem. Phys.*, **12**(45), 15034–15039.

64. Wu, J.J., Reading, M., and Craig, D.Q.A. (2008) Application of calorimetry, sub-ambient atomic force microscopy and dynamic mechanical analysis to the study of frozen aqueous trehalose solutions. *Pharm. Res.*, **25**(6), 1396–1404.

65. Sidebottom, D.L. and Tran, T.D. (2010) Universal patterns of equilibrium cluster growth in aqueous sugars observed by dynamic light scattering. *Phys. Rev. E*, **82**(5).

66. Lerbret, A., et al. (2005) How homogeneous are the trehalose, maltose, and sucrose water solutions? An insight from molecular dynamics simulations. *J. Phys. Chem. B*, **109**(21), 11046–11057.

67. Her, L.-M., Deras, M., and Nail, S.L. (1995) Electrolyte-induced changes in glass transition temperatures of freeze-concentrated solutes. *Pharm. Res.*, **12**(5), 768–772.

68. Izutsu, K., et al. (1996) Effects of sugars and polymers on crystallization of poly(ethylene glycol) in frozen solutions: phase separation between incompatible polymers. *Pharm. Res.*, **13**(9), 1393–1400.

69. Izutsu, K., et al. (1998) Effect of salts and sugars on phase separation of polyvinylpyrrolidone-dextran solutions induced by freeze-concentration. *J. Chem. Soc. Faraday Trans.*, **94**(3), 411–417.

70. Izutsu, K., et al. (2010) Effects of solute miscibility on the micro- and macroscopic structural integrity of freeze-dried solids. *J. Pharm. Sci.*, **99**(11), 4710–4719.

71. Amigo, J.M. (2010) Practical issues of hyperspectral imaging analysis of solid dosage forms. *Anal. Bioanal. Chem.*, **398**(1), 93–109.

72. El-Hagrasy, A.S., et al. (2001) Near-infrared spectroscopy and imaging for the monitoring of powder blend homogeneity. *J. Pharm. Sci.*, **90**(9), 1298–1307.

73. Padilla, A.M., et al. (2011) The study of amorphous phase separation in a model polymer phase-separating system using Raman microscopy and a low-temperature stage: effect of cooling rate and nucleation temperature. *J. Pharm. Sci.*, **100**(4), 1362–1376.

74. Padilla, A.M., et al. (2011) The study of phase separation in amorphous freeze-dried systems. Part I: Raman mapping and computational analysis of XRPD data in model polymer systems. *J. Pharm. Sci.*, **100**(1), 206–222.

75. Padilla, A.M. and Pikal, M.J. (2011) The study of phase separation in amorphous freeze-dried systems, part 2: investigation of Raman mapping as a tool for studying amorphous phase separation in freeze-dried protein formulations. *J. Pharm. Sci.*, **100**(4), 1467–1474.

76. Dong, J., et al. (2010) Spatial distribution of the state of water in a frozen mammalian cell. *Biophys. J.*, **99**(8), 2453–2459.

77. Imamura, K., et al. (2008) Temperature scanning FTIR analysis of interactions between sugar and polymer additive in amorphous sugar–polymer mixtures. *J. Pharm. Sci.*, **97**(1), 519–528.

78. Taylor, L.S. and Zografi, G. (1998) Sugar–polymer hydrogen bond interactions in lyophilized amorphous mixtures. *J. Pharm. Sci.*, **87**(12), 1615–1621.

SECTION 3

PHASE III AND COMMERCIAL DEVELOPMENT

11

LATE-STAGE PRODUCT CHARACTERIZATION: APPLICATIONS IN FORMULATION, PROCESS, AND MANUFACTURING DEVELOPMENT

Christine P. Chan[1] and Li Shi[2]

[1] *Biologics R&D, Genzyme—A Sanofi Company, Boston, MA, USA*
[2] *Shanghai Zerun Biotechnology Co., Ltd., Member of Wison Group, Shanghai, China*

Biophysical Methods for Biotherapeutics: Discovery and Development Applications, First Edition. Edited by Tapan K. Das.
© 2014 John Wiley & Sons, Inc. Published 2014 by John Wiley & Sons, Inc.

11.1 INTRODUCTION

As a product development program progresses to phase III clinical stage, the process development efforts converge and focus on ensuring long-term robustness of the process and high-quality commercial manufacturing operations. An important component in these coordinated activities is to define an optimized formulation for maintenance of drug quality and stability during manufacturing, storage through the end of shelf life, and effective clinical administration. In addition, detailed product characterization data are collected and collated in conjunction with process studies to define the critical quality attributes (CQAs) and critical process parameters (CPPs) as well as establishment of their acceptable ranges. This information is part of the knowledge base crucial in process validation and development of overall control strategies throughout the product life cycle [1].

This chapter summarizes the key considerations in applying biophysical characterization tools on protein therapeutics during phase III/commercial process development. Finalization of the dosage form is discussed in the context of the desired quality target product profile (qTPP) based on disease indication and product supply chain strategy. The importance of gathering product characterization information during different phases of development, including formulation screening, short-term accelerated stability studies, forced degradation studies, real-time stability studies of clinical trial materials, as well as in-process findings during upstream/downstream scale-up studies, is highlighted. Potential issues in protein stability during bioprocess operations are discussed along with the utility of various analytical methods. In most cases, the assay technologies employed during product development are not stage specific, and the detailed principles of individual biophysical methods have been reviewed in earlier chapters in this book. Additional emerging analytical applications such as newer methodologies with mass spectrometry, microscopy, and flow cytometry (FC) for protein aggregation and subvisible particulate characterization are also discussed. The emphasis is on the use of orthogonal methods to derive a comprehensive understanding of product stability and potential issues as the bioprocess operation is finalized.

11.2 STRATEGIES IN USING BIOPHYSICAL METHODS IN LATE-STAGE DEVELOPMENT

11.2.1 Progression from Early- to Late-Stage Development

Biotherapeutic products are large complex molecules often with multiple structural domains contributing to different biological functions for drug efficacy. The broad use of recombinant DNA technology has enabled design and production of a diverse range of protein structures with different functions for the treatment of a variety of disease states. There are over a hundred approved protein therapeutic products on the market and hundreds more in various stages of preclinical and clinical development for many currently unmet medical needs [2].

In most cases, the desired biological and pharmaceutical properties of the product candidates are drafted as part of the TPP at the molecular design and discovery phase of

TABLE 11.1. Quality target product profile examples

Product attribute	A monoclonal antibody product[a]	An enzyme replacement product[b]
Dosage form	Liquid, single use	Lyophilized, single use
Protein content per vial	500 mg	200 or 400 units
Dose	10 mg/kg	2.5–60 U/kg, individualized
Concentration	25 mg/mL	40 U/mL after reconstitution
Mode of administration	IV, diluted with isotonic saline or dextrose	IV, diluted with isotonic saline
Container	20R type 1 borosilicate glass vials, fluoro-resin laminated stopper	Type 1 borosilicate glass vials, siliconized butyl rubber stopper
Shelf life	>24 months at 2–8°C	24 months at 2–8°C
Viscosity	Acceptable for manufacturing, storage, and delivery without the use of special devices (e.g., less than 10 cP at room temperature)	
Compatibility with mfg process	Minimum 14 days at 25°C and subsequent 2 years at 2–8°C, soluble at higher concentration during UF/DF filtration	
Degradants and impurities	Below safety threshold or qualified	
Aggregates	<5%	
Afucosylated glycan	1–10%	
Host cell proteins	<100 ppm (ng/mg)	

Blank/no entry: information not disclosed.
[a]A-mab, from Reference 1.
[b]Cerezyme, from public prescribing information.

development. This description encompasses quality attributes related to purity, integrity, safety, potency, as well as strength and dosing configuration (see Table 11.1 for examples of qTPP) [3]. A collection of characterization assays covering biochemical, biophysical, and biological properties specific for the protein product is thus developed and refined based on prior knowledge. As process development progresses along with clinical development, relevant information on the molecular heterogeneity, formulation, and stability conditions as well as degradation pathways is assembled. Coupled with knowledge from the literature, clinical experience, and manufacturing capabilities assessment, the TPP may be dynamically evolved through late-stage development.

Drug development is a high-cost venture, with relatively low probability of a new pre-IND candidate becoming a commercial product. There are divergent approaches (e.g., front-load, tiered/mixed, back-load) of allocating analytical and formulation development resources [4] for generation of relevant information at appropriate stages to support the most critical needs of individual programs.

Early phase program resources are usually focused on initial clinical assessment of safety and biological activity. Molecular characterization efforts may be minimal, leveraging prior knowledge based on the type of protein domain structure. Formulation development activities tend to be simplified due to knowledge, time, and drug material limitations (see Table 11.2 and below for discussion on stages and respective objectives in formulation development). Nevertheless, the value of defining solution property limits for optimal process recovery is well recognized. Basic information on

TABLE 11.2. Key stages and respective objectives in formulation development

Key stages	Objectives/studies
Preformulation	• Define inherent physicochemical properties of the protein: size, pI, charge variants, major glycan variants • Determine solution property limits, aggregation/oxidation/deamidation at different pH, ionic strengths • Basic stress studies: 25°C/40°C compare to 4°C, cycles of freeze/thaw, shaking, membrane filtration/concentration, chemical oxidation, photo-oxidation • Define core assays for product, investigate any key variants associated with different instability profiles
Formulation design	• Evaluate formulation elements (product concentration, pH, buffer, salts, excipients) on key CQAs in screening DOE to scope the ranges, define main effects • Consider liquid dosage form versus solid • Define excipients • Define stability-indicating assays
Process development	• Given definition of process steps, use scale-down models to investigate effect of processing steps on product stability, use worst-case scenario for failure limits • Verify ranges of excipients
Long-term stability studies and administration protocol	• Conduct real-time stability studies with multiple lots from different stages of process development • Define intended container/closure system, verify compatibility of protein • Define administration protocol and verify stability during administration, compatibility of protein with administration set • Optimized DOE studies to finalize formulation component ranges, define raw material sources and specifications where necessary • Forced degradation study with final formulation configuration using material from final process • Estimate product shelf life
Drug substance/drug product attributes	• Define drug substance stability (hold time) and drug product manufacturing steps • Define drug product configuration, storage conditions, and clinical administration protocol • Define product shelf life and end-of-expiry specifications for CQAs

protein solubility as a function of pH and ionic strength as well as temperature ranges can be obtained employing high-throughput screening formats [5, 6] and using milligram amounts of material. Common analytical tools include various spectroscopic methods (UV absorbance, circular dichroism (CD), intrinsic/extrinsic fluorescence, Fourier transform infrared (FTIR), Raman), calorimetry, light scattering and particle counting, as well as electrophoresis and size-exclusion chromatography. Aggregation of varying sizes, apparent melting point, and significant conformational changes in the overall structure and/or molecular population can be assessed in formulation screens. A rationally designed early formulation consists of (i) an appropriate pH and buffering agent that minimizes various possible degradation mechanisms including deamidation, oxidation, hydrolysis, and aggregation; (ii) salt to control ionic strength; as well as (iii) various excipients (stabilizers, surfactants), which aim to maintain an adequate drug stability profile through the long duration of phase I and II clinical trials. A lyophilized dosage form may be selected over liquid form for early phase studies due to lower risk of stability issues. Quite often, formulation optimization and final dosage form definition are delayed until the later stages, after a safe and efficacious dose, route of administration, and likelihood of success of the product are better defined.

An essential design goal of commercial formulation is optimal delivery of the drug substance in the clinical setting. The final dosage form thus needs to be appropriately designed based on the intended route of administration, which in turn varies depending on the disease application and ease-of-use by the patients. Oncology applications of monoclonal antibodies frequently require high-dose administration as well as extended treatment cycles for maintaining adequate antitumor efficacy [7]. This often necessitates intravenous administration for higher volume tolerance or high-concentration dosage by subcutaneous administration for less invasive delivery. In early phase development, the various parenteral routes of administration (i.e., intravenous, intramuscular, and subcutaneous) with different injection volume limits are often tested in conjunction with the dose-ranging studies. Since ease of use by patients may translate into long-term compliance, more sophisticated dosage configurations such as prefilled syringe, autoinjectors and infusion devices, polymeric controlled-release formulations, as well as other modes of delivery, such as intranasal, transdermal, and transmucosal, may also be evaluated [8]. Overall, based on optimal long-term stability and efficacious clinical delivery considerations, the commercial formulation is locked into a composition and configuration.

The ultimate goal of commercial formulation development for a given program is to define an optimal formulation for the bulk material that confers long-term product stability as well as appropriate product shelf life under robust manufacturing. Thorough evaluation and trend analysis of historical drug product stability data along with characterization information collected throughout product development are important. In addition, a sound understanding of the degradation profiles of the product needs to be developed. Forced degradation studies where pH, heat, light, chemical oxidants, freeze/thaw, and mechanical stress are applied in an incremental manner are particularly useful [9]. Since the purified bulk drug substance undergoes additional unit operations in manufacturing to generate the final drug product, the scope and ranges of these stress studies also require definition of the relevant process equipment configurations and capabilities. In conjunction, the comprehensive set of orthogonal biochemical and biophysical assays [10] established through development can be further assessed for their relative utility in detecting specific molecular or functional details. Additional biophysical methods including analytical ultracentrifugation, field flow fractionation, atomic force microscopy (AFM), as well as various submicron and subvisible particulate analytical tools often provide further details of aggregation pathways and molecular stability [11]. Adding complexity to this effort is the variation of drug product microheterogeneity from one lot to another through the course of process development. In general, the upstream and downstream process improvement and scale-up efforts may result in altered posttranslation modifications as well as impurity profiles of the product preparations [12]. Since these variations may contribute to differences in structural stability as well as functional relevance, detailed comparability studies are necessary to bridge and compare different clinical materials. In this iterative process, further insight into the molecular structure and functional characteristics of the product can be gained as the assays are further refined and optimized. The increased availability of drug substance in late-stage development facilitates various assay qualification and validation efforts to define assay operating characteristics and specifications. In most cases, biophysical assays are qualitative or semiquantitative, and the discriminating capability of a method for particular structural attributes of a given protein needs to be established. Sometimes these efforts may not be feasible in early phase. By late phase, through efforts of process development and execution of coordinated forced degradation studies, multiple batches of materials with differing product qualities are available. This allows more systematic and rigorous assessment of the applicability of the assay methodologies. Optimization of the formulation components can be conducted using a design of experiment (DOE) approach [1, 13]. In these multivariate studies, the effects of a range of product concentration, pH, and excipients are evaluated for their independent effects as well as interactions on various known CQAs based on knowledge gained from earlier studies. In addition to discriminatory potency assays, detailed molecular characterization of altered product population distribution generated in these studies using nondestructive biophysical methods are integral to the overall data interpretations. This type of extended characterization data set allows further understanding of the robustness of the process and supports the setting of shelf life specifications.

11.2.2 Protein Instability and Process/Manufacturing Unit Operations

Protein instability can result from chemical and/or physical perturbations on the structure or conformation. Chemical instabilities involve covalent modifications on the polypeptide chain and/or the glycan components on the glycoprotein structure. These reactions include oxidation, deamidation and isomerization, glycation and altered glycosylation, fragmentation and subunit dissociation, as well as rearrangement of disulfide linkages. Physical instabilities may involve conformational, colloidal, and interfacial mechanisms that affect the structure of the individual molecules as well as the population interactions. These changes alone or in concert can lead to unfolding and denaturation, surface adsorption, aggregation, and precipitation of the biotherapeutic, which in turn may cause loss of activity and efficacy as well as potentially increase the risk of unwanted immunogenicity.

It is well recognized that proteins are only marginally stable with relatively small free energy differences between the folded and partially unfolded states [14]. During commercial manufacturing, the protein is subjected to different environmental stresses that may disrupt the delicate balance of the multiple forces and interactions that stabilize the native (or the optimally active) structure. In addition, chemical modifications may be triggered independently or exacerbated due to unfolding and exposure of reactive functional groups. Likewise, chemical degradation can lead to changes in higher order structures resulting in aggregation. Overall, the different chemical and physical instabilities are often found to be interrelated.

Aggregation is a key issue for biotherapeutics process development as this attribute confers immunogenicity risk which is difficult to assess and often remains unknown. Protein aggregation is widely believed to be initiated from specific association of partially denatured product monomers, progressing to larger oligomeric structures often through hydrophobic interactions. In addition, many proteins exhibit self-association which may or may not be accompanied by conformational changes depending on experimental conditions [15]. Within a given sample, there may be a wide range of aggregation size spanning up to six orders of magnitude, from nanometer to millimeter in diameter. Since no single analytical method could cover this broad range, it is necessary to use a combination of multiple techniques which in turn are based on different physical principles for detection (see Table 11.3 and below). For soluble and insoluble aggregates in the submicron range, methods such as SEC, AUC, FFF, and light scattering are used in conjunction for characterization. In addition, particle analysis techniques such as light obscuration (LO), nanoparticle tracking analysis (NTA), micro-flow imaging, Coulter counter, and microscopy of different modes are used to provide size, count, and morphology information for aggregate sizes from 0.1 μm subvisible range to millimeter visible range. The structural perturbations in the aggregated moieties are also diverse. FTIR, Raman, UV-vis, fluorescence spectroscopies, and hydrophobic dye-binding methods are often used to probe secondary/tertiary structures and exposure of hydrophobic patches. Simple centrifugal separation or filtration capture allows analysis of the insoluble fraction as compared to the soluble aggregates in the supernatant for discernable conformational as well as chemical differences.

It is now common practice to evaluate the effects of processing steps on protein stability during process development while the key manufacturing unit operations and associated equipment configuration and capabilities are being defined. Together with formulation development studies, understanding of the basic properties and stabilization mechanisms as well as instability factors for a protein is progressively established. Each product is usually evaluated on a case-by-case basis using a combination of analytical methods for the extended characterization. The key points are summarized below with application examples and several published reviews are noted [16–19].

TABLE 11.3. Biophysical techniques used in industry for characterization of biotherapeutics

Method	Abbreviation
Visual inspection	
Microscopy-optical, fluorescence	
Light obscuration	LO
Micro-flow digital imaging	MFI
Coulter counter	
Nanoparticle tracking analysis	NTA
Dynamic light scattering	DLS
Multi-angle light scattering	MALS
Turbidity	
Electron microscopy: scanning, transmission	SEM, TEM
Atomic force microscopy	AFM
"Native" mass spectrometry	
Macro-ion mobility spectroscopy	IMS
Hydrogen/deuterium exchange mass spectroscopy	H/D Ex MS
Analytical ultracentrifugation: sedimentation velocity, sedimentation equilibrium	AUC SV, SE
Size exclusion chromatography	SEC
Field flow fractionation	FFF
Polyacrylamide gel electrophoresis: with SDS or native	(SDS)-PAGE
Capillary electrophoresis: with SDS or zone electrophoresis	CE-SDS, CZE
UV-VIS spectroscopy	
Infrared spectroscopy	FTIR
Raman spectroscopy	
Fluorescence spectroscopy	
Circular dichroism spectroscopy	CD
Nuclear magnetic resonance spectroscopy	NMR
Isothermal titration calorimetry	ITC
Differential scanning calorimetry	DSC

11.2.2.1 Upstream Cell Culture and Fermentation Process Impact. During culture of the expression cell lines, the secreted recombinant proteins are accumulated in the culture medium within the bioreactor for long durations before harvest. The conditions optimized for cell growth such as culture temperature (30–37°C), combinations of growth media components, and additives may cause instability in the protein product [16, 20]. Nonenzymatic reactions in conjunction with enzymatic-mediated modifications on the protein often result in a heterogeneous mixture of product variants which may differ in potency as well as stability [21, 22]. Different glycoforms on the Fc region of IgG have been shown to confer varied thermal stability using DSC [23]. In general, glycosylation often stabilizes the domain structures and ample examples can be found with many marketed products [24]. Unpaired thiols, subunit dissociation, high glycation, and aggregation have been observed under different culture conditions using gel electrophoresis, LC/MS, and SEC methods [25–27]. In most cases, the aggregated population can be removed in the downstream purification process.

11.2.2.2 Downstream Primary Recovery and Purification Process

Impact. Purification operations are generally optimized for yield improvements. Maintaining product stability and avoiding aggregation and precipitation are thus very important. However, the chromatographic conditions associated with affinity, ion-exchange, and hydrophobic interaction modes involve significant shifts in pH and ionic strength [28], creating potentially unfavorable solution conditions for protein stability. Low pH conditions (pH 2–4) are commonly used to elute monoclonal antibodies from protein-A affinity columns. Viral inactivation protocols may also involve extended exposure to extreme pH. As much as 25% of aggregates have been reported for a commercial product during this column step [29]. Hence the effect of acidic conditions on antibody stability has been widely studied [30–33]. In a model study, three different monoclonal antibodies were examined under low pH over 24 hours as well as pH titrations to simulate process conditions, i.e. change from neutral to low pH during protein-A column elution and low to neutral pH during neutralization step before loading on the subsequent polishing column [31]. When IgG4-A was analyzed by near-UV CD, upward shifts in the spectra were observed at both pH 3.5 and 2.7 when compared to that obtained at pH 6. Further incubation at both low pH over 24 hours caused little change in the CD spectra while AUC-SV confirmed mainly monomeric distribution. This indicates the acid conformation is stable over this time frame and low pH alone does not induce aggregation. Titration of this IgG4-A from pH 3.5 to 6 resulted in recovery of CD profiles resembling that of the original sample. In contrast, when the IgG4-A was shifted from pH 2.7 to 6.0, the near-UV CD spectra remained altered, and AUC-SV analysis revealed formation of 30% soluble oligomers. The data suggest that subtle conformational changes resulting from exposure to pH 2.7 may persist after titration and these intermediates play a key role in aggregation. There is indication that propensity of instability differs between IgG isotypes, and it has been suggested that IgG2 antibodies are more prone to aggregation [34] due to lower stability in the CH2 domain. However, it often remains unclear which domains are most aggregation prone upon unfolding for individual proteins. Small changes in pH and ionic strength can alter the solubility and/or size of aggregates [35]. In addition, differences in aggregation may be observed under identical pH conditions with different buffer systems (acetic acid vs. citric acid) [33].

Multiple filtration steps are used in protein purification for concentration, buffer exchange, and virus removal. While the low to moderate filtration pressure used is not generally expected to initiate aggregation, high pressure may cause protein unfolding and increased hydrophobic interactions thus enhancing aggregation [36]. During the ultrafiltration/diafiltration (UF/DF) process, protein concentration at the membrane surface may be locally higher than the bulk solution, which may trigger formation of aggregates and lead to membrane fouling [37, 38]. Depending on the filter material and the protein concentration, adsorption on the surface and loss of product may be appreciable [39].

During the purification operation, the intermediate product streams may have variable exposure to air, light, fluctuating room temperature, cycling between cold room storage, as well as various mechanical stress and different types of material contact. Forced degradation and materials-of-construction (MOC) studies performed on purified drug substance may provide insights into the in-process stresses that should be avoided.

However, the purification process intermediates contain impurities that may vary significantly from batch to batch which may impact in-process stability. In most cases, key QC assays such as size by SDS-PAGE, charge variants by ion-exchange chromatography or isoelectric focusing electrophoresis, and aggregation by SEC are appropriate for monitoring product quality during purification operations.

11.2.2.3 Drug Product Formulation, Fill-Finish Operations Process
Impact. In commercial manufacturing, the purified drug substance may be stored as liquid or frozen to facilitate flexibility of operations and transport between manufacturing facilities. The formulation operation may involve dilution of the bulk drug substance, buffer exchange, and additional mixing with excipients. After sterile filtration, the formulated bulk is dispensed through filling machines into the primary containers and stoppered. Filled vials are crimped, inspected visually, and processed through labeling and packaging before transport and distribution. During these unit operations, protein stability may be impacted by a number of factors including temperature (e.g., freeze-thaw, temperature cycling between cold room storage and room temperature operations), interfacial exposure (solid–liquid, liquid–liquid, gas–liquid), photo-oxidation (light exposure in presence of oxygen), as well as interactions with extraneous materials from the process operation such as silicon oil, steel particles shed from stainless steel equipment, extractables/leachables and metal ions from material surfaces, residual hydrogen peroxide from equipment sterilization protocols, and peroxides impurities from excipients.

Different types of agitations such as shaking and stirring may have different impact on protein aggregation depending on the intensity and duration, level of air–liquid interface as well as the protein itself. Shaking creates air–water interfaces and stirring is even harsher with the additional interfacial effects with the stir bar, trapping of air bubbles, and localized heat. In an earlier study where human growth hormone and deoxyribonuclease were exposed to shear stress in the absence of an air–liquid interface [40], the effect of shear alone on aggregation was found to be insignificant. In later studies with monoclonal antibodies where the vial configuration, fill volume, headspace oxygen, and shaking or stirring speed were varied, it was shown that varying types of aggregates were formed at different levels depending on the experimental setup [41, 42]. It is clear that stress conditions need to be incrementally varied to allow detailed comparisons between formulations to be discriminatory. In a study on a marketed IgG, differences in monomer content by SEC between simple phosphate-buffered saline versus the marketed formulation can only be discerned when shaking was done at 200 rpm but not at 250 rpm [42]. SEC results for soluble aggregate tracks differently in the two formulations based on turbidity (indicator for insoluble aggregates), likely due to the presence of polysorbate in the marketed product. Static storage after the shaking stress allows further study of the equilibrium between different types of aggregates [43]. For an IgG1 sample shaken for 168 hours, visible and subvisible particles (insoluble aggregates), turbidity, and Z-average diameter based on DLS were found to decrease while soluble aggregate content increased over storage of 1 month at 25°C. This suggests some depolymerization of large oligomers into smaller aggregates. FTIR second-derivative spectra did not show differences in the shaken supernatant and the pellet as compared to the unstressed sample, suggesting essentially native secondary structures. Spiking an unstressed protein formulation with the shaking-induced aggregates did not lead to further propagation of aggregation. In the same study, heat treatment (85°C for 10 min) induced denatured aggregates, with a size profile different from that generated by shaking.

Given the known disulfide linkage variations in IgG2 antibodies [44], biochemical assays are often used in conjunction with biophysical analytical methods for characterization of aggregation. Unpaired cysteines can be assayed with and without 5 M guanidine HCl denaturation to assess buried versus solvent-exposed sulfhydryl groups and their contributions to aggregation. Covalent aggregates are simply measured by SDS-PAGE. While SEC, AUC, and FTIR revealed no changes in conformation and soluble aggregate profiles, SEM revealed formation of particles with agitation [45]. A recent coordinated study with IgG1 and IgG2 antibodies employed 11 different specific stress conditions and tried to classify the array of aggregates based on size, secondary and tertiary structures (folded/partially folded, mostly unfolded), surface hydrophobicity, reversibility upon dilution, metal content, and presence of chemical modification [46,47]. The study again highlights the complex heterogeneous nature of protein aggregates and suggests that each stress technique may be associated with specific mechanisms of aggregate progression. This classification scheme may then serve as a useful tool to help identify the types of stress that triggers the aggregation based on direct analysis of the antibody aggregates isolated from the treated solutions. The overall information may in turn be helpful in defining remediation steps in the formulation and manufacturing operations.

Protein stability may be affected by interactions with solid surfaces [48]. Adsorption to interfaces or interactions with particles can cause denaturation and aggregation. Metal ions and peroxides may leach from surfaces and exacerbate oxidation. The different MOC to which the protein solution is exposed during bioprocessing thus pose complex risks on product stability due to the influence of multiple factors. For example, the use of stainless steel equipment and vessels may impact aggregation due to protein interactions with the steel surface, the Fe ions leached from the steel equipment causing conformational destabilization [49], as well as steel particles shed from pumping operations acting as heterogeneous nuclei [50] triggering particulate formation. Extractables and leachables from filters, plastic, glass, and other components [51] may cause protein degradation in addition to contaminating the drug product. A recent study with three sterilizing filters indicated that an IgG2 preparation can be destabilized by the extractables, showing increased aggregation, oxidation, and acidic species [52]. Surface tension was determined on the filter extractable solutions to evaluate surface activity in addition to a range of size- and charge-based assays conducted in conjunction with agitation and temperature stress studies on the antibody. The filter extractables were shown to be strongly surface active, behaving analogously to added surfactants and resulted in lower particle formation as determined by DLS and MFI. DSC indicated no change in thermally induced melting transitions. However, the filter extractables were found to have a negative impact on the chemical stability of the IgG2 in the presence of polysorbate-80 in the formulation, as determined by reversed-phase HPLC and imaged capillary electrophoresis. In general, sterile filtration studies need to be conducted with consideration of protein adsorption resulting in loss of product, surfactant additions to ameliorate membrane fouling [53], as well as excipient adsorption to the membrane [54] which in turn may cause the surfactant concentration to drop below the target resulting in product destabilization. Transmembrane pressure as well as temperature of the bulk should also be evaluated for impact on protein stability [55]. Overall, preliminary evaluation on the potential stress issues of a protein product may be obtained by conducting a variety of studies during early phase process development. However, the actual combined destabilization effects encountered during extended large-scale manufacturing remain difficult to cover and characterize before actual scale-up. Development of appropriate scale-down models with considerations of the complex engineering components relevant to the large-scale unit operations is thus very important. In practice, studies are often designed to cover the worst-case scenario of process ranges using orthogonal biochemical and biophysical methods for thorough characterization of product stability.

Surfactants are often used in process operations to stabilize proteins against interface-induced aggregation. In particular, polysorbates (PS20 and PS80) are widely used in various marketed formulations. Their potential interaction with proteins can be evaluated by isothermal titration calorimetry [56]. These chemically diverse mixtures containing mainly sorbitan polyoxyethylene fatty acid esters are also prone to degradation by autoxidation and hydrolysis [57]. The surfactant thermal degradation can be studied by DSC and isothermal thermogravimetric analysis (TGA) and hydrolysis rates can be calculated by following the proton signals of the fatty ester by ^{1}H NMR over time [58]. HPLC with evaporative light scattering detection as well as fluorescence micelle method is used to quantitate PS concentrations while peroxide formation can be followed by simple biochemical methods [59]. A higher level of peroxides in PS80 has been shown to increase oxidation levels of IL-2 mutein [60]. At 0.1%, PS80 has dual effects on IL-2 mutein stability in that shaking-induced aggregation was inhibited, but disulfide- and non-disulfide-linked aggregates were significantly increased during storage at elevated temperature (40°C) [61]. Therefore, the use of surfactants should ideally be kept to a minimum in a product formulation.

Many proteins are sensitive to light/UV exposure. Tryptophan, histidine, phenylalanine, tyrosine residues, as well as cystine can be oxidized in the presence of oxygen [62]. Photostability testing is carried out following ICH Q1B [63] recommendations including the use of a controlled chamber and options for light sources. Proteins typically degrade by a combination of physical and chemical mechanisms. The sensitivity of biopharmaceuticals to light exposure needs to be well characterized for a final definition of the formulation composition as well as the primary container and secondary packaging. Recommendations for duration and light exposure conditions during manufacturing steps and clinical delivery should be made accordingly. Additional oxidation risks in manufacturing may come from exposure to residual vaporized sanitizing agents (hydrogen peroxide and peracetic acid) [64]. Reversed-phase HPLC in conjunction with LC/MS peptide mapping is often used to localize the modified residues [65] while SEC with online MALS and AUC-SV can be used to determine fragment and soluble aggregate size distribution [66].

High-concentration antibody formulation development is now common due to the convenience of high-dose delivery through subcutaneous injections. The increased interactions between solutes and neighboring protein molecules in a crowded environment often have consequences including self-association of native proteins, unfolding of native structure, and reversible or irreversible aggregation [67]. The high-concentration formulation may also exhibit increased viscosity and undesirable opalescence which can arise as a result of Rayleigh scattering [68]. These properties pose additional formulation challenges potentially impacting shelf life, manufacturing process performance (particularly filtration), as well as the clinical administration. Characterization of the species involved in the reversible equilibrium can be done using AUC with computational tools to model the complex sedimentation equilibrium data [69] as well as inclusion of a tracer (radioactive or fluorescent) [70]. In a study with monoclonal antibodies, the concentration dependence of apparent molecular mass of three IgG1 antibodies were studied in different ionic strength buffers using preparative ultracentrifugation with subsequent microfractionation [71]. The self-association appears to be weak and was not detected by AUC-SV and SEC at low protein concentration. For one of the antibodies at concentrations over 100 mg/mL, significant molecular interactions existed in low ionic strength conditions which were diminished in higher ionic strength conditions. This suggests electrostatic interactions of this particular protein contributed to the high viscosity. Overall, self-association is considered an outcome of the net balance between various short-range interactions including hydrophobic and other nonionic interactions as well as electrostatic repulsive and attractive interactions. The strength of protein colloidal interactions is often evaluated by the osmotic second virial coefficient (B22) which describes the nonideal solution behavior from two-body interactions. In addition to sedimentation equilibrium studies by AUC, different methods have been used to experimentally determine B22, including static light scattering, dynamic light scattering, membrane osmometry, self-interaction chromatography, and composition-gradient light scattering [72,73]. The electrostatic potential near the surface of the protein in the formulation can be described by the experimentally derived zeta potential determined through electrokinetic methods [74, 75]. The apparent net charge determined from these measurements is more relevant than the pI determined from IEF. By manipulating solution conditions such as ionic strength, buffer species, and pH, the concentration-dependent reversible self-association, increased viscosity, and opalescence of the protein formulation can be modulated [76–78]. The dissociation/association kinetics of protein self-association phenomenon can vary from seconds to days; thus, a variety of analytical methods should be used in concert to characterize these conditions.

Freeze-thaw of drug substance bulk offers flexibility in the manufacturing operation. However, both freezing and thawing may have a negative impact on protein quality due to different stresses including potential pH changes, cold denaturation, formation/reformation of ice–water interfaces, localized concentration differences of components, and phase separation [79]. Excipients can be added to protect the protein from instability. Beyond the early-stage freeze-thaw studies usually conducted at an uncontrolled rate, large-scale processes can be mimicked with a scale-down model using controlled rate freezer. Characterization studies on an IgG1-indicated aggregation after freeze-thawing (five cycles between −80°C and 25°C) mainly consisted of particulate formation in the micrometer range as determined by LO and also indicated by the increased Z-average diameter and polydispersity index by DLS. Based on CD, attenuated total reflectance (ATR)-FTIR, and extrinsic dye fluorescence spectroscopy, the aggregate contained native-like monomers and were noncovalently linked as determined by SDS-PAGE. SEC and asymmetric flow field flow fractionation (AF4) did not detect soluble aggregates suggesting higher sensitivity of the particulate analysis methods [80]. A similar study with an IgG2 in histidine-based formulations subjected through three cycles of freeze-thaw (−196°C and 5°C) showed subvisible particle levels from 2 to 50 μm as detected by MFI progressively increased after each cycle. SEC did not detect any changes in soluble aggregate which remained at 1% or less. When the formulation was changed to phosphate-buffered saline, soluble aggregates were increased to levels detectable by SEC and a correlation was found with results from MFI [81]. A study with a different antibody showed that storage in particular commercial freezing containers led to higher levels of soluble aggregates by SEC, indicating the significant influence of surface interaction [82].

Lyophilized dosage forms are often used in early stage of a development program and are considered as final formulation for proteins that are not suitably stable in liquid form. This remains a common formulation strategy and almost half of the marketed biopharmaceuticals are lyophilized. During process development, the unit operations employed in the various freezing and drying methods are optimized in conjunction with the formulation excipients which include a bulking agent added to confer acceptable cake and additional stabilizers (see [83, 84] for review). Lyophilized samples can be analyzed after reconstitution using the same set of biochemical and biophysical methods used for the analysis of liquid samples. In addition, residual moisture content is monitored as it generally correlates with degradation such as formation of acidic/basic species, dimerization, and aggregation. Protein secondary structure can be analyzed in the solid state by FTIR and Raman spectroscopy. Physical properties important for storage stability including specific surface area, surface composition, and global and localized molecular mobility are studied in dry form. Characterization methods for the freeze-dried cake include DSC, TGA, near-IR, X-ray diffraction (XRD), neutron scattering, as well as X-ray photoelectron spectroscopy [85].

11.3 ANALYTICAL METHODS APPLICATIONS CONSIDERATIONS

The biophysical methods discussed above all have utility in evaluating the higher order structure of biopharmaceuticals potentially impacted by process, manufacturing, formulation, and storage conditions. However, due to the diversity and complexity of therapeutic protein structures and the heterogeneous nature of a given product, it is well understood that not all the technologies are of equivalent value for a given project. During process and formulation development, the analytical experience gained from the various stress and accelerated stability studies forms the basis of identifying and qualifying the relevant biophysical methods. In addition to finalizing the stability-indicating assays for setting product specifications, the core set of characterization assays that will be used for gaining further process understanding should be defined. For comparability studies, the biophysical characterization methods intended for regulatory submissions need to be qualified for the specific protein along with the formulation. Some commonly used methods are included below with comments on application considerations, along with several emerging techniques and application examples in key areas.

11.3.1 Spectroscopic Methods

Spectroscopic methods are routinely used for secondary and tertiary structure perturbation assessments. In particular, CD, FTIR, and fluorescence spectroscopy are widely used to determine the consistency and/or comparability of the bulk properties of samples. Depending on the magnitude of the perturbation within each molecule and the level of impacted molecules in the solution, the capability of discrimination between samples varies. The signal-to-noise ratio of the assay often depends on the conformation profile of the protein as well as the primary sequence. Buffer components and surfactants that are part of the formulation may affect the assay readout and should be carefully studied. Variability (repeatability and intermediate precision) of the method is assessed by replicate measurements of the same protein sample within the same day and over multiple days with independent sample preparations [86]. Since visual comparisons of spectral data often remain subjective, different quantitative approaches have been employed for objective comparisons and classification of spectra. In one example, a simple customized Excel spreadsheet was constructed to process raw CD spectra to eliminate operational bias [87]. Signals are normalized for concentration and minimum and maximum boundaries are calculated based on the replicate measurements at each data point. For comparability studies, a mean reference CD spectrum was compiled for the bulk protein from a number of historical batches of a representative manufacturing process. Deviations of the sample spectrum from the reference spectrum were then calculated at different wavelengths taking into account the standard deviation at each data point. An acceptance range based on ± 3 standard deviations was set up for a pass/fail evaluation. In other cases, spectral processing software packages [88, 89] and multivariate statistical analysis (principal component analysis, distance to model plots) [90] are used to clarify clusters and trends for comparisons without specific attempt to interpret the underlying protein structure. For evaluation of the capability to discern nonnative structures within a given protein sample, blending/spiking studies are conducted using stressed materials. Depending on the protein and the level of conformational perturbations, spectral changes can be quite sensitive as indicator for aggregation and well correlated with SEC and AUC-SV data. In a study with acid α-glucosidase [91], forced aggregate was generated by dialyzing the commercial product into unfavorable pH condition (phosphate buffer, from 6.2 to 7.5), removing the formulation components including PS80, and heating at 37°C for 18 hours. The high molecular weight aggregates were purified by SEC and combined with purified drug product monomers at different proportions to yield a series of spiked samples at 0–20%. Evaluation of near-UV CD spectra indicated that the method was sensitive to aggregation levels of ~4% for this protein.

For characterization of protein structures of various types, vibrational spectroscopy including FTIR and Raman are particularly useful since these techniques can be applied on samples of all physical states such as liquid, gel, and solid. In recent times, the updated spectral analysis and band assignment techniques along with the development of highly sensitive spectrometers resulted in broad accumulation of spectral databases for proteins and widened applications on biopharmaceutical process development. With modern Raman spectrometers, spectral data can be routinely obtained at concentration of 1 mg/mL which is practical for studies of most protein therapeutics and on par with other spectroscopic methods [92]. Common process stress conditions which lead to aggregation have been studied in IgG1 and IgG2 antibodies using FTIR and Raman [93]. In a case study on IgG1, the antibody was stressed two ways: (i) incubated at pH 5 and 45°C for 3 months and (ii) incubated at pH 8 (unfavorable) under agitation. Second-derivative FTIR spectra were compared for the insoluble aggregates collected by centrifugation and the soluble aggregates in the supernatant, in both cases with the native undisturbed antibody at pH 5. Assignments of amide I bands at different frequencies in the region of 1600–1700 cm^{-1} indicated characteristics of native β-barrel fold at pH 5. While the shear-stressed soluble IgG appeared to maintain most of the native features, the heat-stressed soluble fraction showed both intermolecular β-sheet and irregular structures indicating unfolding. The insoluble fraction from shear stress showed a much broader peak around 1630 cm^{-1}, suggesting more heterogeneous structures with different forms of β-sheet and irregular structures. The heat-stressed insoluble aggregates contained mostly extended antiparallel β-sheet and lacked most of the native β-barrel fold. This case study clearly showed that the aggregate structures formed by heat treatment differ significantly from that formed by agitation. In the same paper, the effect of shifting pH (from 7 to 3) was also investigated by Raman and Raman optical activity (ROA) spectroscopies for characterization of molecular conformations of side chains such as Trp, Tyr, and disulfide bonds. Both IgG1 and IgG2 antibodies showed significant unfolding of tertiary structures at pH 3 while the secondary structures remain largely intact based on ROA results. Raman spectra of isolated Fc aggregates differ from Fc in solution in the 500–560 cm^{-1} region, indicating differences in disulfide conformation. For stress studies to be relevant to practical situations in manufacturing, a common practice is to conduct the study in an incremental manner without destabilizing more than 20–40% of the protein. These case studies exemplify how modern spectroscopic methods could be used to finely gauge the stress studies as well as gain molecular insights into the multiple pathways of degradation and aggregation. Taken together, fine spectroscopic characterization of aggregates formed in different manufacturing stress-related situations can be extremely valuable in the development of overall process understanding and control strategies.

11.3.2 Aggregates and Subvisible Particulate Analysis

Protein aggregation analysis entails characterization and quantitation of product species from dimers to visible particles. The mechanistic basis of this size progression is complex and clearly varies for individual proteins with differing susceptibilities to process parameters. The diverse techniques for studying molecular size distribution have been highlighted in Section 11.1.2.2. Based on different principles of detection and time scale of measurements, the assay data sets provide complementary information for constructing an overall understanding of the complex multi-stage aggregation process. However, many challenges remain including the different levels of sensitivity between assays, the discriminating capabilities (in terms of reproducibility, size resolution, discernable molecular details), and how the conditions for measurements may impact the distribution of aggregates in the sample. Most derived data are expressed in relative terms—each entity is calculated as percentage of total assay signal. A coordinated, quantitative interpretation of the aggregation continuum for a given protein product thus remains challenging.

For routine QC monitoring of product stability, SEC remains the common assay for soluble aggregates. It is widely used due to the simplicity of standardized HPLC operations and overall assay robustness that is adequate for validation and long-term use. However, there are method limitations and data interpretation needs to be carefully qualified for individual products. The dynamic range in size separation is limited and mixed-mode interactions with the resins may generate anomalous results. For example, a monoclonal antibody analyzed under industry standard conditions was found to elute at retention time equivalent to apparent molecular weight of 44 kDa as opposed to the expected 148 kDa, due to secondary interaction with the resins [94]. In other cases, preferential adsorption of aggregates to the column resulted in incomplete protein mass recovery [95], and larger aggregates are often retained in the column frit. Sample dilution and mobile phase composition may also affect the SEC analysis with perturbation of the aggregate distribution under assay protocol [95]. It is now a common regulatory expectation to cross-verify the size determination and quantitation from this method using orthogonal techniques such as AUC-SV [96]. For most IgG programs, SEC works well for analysis of fragments, monomer, dimers, and soluble oligomers. Sensitivity can be increased by using 214 nm in addition to 280 nm detection, and online static light scattering can be added to further determine size range of the early-eluting higher molecular weight aggregates [97, 98]. Denatured SEC can be carried out with inclusion of 3 M guanidine HCl [99] or 0.1% SDS [100] to differentiate between covalently linked and noncovalently linked aggregates. In the latter study, the stability of five IgG1s was followed at 30°C, 40°C, and 50°C over 4 weeks and Fab formation through hinge region fragmentation was analyzed by native SEC, SEC with SDS, and CE-SDS. The upper hinge region of IgG1 has been shown to be solvent exposed and highly flexible, with solution pH and temperature influencing the fragmentation patterns of the same span of amino acid sequence, –SCDKTHTC–. In a detailed study where SEC fractions from an IgG1 incubated at 40°C were analyzed by LC/MS [101], it was shown that the cleavage sites shifted toward the C-terminus (closer to the CH2 domain) as the pH was lowered from 7 to 5. At pH 4, the sole cleavage site occurred within the CH2 domain, entirely outside of the hinge region. These results suggest conformational changes in the structure moving the Fab closer toward the hinge region and affecting the rate of hydrolysis. This interplay of biophysical instability and chemical reaction often confounds detailed characterization of IgG1 degradation. Nevertheless, relative rate behaviors can be assessed without defining relationship of the fragments using statistical cross-correlation analysis. In the study cited above [100], the time-dependent changes of individual fragment species were evaluated by two-dimensional maps and their relative rates of change deconvoluted. The five IgG1s in the study showed very similar apparent fragmentation rates within the limits of the solution conditions used.

AF4 analysis has advantages over chromatography as there is little sample preparation required. The separation is matrix free, employing a crossflow through a membrane channel wall [102]. However, while the techniques show great potential for resolution of larger size range, self-association induced by the concentrating effects of the AF4 technique itself poses great challenge in obtaining good resolution. Reports on comparison of SEC, AF4, and AUC-SV have been published [103, 104] for monoclonal antibody analysis. Both AF4 and AUC provided better recovery and estimates of the aggregates compared with SEC. AUC-SV has better precision than AF4, but consistent quantitation of large oligomers at low levels (below 1–2%) remains difficult. Overall, CE methods remain most robust for size analysis of stable species with relatively high throughput in operation. CE-SDS gives better resolution for all fragment species as well as the covalent higher molecular weight aggregates in general [99, 100]. Furthermore, CZE mode can be used with a variety of background electrolytes for non-denaturing separations. A recent paper showed the utility of CZE with online wavelength-resolved fluorescence detection to evaluate urea unfolding of model proteins such as lysozyme and chymotrypsinogen-A [105]. The unfolded proteins showed a decrease in electrophoretic mobility as well as red shift in the fluorescence emission spectra, suggesting higher exposure of tryptophan residues. This type of assay technology thus yields information on two potentially independent parameters of conformational changes and may be valuable in studying the multiple stages of unfolding and aggregation.

In many cases of manufacturing impact investigations, high-sensitivity detection of aggregates is more important than accurate quantitation and fine definition of size distribution. As such, DLS, fluorescent dye binding, and fluorescence microscopy methods are particularly useful to detect early events in stress studies as well as in long-term stability studies. The mechanism of dye-mediated detection of protein aggregation is very complex and may sometimes give confounded results. However, small extrinsic dyes that interact with proteins noncovalently through hydrophobic interactions have been successfully used to detect low amounts of aggregates against the background of large amounts of folded monomer [106, 107], even in the presence of polysorbates [108]. One study used Nile Red to stain an IgG1 and evaluated aggregate formation using fluorescence microscopy [109, 110]. In conjunction with AF4, static, and dynamic light scattering and spectroscopic methods, the range of aggregates formed in different solutions and the kinetics of formation over hours were characterized. However, whether the addition of the fluorescence probe affects aggregate distribution in the sample remains a question that needs to be addressed by orthogonal methods. For relative trending in a semiquantitative fashion, a study was conducted on three monoclonal antibodies and fluorescence dye binding was evaluated along with other more established assays [107]. Aggregates were first created by heating at 50°C for up to 8 hours, and the fluorescence signal obtained from SYPRO Orange binding was shown to correlate well with SEC data at 1–3% higher molecular weight soluble aggregates. In addition, particulate samples were generated by stirring at room temperature for up to 24 hours and the dye fluorescence intensity correlated with particle counting data from MFI. In a case study with several commercial products with polysorbate-containing formulations [108], the dyes DCVJ and CCVJ appeared to be effective tools for analysis even at relatively high protein and PS80 concentrations. However, it is recognized that the signal intensity generated is dependent on dye binding to altered regions of the protein. Accessibility of particular hydrophobic sites in a micron-sized particle may not be comparable to that on a soluble dimer. Given the diverse heterogeneous nature of an aggregated sample, it is not possible to adequately standardize the fluorescence signal for proper quantitation of aggregate content.

Subvisible particle analysis is a major concern for commercial process development due to difficulties in defining the nature and risk of unwanted immunogenicity. There are as yet no reliable predictive theoretical models or relevant *in vitro* assays to assess immunogenicity in the clinical setting. To date, many laboratories in the industry have made significant strides in establishing methods for extending the particle size detection limit below 10 μm. This is the limit stipulated in USP<788> based on concerns with capillary occlusion. The major methods and considerations have been reviewed [111, 112]. The key point is to use orthogonal methods for coverage and optics-based detection techniques remain most common. These may include LO, MFI, and FlowCAM instruments with detection size limit down to 2 μm. The NTA system further covers size range lower than 1000 nm and has been shown to give better size resolution than DLS [46, 113]. LO usually gives lower counts as translucent particles may not be detected. The associated higher RSD presents additional challenge for statistically meaningful comparisons between samples. The flow-imaging techniques provide size sorting, counts, as well as particle shape and optical properties information. This is of particular value since air bubbles and silicone oil can be categorized differently from nonspherical proteinaceous material based on aspect ratio calculation or manual review of the acquired images. For the second (orthogonal) method, a different optics-based instrument or an entirely different detection principle such as electrical impedance, for example, the Coulter counter [114], is recommended. The Coulter counter method is of most interest for high-concentration formulation analyses where the contrast is diminished, i.e. smaller difference in refractive index between the particles and surrounding medium with increasing monomeric protein concentrations. However, there is no morphology information derived, and the samples need to contain electrolytes for conductivity. Many formulation buffers are actually adequate but sugar-based solutions often cannot be used directly. Dilution in stronger electrolytes is possible but then may change the size distribution of the particles and should be evaluated on a case-by-case basis. With the different instrumentation setup and data-processing algorithms, the absolute counts obtained from the different techniques are not expected to be identical. Nevertheless, the relative trending among a set of samples is often quite similar between methods, allowing comparative conclusions for process and formulation development. There are many published reports on the combined use of particle counting methods along with additional biophysical characterization. One paper [115] discussed the findings on the stability profile of a 90 mg/mL IgG formulation stored at 2–8°C for up to 18 months. MFI was demonstrated to have improved sensitivity for subvisible particulates compared to conventional LO detection. In the course of the long-term stability study, there was progressive increase in the overall particle size distribution (2–100 μm) over time across all four batches analyzed when compared to the frozen retain stored at −70°C. The equivalent weight percentage of protein particulates was estimated from the MFI data by assuming a spherical shape and using a protein partial specific volume value of 0.7 cm^3/g to calculate the total weight of particles. In addition, particles were collected on gold-coated membranes from a total of 85 vials and weight content determined directly. For the samples stored for 16 months, these two approaches gave very similar results, at 0.018% and 0.022%, respectively, indicating very low level of protein particulates by weight. The nature and composition of the protein particulates were characterized by a combination of inverted microscopy, FTIR microscopy, and SEM-EDX (energy-dispersive X-ray spectrometer). Particulates were identified as protein with silicone, although some particles also contained other elements such as aluminum.

11.3.3 Emerging Applications

Several well-proven technologies in molecular characterization have gained broader applications in recent years in the studies of aggregation and particulate analysis. FC is well known for its high throughput and with a detection limit of 0.5 μm has shown great potential in subvisible particle analysis [11]. Typically, forward and side scattering signals and a number of fluorescence signals can be collected. The effect of piston pump filling of monoclonal antibody samples into vials have been analyzed using FC and compared to that from MFI [116]. SYPRO Orange was used to differentiate protein and nonprotein particles. Total particle count was found to increase with increasing numbers of cycles through the pump, from 1 to 10 passes, while the proportion of protein and nonproteinaceous particles was constant. The trends were similar for FC and MFI, but the latter is less sensitive in detecting changes in the early passes due to size detection limit of 2 μm. Given the relatively low sample volume required (100–200 μL), the volume actually scanned (30–100 μL), the time required for one scan (1–2 min) and the 96-well autosampler capability, this method is a very promising technique for general subvisible particulate enumeration. Another recent study showed application of FC on the analysis of silicone oil-induced particulate formation in protein formulations [117].

For characterization of the continuum of aggregation growth during the shelf life of a protein product, molecular-scale imaging technique can provide valuable information. Transmission electron microscopy (TEM) offers excellent resolution but the staining method often distorts delicate structures. AFM has nanometer resolution and can image single biomolecules in liquid medium. In addition, there are different modes of AFM force spectroscopy for probing molecular interactions [118]. It is thus possible to visualize individual IgG molecules and examine the structural organization of various aggregates. In a recent study using the simple dry tapping mode, several different antibodies were shown to exhibit similar aggregation intermediates consisting of several monomers [119]. There was indication that the oligomers then coalesce into larger assemblies, but the progression differs between antibodies. In some cases where the protein was stressed by increased temperature, oligomers were not found and only large structures were observed presumably from completely unfolded molecules. These results suggest that AFM can be used to evaluate the propensity of condensation of protein products from oligomers to subvisible particles.

Mass spectrometry has been widely used to characterize glycoprotein covalent structure. In addition, the technique can be used in the native mode under conditions that aim to preserve quaternary protein structures to investigate complex topology and dynamics. Analysis of charge-state distribution and ion mobility separation coupled with MS [120–122] are common approaches for monitoring conformational changes and aggregation. A recent paper describes the use of temperature-controlled electrospray ionization (ESI), where the protein is heated at a fixed temperature for a period of time and followed immediately by MS analysis [123]. This allows the monitoring of both oligomers formation and conformational changes. Another major technique is hydrogen/deuterium exchange MS [120, 124, 125], which probes protein structures by monitoring the rate and extent of deuterium exchange with backbone amide hydrogen using MS. The level of H/D exchange depends on the solvent accessibility of backbone amide and the conformation of the protein. The exchange process can be monitored by MS at different time points either globally (on the intact protein) or combined with enzymatic digestion (typically pepsin) to categorize solvent exposure of different domains of the protein and enable structural changes to be resolved at the peptide level. Coexisting populations with different conformations can be revealed at levels of ~10% [125].

11.4 CONCLUDING REMARKS

11.4.1 Accelerated Stability Studies Considerations

Accelerated stability studies are often conducted during development due to limited time available for completing studies and making formulation choices. In support of clinical dating, front-run accelerated studies are often performed to predict stability of the protein product under the real-time condition. However, protein stability at low-temperature storage is not always predictable based on studies at higher temperature, particularly when the degradation mechanism is very complex. While the overall protein stability would follow the combined kinetics of the individual molecular instabilities, the relative contribution of the different degradation mechanisms may change at different temperatures. This complicated interrelationship of the various changes often means poor predictability of the Arrhenius plots. For example, a study with an IgG2 [126] showed aggregate formation lowest at pH 5 under normal storage conditions (4°C and 29°C) while at 37°C, the rate of aggregation was highest at pH 5. Meanwhile, DSC indicated the onset of thermal melting does not begin until beyond 50°C. Since IgG2 is well known to consist of at least two structural isoforms with differing disulfide linkages [44, 127], the authors used a redox refolding approach to generate solutions enriched in the IgG2-A and IgG2-B forms. After 12 months of storage, there was more aggregate found in the IgG2-B-enriched solution than in the IgG2-A solution at 4°C. In contrast, more aggregate was found in the IgG2-A-enriched solution at 37°C. While the IgG2 isoforms exhibited significantly different stability profiles, the Arrhenius kinetics of aggregation for each of the individual IgG2 isoforms was also nonlinear. In addition, denatured SEC analysis showed differences in hinge region clipping with the IgG2-B form more stable. It is suggested that the IgG2-B isoforms possess a more compact, less flexible structure and the higher resistance to clipping may be related to conformation of the hinge which is mediated by disulfide connectivity. Due to its steep temperature dependence, clipping-mediated aggregation in lower pH formulations was more prominent at elevated temperature. This complex pH dependence of aggregation as a function of temperature thus makes formulation decisions based on accelerated data difficult. The study illustrates the complexity of antibody aggregation where multiple pathways are in play and the different product variants may have different stability properties.

Taken together, the most stable formulation at elevated temperature may not be the optimal formulation under normal storage conditions. As such, only real-time stability studies data should be used to define the expiry dating for the final product. Keeping in mind that early-stage material may be different in product microheterogeneity from material generated in a late-stage process, it is important that all stability data, from real-time, accelerated, and various stress studies, be collated for concerted understanding of the product molecular instabilities. Bridging or linking studies should be conducted as part of the overall effort to define the specification limits of the protein concentration and formulation components for long-term robustness.

11.4.2 Conclusions

Selection of the final formulation for a protein product involves a balancing act. The considerations in providing ease of use in the clinic and maintaining stability during manufacturing operations as well as stability during long-term storage have been discussed here. It is recognized that chemical and physical instabilities of a protein have different sensitivities to solution conditions and environmental factors. The choice of the formulation conditions may thus be optimized for maintaining key product CQAs while perhaps suboptimal for some individual degradation reactions. As such, a well-defined drug product shelf life supported by thorough stability and forced degradation studies in conjunction with adequate development of process control strategies is important. In some cases, process and engineering modifications are necessary along with appropriate formulation for optimal product quality; for example, process tubing with E/L leaching problems needs to be substituted. Based on process characterization studies, the expected/acceptable ranges of variations in product heterogeneity and component raw materials are specified. In all cases, fine characterization by multiple analytical techniques is essential in providing confirmatory and complementary information on various aspects of the product and its instability. In particular, biophysical methods that provide nondestructive molecular characterization of the drug product and the high-sensitivity analysis of aggregates and particulates formation are key areas of interest in the development of robust commercial formulation and manufacturing. Overall, a rigorous science-based approach is used in conjunction with risk-based evaluations in the quality-by-design (QbD) approach to drug product development [128].

REFERENCES

1. CMC Biotech Working Group. (2009) *A-Mab: A Case Study in Bioprocess Development.* CASSS and ISPE website.
2. Walsh, G. (2010) Biopharmaceutical benchmarks 2010. *Nat. Biotechnol.*, **28**, 917–924.

3. International Conference on Harmonization. (2009) *Guidance for Industry. Q8 (R2) Pharmaceutical Development.*

4. den Engelsman, J., et al. (2011) Strategies for the assessment of protein aggregates in pharmaceutical biotech product development. *Pharm. Res.,* **28**, 920–933.

5. Goldberg, D.S., Bishop, S.M., Shah, A.U., and Sathish, H.A. (2011) Formulation development of therapeutic monoclonal antibodies using high-throughput fluorescence and static light scattering techniques: role of conformational and colloidal stability. *J. Pharm. Sci.,* **100**, 1306–1315.

6. Li, Y., Mach, H., and Blue, J.T. (2011) High throughput formulation screening for global aggregation behaviors of three monoclonal antibodies. *J. Pharm. Sci.,* **100**, 2120–2135.

7. Satoh, M., Iida, S., and Shitara, K. (2006) Non-fucosylated therapeutic antibodies as next-generation therapeutic antibodies. *Expert Opin. Biol. Ther.,* **6**, 1161–1173.

8. Daugherty, A.L. and Mrsny, R.J. (2006) Formulation and delivery issues for monoclonal antibody therapeutics. *Adv. Drug Deliv. Rev.,* **58**, 686–706.

9. Hawe, A., et al. (2012) Forced degradation of therapeutic proteins. *J. Pharm. Sci.,* **101**, 895–913.

10. Chan, C.P. (2008) Biochemical and biophysical methods currently used for therapeutic protein development. *Am. Pharm. Rev.,* **11**(7), 24–32.

11. Mach, H. and Arvinte, T. (2011) Addressing new analytical challenges in protein formulation development. *Eur. J. Pharm. Biopharm.,* **78**, 196–207.

12. Lubiniecki, A., et al. (2011) Comparability assessments of process and product changes made during development of two different monoclonal antibodies. *Biologicals,* **39**, 9–22.

13. Grillo, A.O., Kane, M., Penn, N., and Perkins, M. (2010) Characterizing the formulation design space. *BioPharm. International,* **23**(3), 30–39.

14. Dill, K.A. (1990) Dominant forces in protein folding. *Biochemistry,* **29**, 7133–7155.

15. Chi, E.Y., Krishnan, S., Randolph, T.W., and Carpenter, J.F. (2003) Physical stability of proteins in aqueous solution: mechanism and driving forces in nonnative protein aggregation. *Pharm. Res.,* **20**, 1325–1336.

16. Cromwell, M.E., Hilario, E., and Jacobson, F. (2006) Protein aggregation and bioprocessing. *The Am. Assoc. Pharm. Sci. J.,* **8**, E572–E579.

17. Shire, S.J., (2009) Formulation and manufacturability of biologics. *Curr. Opin. Biotechnol.,* **20**, 708–714.

18. Manning, M.C., et al. (2010) Stability of protein pharmaceuticals: an update. *Pharm. Res.,* **27**, 544–575.

19. Wang, W., Nema, S., and Teagarden, D. (2010) Protein aggregation-pathways and influencing factors. *Int. J. Pharm.,* **390**, 89–99.

20. Jenkins, N., Murphy, L., and Tyther, R. (2008) Post-translational modifications of recombinant proteins: significance for biopharmaceuticals. *Mol. Biotechnol.,* **39**, 113–118.

21. Walsh, G. and Jefferis, R. (2006) Post-translational modifications in the context of therapeutic proteins. *Nat. Biotechnol.,* **24**, 1241–1252.

22. Liu, H., et al. (2008) Heterogeneity of monoclonal antibodies. *J. Pharm. Sci.,* **97**, 2426–2447.

23. Ghirlando, R., Lund, J., Goodall, M., and Jefferis, R. (1999) Glycosylation of human IgG-Fc: influences on structure revealed by differential scanning micro-calorimetry. *Immunol. Lett.,* **68**, 47–52.

24. Sola, R.J. and Grienenow, K. (2009) Effects of glycosylation on the stability of protein pharmaceuticals. *J. Pharm. Sci.,* **98**, 1223–1245.

25. Zhang, W. and Czupryn, M.J. (2002) Free sulfhydryl in recombinant monoclonal antibodies. *Biotechnol. Prog.*, **18**, 509–513.

26. Lim, A., Reed-Bogan, A., and Harmon, B.J. (2008) Glycosylation profiling of a therapeutic recombinant monoclonal antibody with two N-linked glycosylation sites using liquid chromatography coupled to a hybrid quadrupole time-of-flight mass spectrometer. *Anal. Biochem.*, **375**, 163–172.

27. Trexler-Schmidt, M., et al. (2010) Identification and prevention of antibody disulfide bond reduction during cell culture manufacturing. *Biotechnol. Bioeng.*, **106**, 452–461.

28. Xiao, Y., et al. (2006) Protein instability during HIC: describing the effects of mobile phase conditions on instability and chromatographic retention. *Biotechnol. Bioeng.*, **93**, 1177–1189.

29. Philips, J., et al. (2001) Manufacture and quality control of CAMPATH-1 antibodies for clinical trials. *Cytotherapy*, **3**, 233–242.

30. Buchner, J., et al. (1991) Alternatively folded states of an immunoglobulin. *Biochemistry*, **30**, 6922–6929.

31. Ejima, D., et al. (2007) Effects of acid exposure on the conformation, stability, and aggregation of monoclonal antibodies. *Proteins. Struct. Funct. Bioinform.*, **66**, 954–962.

32. Sahin, E., Grillo, A.O., Perkins, M.D., and Roberts, C.J. (2010) Comparative effects of pH and ionic strength on protein-protein interactions, unfolding, and aggregation for IgG1 antibodies. *J. Pharm. Sci.*, **99**, 4830–4848.

33. Hari, S.B., et al. (2010) Acid-induced aggregation of human monoclonal IgG1 and IgG2: molecular mechanism and the effect of solution composition. *Biochemistry*, **49**, 9328–9338.

34. Franey, H., Brych, S.R., Kolvenbach, C.G., and Rajan, R.S. (2010) Increased aggregation propensity of IgG2 subclass over IgG1: role of conformational changes and covalent character in isolated aggregates. *Protein Sci.*, **19**, 1601–1615.

35. Sahin, E., et al. (2012) Aggregation and pH-temperature phase behavior for aggregates of an IgG2 antibody. *J. Pharm. Sci.*, **101**, 1678–1687.

36. Seefeldt, M.B., et al. (2005) High-pressure studies of aggregation of recombinant interleukin-1 receptor antagonist: thermodynamics, kinetics, and application to accelerated formulation studies. *Protein Sci.*, **14**, 2258–2266.

37. Stoner, M.R., et al. (2004) Protein-solute interactions affect the outcome of ultrafiltration/diafiltration operations. *J. Pharm. Sci.*, **93**, 2332–2342.

38. Kim, K.J., Chen, V., and Fane, A.G. (1993) Some factors determining protein aggregation during ultrafiltration. *Biotechnol. Bioeng.*, **42**, 260–265.

39. Mahler, H.C, et al. (2010) Adsorption behavior of a surfactant and a monoclonal antibody to sterilizing-grade filters. *J. Pharm. Sci.*, **99**, 2620–2627.

40. Maa, Y.F. and Hsu, C.C. (1997) Protein denaturation by combined effect of shear and air-liquid interface. *Biotechnol. Bioeng.*, **54**, 503–512.

41. Kiese, S., Pappenberger, A., Friess, W., and Mahler, H.C. (2008) Shaken, not stirred: mechanical stress testing of an IgG1 antibody. *J. Pharm. Sci.*, **97**, 4347–4366.

42. Eppler, A., Weigandt, M., Hanefeld, A., and Bunjes, H. (2010) Relevant shaking stress conditions for antibody preformulation development. *Eur. J. Pharm. Biopharm.*, **74**, 139–147.

43. Kiese, S., Pappenberger, A., Friess, W., and Mahler, H.C. (2010) Equilibrium studies of protein aggregates and homogeneous nucleation in protein formulation. *J. Pharm. Sci.* **99**, 632–644.

44. Wypych, J., et al. (2008) Human IgG2 antibodies display disulfide-mediated structural isoforms. *J. Biol. Chem.*, **283**, 16194–16205.

45. Brych, S.R., et al. (2010) Characterization of antibody aggregation: role of buried, unpaired cysteines in particle formation. *J. Pharm. Sci.*, **99**, 764–781.

46. Joubert, M.K., et al. (2011) Classification and characterization of therapeutic antibody aggregates. *J. Biol. Chem.*, **286**, 25118–25133.

47. Luo, Q., et al. (2011) Chemical modifications in therapeutic protein aggregates generated under different stress conditions. *J. Biol. Chem.*, **286**, 25134–25144.

48. Bee, J.S., et al. (2011) Effects of surfaces and leachables on the stability of biopharmaceuticals. *J. Pharm. Sci.*, **100**, 4158–4170.

49. Zhou, S., et al. (2010) Comparative evaluation of disodium edetate and diethylenetriaminepentaacetic acid as iron chelators to prevent metal-catalyzed destabilization of a therapeutic monoclonal antibody. *J. Pharm. Sci.*, **99**, 4239–4250.

50. Tyagi, A.K., et al. (2009) IgG particle formation during filling pump operation: a case study of heterogeneous nucleation on stainless steel nanoparticles. *J. Pharm. Sci.*, **98**, 94–104.

51. Allain, L. and Wang, Q. (2007) Impact of package leachables on the stability of pharmaceutical products. *Am. Pharm. Rev*, **10**(4), 38–44.

52. Huang, M., Horwitz, T.S., Zweiben, C., and Singh, S.K. (2011) Impact of extractables/leachables from filters on stability of protein formulations. *J. Pharm. Sci.*, **100**, 4617–4630.

53. Maa, Y.F. and Hsu, C.C. (1998) Investigation on fouling mechanisms for recombinant human growth hormone sterile filtration. *J. Pharm. Sci.*, **87**, 808–812.

54. Mahler, H.C., et al. (2008) Behavior of polysorbate 20 during dialysis, concentration and filtration using membrane separation techniques. *J. Pharm. Sci.*, **97**, 764–774.

55. Bee, J.S., et al. (2009) Response of a concentrated monoclonal antibody formulation to high shear. *Biotechnol. Bioeng.*, **103**, 936–943.

56. Hoffmann, C., Blume, A., Miller, I., and Garidel, P. (2009) Insights into protein-polysorbate interactions analysed by means of isothermal titration and differential scanning calorimetry. *Eur. Biophys. J.*, **38**, 557–568.

57. Kerwin, B.A. (2008) Polysorbates 20 and 80 used in the formulation of protein biotherapeutics: structure and degradation pathways. *J. Pharm. Sci.*, **97**, 2924–2935.

58. Kishore, R.S.K., et al. (2011) Degradation of polysorbates 20 and 80: studies on thermal autoxidation and hydrolysis. *J. Pharm. Sci.*, **100**, 721–731.

59. Towne, V., et al. (2009) Measurement and decomposition kinetics of residual hydrogen peroxide in the presence of commonly used excipients and preservatives. *J. Pharm. Sci.*, **98**, 3987–3996.

60. Ha, E., Wang, W., and Wang, Y.J. (2002) Peroxide formation in polysorbate 80 and protein stability. *J. Pharm. Sci.*, **91**, 2252–2264.

61. Wang, W., Wang, Y.J., and Wang, D.Q. (2008) Dual effects of Tween 80 on protein stability. *Int. J. Pharm.*, **347**, 31–38.

62. Kerwin, B.A. and Remmel, R.L. (2007) Protect from light: photodegradation and protein biologics. *J. Pharm. Sci.*, **96**, 1468–1479.

63. International Conference on Harmonisation (ICH) (1997) *Guideline for the Photostability Testing of New Drug Substances and Products, (Q1B).*

64. Wang, W., Cui, T.Y., Wang, Y.J., and Martin-Moe, S. (2004) Oxidation of protein by vaporized sanitizing agents. *PDA J. Pharm. Sci. Technol.*, **58**, 121–129.

65. Yang, J., Wang, S., Liu, J., and Raghani, A. (2007) Determination of tryptophan oxidation of monoclonal antibody by reversed phase high performance liquid chromatography. *J. Chromatogr. A*, **1156**, 174–182.

66. Qi, P., et al. (2009) Characterization of the photodegradation of a human IgG1 monoclonal antibody formulated as a high-concentration liquid dosage form. *J. Pharm. Sci.*, **98**, 3117–3130.

67. Minton, A.P. (2005) Influence of macromolecular crowding upon the stability and state of association of proteins: predictions and observations. *J. Pharm. Sci.*, **94**, 1668–1675.

68. Sukumar, M., Doyle, B.L., Combs, J.L., and Pekar, A.H. (2004) Opalescent appearance of an IgG1 antibody at high concentrations and its relationship to noncovalent association. *Pharm. Res.*, **21**, 1087–1093.

69. Scott, D.J. and Winzor, D.J. (2009) Comparison of methods for characterizing nonideal solute self-association by sedimentation equilibrium. *Biophys. J.*, **97**, 886–896.

70. Rivas, G. and Minton, A.P. (2011) Beyond the second virial coefficient: sedimentation equilibrium in highly non-ideal solutions. *Methods*, **54**, 167–174.

71. Liu, J., Nguyen, M.D.H., Andya, J.D., and Shire, S.J. (2005) Reversible self-association increases the viscosity of a concentrated monoclonal antibody in aqueous solution. *J. Pharm. Sci.*, **94**, 1928–1940.

72. LeBrun, V.L., et al. (2010) A critical evaluation of self-interaction chromatography as a predictive tool for the assessment of protein-protein interactions in protein formulation development: a case study of a therapeutic monoclonal antibody. *Eur. J. Pharm. Biopharm.*, **75**, 16–25.

73. Attri, A.K. and Minton, A.P. (2005) Composition gradient static light scattering: a new technique for rapid detection and quantitative characterization of reversible macromolecular hetero-associations in solution. *Anal. Biochem.*, **346**, 132–138.

74. Yadav, S., Liu, J., Shire, S.J., and Kalonia, D.S. (2010) Specific interactions in high concentration antibody solutions resulting in high viscosity. *J. Pharm. Sci.*, **99**, 1152–1168.

75. Lehermayr, C., Mahler, H.C., Mader, K., and Fischer, S. (2011) Assessment of net charge and protein-protein interactions of different monoclonal antibodies. *J. Pharm. Sci.*, **100**, 2551–2562.

76. Kanai, S., Liu, J., Patapoff, T.W., and Shire, S.J. (2008) Reversible self-association of a concentrated monoclonal antibody solution mediated by Fab-Fab interaction that impacts solution viscosity. *J. Pharm. Sci.*, **97**, 4219–4227.

77. Wang, N., et al. (2009) Opalescence of an IgG1 monoclonal antibody formulation is mediated by ionic strength and excipients. *BioPharm. Int.*, **22**(4), 36–47.

78. Salina, B., et al. (2010) Understanding and modulating opalescence and viscosity in a monoclonal antibody formulation. *J. Pharm. Sci.*, **98**, 82–93.

79. Bhatnagar, B.S., Bogner, R.H., and Pikal, M.J. (2007) Protein stability during freezing: separation of stresses and mechanisms of protein stabilization. *Pharm. Dev. Technol.*, **12**, 505–523.

80. Hawe, A., Kasper, J.C., Friess, W., and Jiskoot, W. (2009) Structural properties of monoclonal antibody aggregates induced by freeze-thawing and thermal stress. *Eur. J. Pharm. Sci.*, **38**, 79–87.

81. Barnard, J.G., Singh, S., Randolph, T.W., and Carpenter, J.F. (2011) Subvisible particle counting provides a sensitive method of detecting and quantifying aggregation of

monoclonal antibody caused by freeze-thawing: insights into the roles of particles in the protein aggregation pathway. *J. Pharm. Sci.*, **100**, 492–503.

82. Kueltzo, L.A., Wang, W., Randolph, T.W., and Carpenter, J.F. (2008) Effects of solution conditions, processing parameters, and container materials on aggregation of a monoclonal antibody during freeze-thawing. *J. Pharm. Sci.*, **97**, 1801–1812.

83. Kasper, J.C. and Friess, W. (2011) The freezing step in lyophilization: physical-chemical fundamentals, freezing methods and consequences on process performance and quality attributes of biopharmaceuticals. *Eur. J. Pharm. Biopharm.*, **78**, 248–263.

84. Abdul-Fattach, A.M., Kalonia, D.S., and Pikal, M.J. (2007) The challenge of drying method selection for protein pharmaceuticals: product quality implications. *J. Pharm. Sci.*, **96**, 1886–1916.

85. Johnson, R.E., Teagarden, D.L., Lewis, L.M., and Gieseler, H. (2009) Analytical accessories for formulation and process development in freeze-drying. *Am. Pharm. Rev.*, **12**(5), 54–60.

86. International Conference on Harmonization (2005) *Guidance for Industry Q2(R1) Validation of Analytical Procedures: Text and Methodology.*

87. Bierau, H., et al. (2008) Higher-order structure comparison of proteins derived from different clones or processes. unbiased assessment of spectra by quantitative CD. *BioProcess Int.*, **6**, 52–59.

88. Jiang, Y., et al. (2008) Meeting new regulatory expectations of characterization methods qualification of biophysical analyses. *Am. Pharm. Rev.*, **11**, 1–7.

89. Li, C.H., et al. (2011) Applications of circular dichroism (CD) for structural analysis of proteins: qualification of near- and far-UV CD for protein higher order structural analysis. *J. Pharm. Sci.*, **100**, 4642–4654.

90. Ravi, J., et al. (2010) International comparability in spectroscopic measurements of protein structure by circular dichroism: CCQM-P59. 1. *Metrologia*, **47**(6), 631–641.

91. Hughes, H., et al. (2009) A multi-tiered analytical approach for the analysis and quantitation of high-molecular-weight aggregates in a recombinant therapeutic glycoprotein. *Am. Assoc. Pharm. Sci. J.*, **22**, 335–341.

92. Wen, Z.Q. (2007) Raman spectroscopy of protein pharmaceuticals. *J. Pharm. Sci.*, **96**, 2861–2878.

93. Li, C.H. and Li, T. (2009) Application of vibrational spectroscopy to the structural characterization of monoclonal antibody and its aggregate. *Curr. Pharm. Biotechnol.*, **10**, 391–399.

94. Yang, J., Wang, S., Liu, J., and Raghani, A. (2007) Determination of tryptophan oxidation of monoclonal antibody by reversed phase high performance liquid chromatography. *Journal of Chromatography A*, **1156**, 174–182.

95. Arakawa, T., Ejima, D., Li, T., and Philo, J.S. (2010) The critical role of mobile phase composition in size exclusion chromatography of protein pharmaceuticals. *J. Pharm. Sci.*, **99**, 1674–1692.

96. Carpenter, J.F., et al. (2010) Potential inaccurate quantitation and sizing of protein aggregates by size exclusion chromatography: essential need to use orthogonal methods to assure the quality of therapeutic protein products. *J. Pharm. Sci.*, **99**, 2200–2208.

97. Bond, M.D., et al. (2010) Evaluation of a dual-wavelength size exclusion HPLC method with improved sensitivity to detect protein aggregates and its use to better characterize degradation pathways of an IgG1 monoclonal antibody. *J. Pharm. Sci.*, **99**, 2582–2597.

98. Philo, J.S., (2009) A critical review of methods for size characterization of non-particulate protein aggregates. *Curr. Pharm. Biotechnol.*, **10**, 359–372.

99. Michels, D.A., Brady, L.J., Guo, A., and Balland, A. (2007) Fluorescent derivatization method of proteins for characterization by capillary electrophoresis-sodium dodecyl sulfate with laser-induced fluorescence detection. *Anal. Chem.*, **79**, 5963–5971.

100. Kamerzell, T.J., et al. (2011) The relative rate of immunoglobulin gamma-1 fragmentation. *J. Pharm. Sci.*, **100**, 1341–1349.

101. Gaza-Bulseco, G. and Liu, H. (2008) Fragmentation of a recombinant monoclonal antibody at various pH. *Pharm. Res.*, **25**, 1881–1890.

102. Cao, S., Pollastrini, J., and Jiang, Y. (2009) Separation and characterization of protein aggregates and particles by field flow fractionation. *Curr. Pharm. Biotechnol.*, **10**, 382–390.

103. Liu, J., Andya, J.D., and Shire, S.J. (2006) A critical review of analytical ultracentrifugation and field flow fractionation methods for measuring protein aggregation. *Am. Assoc. Pharm. Sci. J.*, **8**(3), E580–E589.

104. Gabrielson, J.P., et al. (2007) Quantitation of aggregate levels in a recombinant humanized monoclonal antibody formulation by size-exclusion chromatography, asymmetrical flow field flow fractionation, and sedimentation velocity. *J. Pharm. Sci.*, **96**, 268–279.

105. de Kort, B.J., ten Kate, G.A., de Jong, G.J., and Somsen, G.W. (2011) Capillary electrophoresis with lamp-based wavelength-resolved fluorescence detection for the probing of protein conformational changes. *Anal. Chem.*, **83**, 6060–6067.

106. Hawe, A., Sutter, M., and Jiskoot, W. (2008) Extrinsic fluorescent dyes as tools for protein characterization. *Pharm. Res.*, **25**, 1487–1499.

107. He, F., et al. (2010) Detection of IgG aggregation by a high throughput method based on extrinsic fluorescence. *J. Pharm. Sci.*, **99**, 2598–2608.

108. Hawe, A., Filipe, V., and Jiskoot, W. (2010) Fluorescent molecular rotors as dyes to characterize polysorbate-containing IgG formulations. *Pharm. Res.*, **27**, 314–326.

109. Demeule, B., Gurny, R., and Arvinte, T. (2007) Detection and characterization of protein aggregates by fluorescence microscopy. *Int. J. Pharm.*, **329**, 37–45.

110. Demeule, B., et al. (2007) Characterization of protein aggregation: the case of a therapeutic immunoglobulin. *Biochim. Biophys. Acta*, **1774**, 146–153.

111. Singh, S.K., et al. (2010) An industry perspective on the monitoring of subvisible particles as a quality attribute for protein therapeutics. *J. Pharm. Sci.*, **99**, 3302–3321.

112. Narhi, L.O., et al. (2009) A critical review of analytical methods for subvisible and visible particles. *Curr. Pharm. Biotechnol.*, **10**, 373–381.

113. Filipe, V., Hawe, A., and Jiskoot, W. (2010) Critical evaluation of nanoparticle tracking analysis (NTA) by nanosight for the measurement of nanoparticles and protein aggregates. *Pharm. Res.*, **27**, 796–810.

114. Demeule, B., Messick, S., Shire, S.J., and Liu, J. (2010) Characterization of particles in protein solutions: reaching the limits of current technologies. *Am. Assoc. Pharm. Sci. J.*, **12**, 708–715.

115. Wuchner, K., et al. (2010) Development of a microflow digital imaging assay to characterize protein particulates during storage of a high concentration IgG1 monoclonal antibody formulation. *J. Pharm. Sci.*, **99**, 3343–3361.

116. Mach, H., et al. (2011) The use of flow cytometry for the detection of subvisible particles in therapeutic protein formulations. *J. Pharm. Sci.*, **100**, 1671–1678.

117. Ludwig, D.B., et al. (2011) Flow cytometry: a promising technique for the study of silicone oil-induced particulate formation in protein formulations. *Anal. Biochem.*, **410**, 191–199.

118. Raab, A., et al. (1999) Antibody recognition imaging by force microscopy. *Nat. Biotechnol.*, **17**, 902–905.

119. Lee, H., Kirchmeier, M., and Mach, H. (2011) Monoclonal antibody aggregation intermediates visualized by atomic force microscopy. *J. Pharm. Sci.*, **100**, 416–423.

120. Bobst, C.E., et al. (2008) Detection and characterization of altered conformations of protein pharmaceuticals using complementary mass spectrometry-based approaches. *Anal. Chem.*, **80**, 7473–7481.

121. Heck, A.J.R. (2008) Native mass spectrometry: a bridge between interactomics and structural biology. *Nat. Methods*, **5**, 927–933.

122. Pease, L.F., et al. (2008) Determination of protein aggregation with differential mobility analysis: application to IgG antibody. *Biotechnol. Bioeng.*, **101**, 1214–1222.

123. Wang, G., Abzalimov, R.R., and Kaltashov, I.A. (2011) Direct monitoring of heat-stressed biopolymers with temperature-controlled electrospray ionization mass spectrometry. *Anal. Chem.*, **83**, 2870–2876.

124. Houde, D., Peng, Y., Berkowitz, S.A., and Engen, J.R. (2010) Post-translational modifications differentially affect IgG1 conformation and receptor binding. *Mol. Cell. Proteom.*, **9**, 1716–1728.

125. Houde, D., Berkowitz, S.A., and Engen, J.R. (2011) The utility of hydrogen/deuterium exchange mass spectrometry in biopharmaceutical comparability studies. *J. Pharm. Sci.*, **100**, 2071–2086.

126. Perico, N., Purtell, J., Dillon, T.M., and Ricci, M.S. (2009) Conformational implications of an inversed pH-dependent antibody aggregation. *J. Pharm. Sci.*, **98**, 3031–3042.

127. Dillon, T.M., et al. (2008) Structural and functional characterization of disulfide isoforms of the human IgG2 subclass. *J. Biol. Chem.*, **283**, 16206–16215.

128. Rathore, A.S. and Winkle, H. (2009) Quality by design for biopharmaceuticals. *Nat. Biotechnol.*, **27**, 26–34.

12

BIOPHYSICAL ANALYSES SUITABLE FOR CHEMISTRY, MANUFACTURING, AND CONTROL SECTIONS OF THE BIOLOGIC LICENSE APPLICATION (BLA)

Zahra Shahrokh, Nazila Salamat-Miller, and John J. Thomas

Rare Disease Business Unit, Shire, Department of Pharmaceutical and Analytical Development, Lexington, MA, USA

Biophysical Methods for Biotherapeutics: Discovery and Development Applications, First Edition. Edited by Tapan K. Das.
© 2014 John Wiley & Sons, Inc. Published 2014 by John Wiley & Sons, Inc.

12.1   INTRODUCTION

The development of biological therapeutics is challenged by the need to control their chemical, structural, and biological properties during manufacturing, handling, distribution, storage, and usage of the product. Detailed knowledge of a product's characteristics and understanding of the process–product relationship are essential in managing the complexity and controlling the microheterogeneity of biotherapeutics. Biophysical methods are needed to provide insight into structural properties as supporting information for the selection of product formulation, manufacturing/handling/storage conditions, and comparability following process changes. Technologies are selected from the biophysical tool box to answer specific questions during the development phase through product licensure.

Inclusion of biophysical methods in regulatory submissions is intended to demonstrate knowledge and control of the product as related to its quality. The overall expectation of regulatory agencies in the license applications is the documentation of adequate biophysical data that complement the biochemical and process knowledge to provide an integrated view of product quality. The specific expectations may vary among regulatory agencies and the types of products; in the United States, there is an increasing expectation to include state-of-the-art technologies and orthogonal methodologies for the characterization of therapeutic protein products. To meet these expectations, a company's challenge is to have enough know-how about newer technologies to interpret the biological significance of the data, and to balance this significance with the resources, time, and cost of conducting biophysical studies.

Generally, three Chemistry, Manufacturing, and Control (CMC) sections of the license applications contain biophysical data: Elucidation of Structure, Comparability, and Pharmaceutical Development. Biophysical data provided in the "Elucidation of Structure" section is used to define folded structure, association states, and ligand or receptor interaction that would affect function. Some of these characterization methods are also included in the "Comparability" section to demonstrate maintenance of the structural properties following process changes. A subset of biophysical characterization methods is used for the evaluation of molecular structural stability under different processing, storage, or clinical use conditions. These data may be included in the "Pharmaceutical Development" section of a license application to support the formulation selection and drug product manufacturing process. The collective information in these sections should provide an integrated view of product quality and its maintenance during manufacture, storage, and usage.

This chapter provides a philosophical and practical approach to biophysical studies performed during the product life cycle from molecule selection to licensure. Case examples are included to demonstrate the use of different types of biophysical methods in addressing product-specific questions. The limitations and advantages of the methods are outlined as the basis for their selection and for inclusion in the license application. Since significant expertise and cost are involved in conducting and interpreting most biophysical methods and data, the effort to gain biophysical knowledge needs to be balanced with the value of the information obtained in advancing products to the market.

12.2 THE BIOPHYSICAL TOOL BOX

Many spectroscopic-, calorimetric-, electrical-, and centrifugal, interaction-based technologies constitute the biophysical tool box for assessing a protein's folded structures, association states, structural stability, and function. Table 12.1 lists the most commonly used methods along with their key benefits and limitations and aspects that determine the use of orthogonal methods. Table 12.2 follows with a summary of the typical application of these methods during product development. In addition to the most commonly used biophysical methods, the emerging technologies (listed in Table 12.3) have the potential to provide further insight into protein's structure. Some of these emerging methods are complex and not robust enough to demonstrate small differences, and may also not be readily available for common use outside of the academic centers.

The selection of the technologies depends on the complexity of the question being addressed. Several factors contribute to the choice of biophysical characterization tools:

- The protein's attributes
- The method's ease of use
- The availability of expert resources
- The method's advantages and limitations
- The material limitations

TABLE 12.1. Attributes of common biophysical methods for analysis of therapeutic proteins

Biophysical properties	Analytical methodology (amounts needed)	Key benefits	Key limitations
Folded structure (secondary and tertiary)	Circular dichroism spectroscopy (μg to mg)	• Rapid assessment of protein folding • Reproducible	• Resolution • Requires data deconvolution to estimate the content of structural components
	Fourier transform infrared spectroscopy (μg to mg)	• Assessment of native protein structure in solid state • Rapid and reproducible • Minimal interference from light scattering phenomena	• Resolution and complexity of spectra • Interpretation of structural significance
	Raman spectroscopy (mg)	• Assessment of native protein structure in solid state	• Low signal-to-noise ratio • Resolution and complexity of spectra
	Differential scanning calorimetry (mg)	• Rapid assessment of unfolding temperature under different solution conditions • Precision	• Specificity • T_m does not always correlate to the physical stability
	Intrinsic fluorescence spectroscopy (μg)	• Sensitivity • Assessment of native protein structure without modification	• Resolution (mix of local and global structural information) • Signal sensitive to environmental conditions not related to protein structure
	X-ray crystallography NMR spectroscopy (tens of mg)	• High resolution • High content of molecular-level information	• Difficult to generate appropriate crystals • Low throughput • Instrumentation complexity

TABLE 12.1. (Continued)

Biophysical properties	Analytical methodology (amounts needed)	Key benefits	Key limitations
	Secondary-derivative ultraviolet spectroscopy (μg)	• Sensitivity • Ease of use • Detection of subtle structural changes	• Interpretation of structural significance
Association states/ aggregates	Analytical ultracentrifugation (mg)	• Solution phase (native condition) • No matrix effect	• Instrumentation complexity and data interpretation requires expertise • Reproducibility (may show artifact peaks at low levels) • Sensitivity (low accurate quantitation <1% aggregates)
	Dynamic light scattering (μg)	• Solution phase (native condition) • No matrix effect • Large dynamic range	• Resolution (cannot distinguish less than ~eightfold difference in mass for DLS) • Quantitative estimates are based on shape assumptions and different algorithms • Observation of artifacts at very high concentration solutions due to multiple scattering effect • Not quantitative
Subvisible particles	Light obscuration (hundreds of mg)	• Assessment of particles in the range 2–150 μm • Assessment of particle count in the range ~10–18,000 particles/mL	• Requires large quantity of material (~25 mL) • False-positive signals • Cannot distinguish between proteins and other particles

(Continued)

TABLE 12.1. (Continued)

Biophysical properties	Analytical methodology (amounts needed)	Key benefits	Key limitations
	Imaging techniques (mg to g)	• Assessment of a wide range of particles (from 1 to more than 25 µm) • Visualization of particle morphology • Characterization of particles • Analysis algorithms for data visualization	• Sample preparation and handling artifacts, resulting in poor robustness • Throughput • Lack of protein standards for calibration • Superior image quality in some techniques (FlowCam) compared to the others
Ligand or receptor binding interactions	Surface plasmon resonance (µg)	• Specificity • Provides binding kinetics	• Low throughput (compared to ELISA) • Robustness
	Bio-layer interferometry (µg)	• Specificity • Provides binding kinetics • High throughput	• Robustness

Often the industry experience with a family of products (e.g., monoclonal antibodies), or the history of a product, and the development phase may dictate the methodologies selected. The intent is to produce well-characterized and -controlled therapeutic products.

A combination of techniques is typically used to thoroughly characterize a protein product. In some cases an orthogonal method is combined with a routine method to verify the information obtained by the routine method. Such is the case with size exclusion HPLC (SEC) that is routinely performed for lot release testing and process development. It is a simple method and has shown to have appropriate precision and sensitivity for its intended use. As an orthogonal method, analytical ultracentrifugation (AUC) is used to correlate solution phase data with SEC results to confirm the suitability of the SEC method for determination of aggregate levels. In other cases, an established method may be combined with an emerging technology to answer a more complex question.

Despite the value of biophysical methods throughout product development, these techniques are not validated for reasons such as poor robustness, lack of expertise, inconsistencies in the experimental parameters, and sample preparation/handling. Additionally, extensive knowledge and experience with the technology may be required to

TABLE 12.2. Applications of commonly used biophysical methods during development of therapeutic proteins

Biophysical properties	Analytical methodology	Applications of methods per development phase		
		Pre-IND	Clinical development	BLA filing
Folded structure (secondary and tertiary)	Circular dichroism spectroscopy	Preformulation	Structural characterization Comparability	
	Differential scanning calorimetry	Preformulation Formulation development		
	Fluorescence spectroscopy	Preformulation characterization		
	Fourier transform infrared spectroscopy	Preformulation	Lyophilized formulation development Structure characterization Surface interaction	As applicable
	NMR spectroscopy (small proteins and peptides)	Structural characterization Device interaction using solid-state NMR		
	X-ray crystallography	Structural characterization when crystal structure data are available		
Association states/ aggregates	Analytical ultracentrifugation	Preformulation	Formulation development, stability, size distribution, characterization, comparability	
Subvisible particulates	Light scattering	Preformulation		
	Light obscuration	Not applicable	Formulation development Lot release	Lot release
	Imaging techniques	Preformulation	Formulation development, particle characterization, comparability	Particle characterization, comparability
Ligand/receptor binding kinetics	Surface plasmon resonance, bio-layer interferometry	Bio-characterization[a]	Bio-characterization,[a] comparability	

[a] Bio-characterization relates to assessing bioactivity of the product.

TABLE 12.3. Emerging technologies for protein biophysical characterization

Biophysical properties	Analytical methodology	Key benefits	Key limitations
Folded structure (secondary and tertiary)	Hydrogen/ deuterium exchange—mass spectrometry	• Specificity • Robustness • Range of structural information (global and local structure)	• Low throughput (sample handling and data interpretation) • Sensitivity to detect small changes
	Ion mobility mass spectrometry	• Direct structural information from mixtures	• Resolution • Relevance to solution structure
	NMR (large proteins)	• Resolution • High structural content	• High magnetic fields not commonly available • Low throughput • Sample requirements
Association states/ aggregates	Asymmetric field-flow fractionation	• High throughput • Robustness (applicable to formulated protein solutions)	• Potential for artifactual results or sample loss
Ligand or receptor binding	Microscale thermophoresis	• Ease of use	• Specificity

interpret the significance of any differences observed in the experimental results. Furthermore, the use of the instruments and the data interpretation algorithms may require expertise not readily available outside of academic centers or specialized contract research organizations. Finally, the relevance of differences in biophysical characteristics to the biological properties or *in vivo* effects is often speculative.

12.3 COMMON BIOPHYSICAL METHODS FOR ASSESSING FOLDED STRUCTURE

12.3.1 Circular Dichroism Spectroscopy

Circular dichroism (CD) is a common characterization method included in the license application to describe the secondary and tertiary structural elements of a protein product. CD spectroscopy utilizes differential absorption of left and right circularly polarized light to highlight the asymmetry of peptide bonds. Far-UV CD spectra denote secondary structural features in the range of 180–250 nm. Near-UV CD spectra, in the range of 250–350 nm, provide insights into tertiary structure. Changes in the aromatic amino acid band wavelength and intensities may indicate variations in their local microenvironment. To infer a global understanding of solution phase tertiary structure of a product, the near-UV CD spectra are often complemented with data from other biophysical methods such as fluorescence or second-derivative UV spectroscopy.

CD spectroscopy is useful for assessing the structural effects of post-translational protein modifications and degradation [1–3], but generally large structural changes should occur to detect any changes in the far-UV CD spectral profile. In one example, however, the combination of CD spectroscopy and other biophysical methods was used to conclude minor structural differences following manufacturing process changes. Far-UV CD spectroscopic analysis of Epogen® and Eprex®, the injectable formulations of recombinant human epoetin alfa made by different manufacturing processes, showed 5% less helical content for Eprex® than for Epogen® [4]. Moreover, the near-UV CD spectral findings combined with fluorescence spectroscopy demonstrated a likely increase in the flexibility of tryptophan residues in Eprex® compared to Epogen®. The data collectively indicated that there were slight structural differences between the two molecules.

CD spectroscopy is also useful for comparing structural properties of product variants. Such analysis combined with other biochemical and biological assessments is used to categorize variants as either product-related substances (with desirable structure and function), or as impurities (lacking the desired quality) as defined in ICH Guideline Q6B. For example, C-terminal truncation of monoclonal IgG antibodies is a common modification that was shown to have no effect on several biophysical properties (including CD spectra), as well as no effect on biological activity [5, 6]. In another example, proteolytically fragmented species were observed by SDS-PAGE during scale-up production of a monomeric 80 kDa sulfatase. In the case of this sulfatase, peptide mapping revealed clipping at several sites, yet the CD spectra (200–260 nm) and the apparent size under non-denaturing conditions by SEC were indistinguishable between the native and the "nicked" species, making it difficult to separate them from the native species during purification. Interestingly, the cell-based biological activity and enzymatic activity of the "nicked" form were identical to the native species despite extensive "nicking" observed by peptide mapping. Since removal of cell culture-derived proteases, which typically give rise to species with altered structural and biochemical properties, poses a challenge to purification of protein therapeutics. Subsequently, the manufacturing process was improved in such a way to control this proteolytic fragmentation that occurred in the early steps. The lack of impact of such nicking on structure and bioactivity was used in the license application to claim the fragmented species as a "product-related variant" rather than an impurity.

Estimates of the level of secondary structural features (e.g., α-helix, β-sheet) can be made following deconvolution of the far-UV CD spectra, although the lack of consistency in these algorithms often limits quantitation with this method. Likewise, it is difficult to obtain quantitative information about tertiary structure by near-UV CD [7] where changes in the aromatic amino acid wavelength band and intensities may indicate variations in their local microenvironment. Thus, while useful in comparability studies, such limitations have restricted CD methods to a semiquantitative fingerprint method. Another limitation of the methods is material requirements, where near-UV CD requires a higher concentration of protein compared with far-UV CD to produce a clear signal.

To infer a global understanding of solution phase tertiary structure of a product, the near-UV CD spectra are often complemented with data from other biophysical methods such as fluorescence or second-derivative UV spectroscopy.

12.3.2 Vibrational Spectroscopy

Vibrational spectroscopy (i.e., infrared (IR) and Raman) is another long-established methodology for characterizing the secondary structure of proteins and peptides. The vibrational modes of specific chemical moieties such as amide carbonyls, cysteine sulfhydryls, and acidic amino acid side chains are well defined. The vibrational states of these abundant moieties in peptides and proteins are greatly influenced by their environment, making IR a sensitive tool to monitor changes in the secondary structure. However, the vast number of vibrational modes in proteins also results in complex spectra of multiple overlapping bands, requiring mathematical data manipulations such as Fourier transform and second-order derivative spectral analysis to enhance resolution [8–10]. Previously, the presence of water bands added complexity to the spectra, but with the new instrumentation, the interference from water is minimal [11]. Alternatively, for comparability purposes, quantitative spectral analysis is not necessarily needed.

A useful application of FTIR is in the development of lyophilized formulation processes. Using attenuated total reflectance (ATR)-FTIR, changes in the secondary structure can be monitored before and after lyophilization [12]. ATR-FTIR takes advantage of a total internal reflection where a beam of IR light enters a transparent crystal and undergoes several internal reflections before reaching a detector. The crystal is in contact with the sample where the IR beam, at the reflection points, penetrates a few micrometers beyond the crystal surface into the sample [13–15]. In addition to its versatility in assessing secondary protein structure in liquid and solid states, an advantage of this technique is its compatibility with high protein concentrations at different states (aqueous, frozen, lyophilized powder). The introduction of ATR-FTIR revolutionized the application of FTIR to a much faster method where liquids, solids, dispersion samples, as well as surface adsorption phenomena can be studied.

FTIR is also valuable for monitoring changes in the higher order structures of small proteins and peptides. It provides a rapid assessment of the structural integrity for comparability studies. For example, this method has been used to determine the effect of deamidation on the folded structure of the therapeutic peptide, amylin [16]. The data show the appearance of new absorption bands and a shift in the profile related to unfolding following deamidation. Overall, the use of FTIR is product-specific and its inclusion in regulatory filings is usually tied to situations where a solid-state characterization of a protein or peptide structure is needed. Depending on the company's development history and expertise, FTIR may be used for comparability studies.

UV resonance Raman (UVRR) spectroscopy is another technique used to probe the higher order structure of proteins. Raman spectroscopy can provide similar information as FTIR; however, its lower sensitivity has limited its application to high-concentration products (e.g., monoclonal antibody formulations). Further developments in more sensitive UVRR techniques have enabled the use of this method to assess the secondary and tertiary structures of therapeutic proteins by monitoring the response from specific side chain excitation [17,18]. Quantitation of proteins' secondary structure has been obtained by utilizing specific amide vibrational modes and reference data sets containing pure secondary structural elements [19, 20]. A similar common approach of using specific wavelengths for quantifying secondary elements from far- and near-CD data has drawn comparisons to UVRR analysis. However, nonlinear spectral resolution across the wavelength range makes Raman spectroscopic data set comparisons problematic. For this reason Raman spectroscopy has been restricted to research and discovery applications [21].

12.3.3 X-Ray Crystallography

X-ray crystallography provides the most detailed understanding of protein structure. The X-ray wavelength ($\sim$1.5 Å) is similar to the inter-atomic distance found in protein molecules; thus, it is well suited for atomic-level characterization of proteins. As such, it is most desirable to have crystal structure data early in development of a therapeutic protein and to include in the licensure as part of characterization of the product.

While this technique provides a high-resolution view of the protein structure, it is often difficult to obtain crystal structure of proteins. Significant amounts of time are spent obtaining high-quality protein crystals to produce the necessary strong in-phase X-ray diffraction pattern. Recent attempts to develop high-throughput methods for obtaining quality crystals may eventually lead to greater use of X-ray crystallography during product development and its greater inclusion in regulatory filings [22–24].

Glycosylation remains a challenge for crystallization of glycoproteins, resulting in less defined data even if high-quality crystals are obtained. Exceptions are observed with IgG monoclonal antibodies where the N-glycans often bridge two unit cells and thus fill the intra-molecular void. The common approach toward many glycoproteins is to truncate the glycans to produce quality X-ray data while retaining the location of the glycosylation. Another limitation of X-ray crystallography is that it provides little information regarding protein dynamics. Partial unfolding and refolding are natural mechanisms that enable protein binding to ligands/receptors and its biological activity. Methods have been developed to show protein mobility based upon observed crystal heterogeneity, but information on large-scale protein dynamics remains limited due to crystal packing forces incurred during sample preparation [25].

While X-ray crystallographic analysis has been primarily used as a discovery tool [26,27], the increased availability of X-ray structural analysis is now providing information in support of protein therapeutic development. In a case study, X-ray crystallographic data demonstrated the structural comparability of acid beta-glucocerebrosidase produced from different expression systems. A recombinant form of acid beta-glucocerebrosidase was expressed in CHO cells and exposed to exoglycosidases to produce phosphorylated high mannose glycoforms. The sequence also had one amino acid substitution (Arg -> His) different from the endogenous human sequence. X-ray crystallographic data from this enzyme showed that this substitution had no impact on the crystal structure, as it was similar to the crystallographic data obtained from the natural form derived from placenta [28, 29]. Another version of this enzyme was produced through gene-activated expression in human cell lines with cell culture media additives that directly produce high mannose glycosylation. The X-ray crystal data demonstrated the structural comparability between the recombinant and the gene-activated forms of the enzymes produced by different processes [30]. Despite the comparable protein structure at the atomic level, the biological functions of the two proteins were found to be different [31]. The X-ray crystallographic data provided detailed structural information in support of the biologic license application (BLA).

12.3.4 Nuclear Magnetic Resonance Spectroscopy

The specificity of NMR as a structural characterization tool comes from its ability to measure different frequencies for each atom within the molecule that are dependent on the linkage to nearby atoms. Unlike the common use of one-dimensional proton NMR for structural identification of small molecules, signal overlap and peak broadening with increasing number of atoms in a molecule require the use of 2D-NMR spectroscopic analysis and limits its application to proteins smaller than 50 kDa. The complexity of 2D-NMR analysis of proteins requires resources not readily available in most biotechnology companies or contract laboratories. NMR analysis also requires millimolar concentrations of proteins, which may be limiting for development purposes. Although NMR can provide valuable insight into protein's folded structure under various buffer and excipient conditions, these practical limitations restrict the use of NMR to a one-time characterization that might be included in regulatory filings.

NMR analysis has been utilized in the development of some small protein or peptide therapeutics. Insulin was one of the first therapeutic proteins structurally characterized by NMR. The limited complexity of insulin is well within the capabilities of NMR, allowing for the use of this method toward evaluating the association state of insulin under various solution conditions [32–34]. EMEA has also documented the use of NMR (orthogonal to CD) analysis for characterization and comparability of Exenatide®, a peptide for the treatment of diabetes [35, 36]. For a larger protein, the initial structural attributes of human erythropoietin (EPO) were obtained using NMR studies [37]. The data have indicated that the 34 kDa epoetin alfa adopts a four-helix bundle structure connected through large loops. By detecting the large helical content of epoetin alfa, NMR spectra allowed interpretation of far-UV CD studies of potential changes in the helicity of a follow-on product. Additionally, the proximity of three tryptophan residues deduced by NMR has shown the potential for detecting changes in the local environment by other biophysical methods such as fluorescence or near-UV CD spectroscopy. NMR has also been cited in a regulatory comparison of a biosimilar protein therapeutic for the treatment of neutropenia [38]. In comparison to an innovator product, Neupogen®, the ^{1}H-NMR data from the follow-on molecule, Zarzio®, showed a high degree of similarity with no unexpected shifts or broadening compared to the original product.

The NMR spectral fingerprint allows for a higher throughput comparison of spectra rather than the complete assignment of the entire structure. The role of excipients on

peptide folding has been evaluated by NMR profile analysis without intensive structural assignment [39]. By utilizing fingerprint comparisons, NMR can demonstrate the
structural integrity of a protein during comparability studies. In addition to its utility
as a high-resolution characterization tool, NMR spectroscopy has been utilized as an
orthogonal approach for protein identification and comparability [38].

12.3.5 Differential Scanning Calorimetry

Calorimetric measurements provide a reproducible and practical means for rapid assessment of the thermal stability of protein therapeutics. In differential scanning calorimetry
(DSC), the heat capacity of a protein linearly increases with increasing temperature
until the protein begins to unfold, which results in a large increase in heat capacity.
DSC is capable of measuring changes in protein folding thermodynamic parameters
(e.g., enthalpy) as a function of temperature, provided that the unfolding transition is
reversible and the change in transition heat capacity is independent of scan rate [40].
The thermogram also offers insight into the molecular domains within a protein, but the
most commonly reported information is the melting temperature (T_m) as a measure of a
protein's thermal stability.

The ease of use, relatively low material requirements, and simple data outputs have
made DSC the method of choice from preformulation studies in the early development
phase to comparability studies in the license application. For example, DSC was used to
establish the comparability of a 51 kDa glycoprotein before and after a manufacturing
process change. The process change was implemented to increase the yield as well as to
remove any animal-derived components in the cell culture media. The thermograms were
obtained for three lots manufactured by each manufacturing process, using 1 mg/mL
protein concentration and a scan rate of 1°C/min. The irreversible melting profiles for
all lots remained superimposable and consisted of one single symmetrical peak at ~60.7
$\pm$ 0.1°C. The results demonstrated no change in the global conformational stability of
this protein following changes in the manufacturing process.

The information on the global unfolding event(s) obtained by DSC may or may
not be corroborated by other biophysical techniques. For example, DSC was found to
be less sensitive to the partial unfolding of interferon-α2a compared to fluorescence
and near-UV CD spectroscopy [41]. Consequently, the use of orthogonal methods is
recommended for characterizing a protein's folded structure during characterization and
comparability studies.

12.3.6 Fluorescence Spectroscopy

Fluorescence spectroscopy is a well-established biophysical method to assess the folded
state(s) of proteins [42, 43]. Given the relatively low sample amount requirements and
the universal application to proteins, fluorescence spectroscopy is often used to assess
structural integrity of proteins such as commercialized monoclonal antibodies [44, 45].
Fluorescence spectroscopy can also be applied to evaluate thermal denaturation of
proteins as an orthogonal technique to DSC [46]. The fluorescence studies are done in

two major categories: intrinsic and extrinsic protein fluorescence. Intrinsic fluorescence arises from the aromatic residues and does not involve any protein modification or adducts, while the extrinsic fluorescence utilizes different types of dyes or probes that fluoresce upon interaction with the protein. Intrinsic fluorescence often produces a much weaker signal compared with the fluorescent probes.

Intrinsic protein fluorescence measurements depend on the microenvironment of the aromatic amino acids. The signal arises predominantly from tryptophans; other aromatic amino acids (i.e., Tyr, Phe) can also contribute to the fluorescence signal, but to a much lesser degree. A shift in the peak maximum wavelength or the signal intensity can be detected if changes in the tryptophan(s) microenvironment occur, for example, upon protein denaturation. These traits, together with the simplicity and high throughput of intrinsic protein fluorescence methodology, make it a useful tool in preformulation and formulation development studies. In one example, fluorescence spectroscopy was used to evaluate the impact of lyophilization on the tertiary structure of interferon-α2a, showing a slight blue shift in the solid state compared to solution phase [47].

An established orthogonal approach to intrinsic protein fluorescence measurements is the use of extrinsic fluorescent dyes [44, 48, 49]. This approach utilizes dyes such as 8-anilino-1-naphthalene sulfonate (ANS) that have weak fluorescence in water or hydrophilic environments and strong fluorescence in hydrophobic environments. Such methods are used for assessing global conformational changes involving exposure of hydrophobic patches.

One limitation of extrinsic probes is their sensitivity to formulation excipients such as polysorbates [50]. Moreover, extrinsic fluorescence methods depend on exposure of hydrophobic regions of proteins; however, these regions may not always be exposed following structural changes. Such limitations have led to the recent development of molecular fluorescence rotor probes, such as 9-(2,2-dicyanovinyl) julolidine (DCVJ) that detects changes in the local viscosity of the protein microenvironment [50,51]. The flexible molecular structure of rotor probes results in fluorescence anisotropy, that is, unequal fluorescence along the probe's axes of rotation [52,53]. The binding of these probes to multimeric aggregates with greater local viscosity constrains their rotation and results in a change in their fluorescence anisotropy. The use of fluorescent molecular rotor probes is an evolving methodology that may allow for monitoring protein aggregation independent of solution composition.

Protein fluorescence spectra may be included in the "Elucidation of Structure" section of the BLA to complement data from other measures of protein conformation. The dependence of the fluorescence signal on environmental conditions requires collecting adequate historical data to be able to delineate analytical variation from significant structural changes when interpreting comparability data. With adequate understanding of the fluorescence data, this information may be included in the "Comparability" section of BLA.

12.3.7   Ultraviolet Spectroscopy

UV spectroscopy is a low-resolution technique for monitoring the global structure of proteins by measuring the absorbance of the aromatic residues that are affected by their microenvironment. This method can be used as an orthogonal biophysical method for demonstrating comparability following manufacturing process changes. A mathematical conversion of UV spectra to second-derivative spectra creates a higher resolution method for investigating subtle global structural changes around the microenvironment of aromatic residues in proteins [54]. This method has been applied in preformulation studies and in constructing an empirical pH–temperature stability phase diagram for bovine colony-stimulating factor [55, 56]. Given the higher resolution of second-derivative UV spectroscopy, it is potentially more useful than the zero-order UV analysis for comparability studies in the regulatory submissions.

12.4 COMMON BIOPHYSICAL METHODS FOR ASSESSING SIZE HETEROGENEITY, ASSOCIATION STATE, AGGREGATION

Concerns with immunogenicity of protein aggregates have triggered an increase in the use and development of a number of orthogonal sizing methods and inclusion of data generated by these methods in the license application to demonstrate product characterization and control. SEC remains the routine method for monitoring aggregates due to its ease of use, high precision, and quantitative nature of the data. When combined with online light scattering (LS), SEC also provides molecular weight determinations of the chromatographically resolved species. However, SEC analysis may result in dilutional dissociation of reversible association states (and meta-stable aggregated species), and unwanted interactions with the column stationary phase. Such limitations necessitate the confirmation of the aggregate levels by orthogonal biophysical methods that operate under solution phase conditions as described below.

12.4.1 Analytical Ultracentrifugation

AUC is a valuable tool for understanding the molecular association and conformation of proteins including the affinity and rate of reversible protein self-association. Given the length of AUC procedure (i.e., more than several hours) compared to SEC analysis (approximately 15–30 minutes), the detection of self-association by AUC method implies that the lifetime of the associated species is in the order of hours [57].

AUC is used throughout product development, from pre-IND characterization studies to clinical development, and comparability assessment; hence the data are often included in regulatory submissions. Two types of AUC techniques, sedimentation velocity (SV) and sedimentation equilibrium (SE), are used to understand different aspects of the protein conformation or association.

SV-AUC is generally used to verify the level of aggregates detected by SEC [58]. An example of such application is the case of SEC and AUC (SV) comparison for a glycoprotein where aggregate levels changed during freeze-thaw cycling. The SEC method resolved the high molecular weight species as two earlier eluting peaks in addition to the main peak. Similarly, AUC (SV) method demonstrated the presence of a main as well as faster sedimenting species. A linear correlation

TABLE 12.4. SEC and AUC (SV) data for orthogonal verification of the amount of protein aggregates

Sample preparation	% Aggregates by SEC	% Aggregates by sedimentation velocity[a]
1	1.9	2.1
2	2.0	2.0
3	3.0	5.7
4	5.2	5.2
5	7.1	7.1
6	11.1	10.3

[a]Data were analyzed using $c(s)$ method in SEDFIT [58]. The peak areas of all species sedimenting faster than the main protein were summed to report the aggregate content for each sample.

was observed between the percentage of total aggregates determined by AUC (SV) and SEC, demonstrating the suitability of SEC to quantify aggregates in this product (Table 12.4).

SE-AUC has been used to characterize protein association states and to calculate the weighted-average molecular weight of proteins. In an example of a therapeutic enzyme, SE data showed a systematic trend of increasing MW with increasing protein concentration, suggesting a reversible self-association of monomer and dimer in equilibrium. Fitting the AUC data to a model for monomer–dimer equilibrium suggested a broad range of dissociation constants, which may be indicative of heterogeneity in glycosylation and/or conformation of the protein. The observed SE trend was consistent with the SV data, which demonstrated an increasing sedimentation coefficient with an increasing protein concentration. The orthogonal SEC-multi-angle light scattering (MALS) method also demonstrated an increasing MW of the main peak with an increasing amount of protein injected on the column.

Novel applications of AUC have expanded the role of this method to potentially understand the *in vivo* fate of protein therapeutics. A recent example is a feasibility study of the structural properties of a fluorescently labeled antibody (omalizumab) in serum samples from several healthy volunteers [59]. In this study, SV demonstrated a comparable sedimentation coefficient for labeled omalizumab in serum and in phosphate buffer. Interestingly, a more compact omalizumab–IgE complex was formed in serum than in PBS. This new application can potentially be used to compare the behavior of molecules in a clinical setting provided that the instrumentation and the expertise are available for interpretation of the results.

12.4.2 Light Scattering

Static or Multi angle light scattering (LS or MALS) light scattering provides a measure of the molecular weight of proteins. The detected scattered light is directly dependent on the molecular weight of protein species as well as their concentration [60]. The moderate sensitivity of MALS can limit its use for evaluating minor product variants.

Dynamic light scattering (DLS) or quasi-elastic LS measures fluctuations of light intensity occurring on a microsecond or millisecond scale. These fluctuations are a function of the Brownian motion of the molecules in solution, which in turn is related to the size and shape of the protein. The high-throughput analysis in the formulation matrix with no dilution and no column interaction (compared to SEC-MALS) is among the advantages of this method.

DLS has been useful in detecting distinct molecular population(s) that may exist in a protein solution and in determining the size (hydrodynamic radius) and molecular heterogeneity profile. It has also been valuable in detecting aggregation "nuclei" [57]. Although applicable for rapid analysis, DLS has a number of limitations, including the low resolution between molecular species (a difference of eightfold in MW is required to differentiate in a mixture), lack of quantitative data on the amount of each species, dependency of results on data reduction algorithms, and the high scattering signal from species other than the target protein species.

12.5 METHODS FOR ASSESSING SUBVISIBLE PARTICULATES

12.5.1 Light Obscuration

The standard technique for measuring subvisible particles has been light obstruction (LO), based on US and European Pharmacopeia's guidelines for parenteral products, allowing ≤ 6000 particles at ≥ 10 μm and ≤ 600 particles at ≥ 25 μm for small volume preparations (≤ 100 mL). As indicated by the method's name, the obstruction of light by flowing particles is detected and calculated based on particle counting standards (polystyrene latex spheres). The technique is reproducible and is capable of counting the amount of particles in a wide size and concentration range. This method is required as a lot release test for parenteral products based on USP <788>. The large volume ($\sim$25 mL) required for the USP method makes it unsuitable as a characterization tool during development; smaller sample volumes ($\sim$2 mL) can be used, but the data are not as reproducible.

This method does not distinguish between extrinsic (nonprotein) particles and proteinaceous particles since it lacks detection for qualities such as shape characteristics and/or differences in refractive index. The LO technique is also highly dependent on the sample handling prior to the analysis. These factors can cause inaccuracy in particle count since extrinsic particles and proteins can be mistakenly counted as the same species [61,62].

LO is usually the first method used in assessing subvisible particles, but it can be supplemented with microscopy as a secondary technique. In the latter method, samples are filtered and isolated for manual particle counting using a light microscope. As is common with most of the microscopy techniques, this method suffers from low throughput as well as the potential for protein–membrane interaction that can cause erroneous particle counting.

12.5.2 Imaging Techniques

Microscopy-based techniques capable of imaging the structure and number of particulates in flowing solutions have recently been developed. Examples of such techniques (FlowCam, Flow Particle Imaging Analyzer, and ParticleInsight) are described in several reviews [63, 64].

One example from this category is microflow imaging (MFI). This technique has recently emerged as a technical advancement in particle counting. This method captures and analyzes digital images of particles as they pass through a flow cell providing data such as particle count, size, concentration, and shape [65]. Through its software capabilities, MFI is able to differentiate proteins from other particles, for instance silicon oil or air micro-bubbles, by allowing for shape analysis of the captured digital images [66].

MFI has a few advantages when compared with LO. It has an increased sensitivity compared to LO, where it can count the particles not only in the range of ≥ 10 and ≥ 25 µm, but also in a much smaller particle range of 2–10 µm. It has been documented that LO underestimates the number of particles below 10 µm when compared with MFI [63]. Also, MFI requires less material for particle analysis (~ 1 mL for non-comendial signal measurement) compared with LO (~ 2 mL for a non-compendia signal measurement). Overall, MFI provides information about particle count and type present at any stage of development, and it is particularly informative during formulation development.

Despite these advantages, there are some inherent limitations associated with MFI. The technique still requires quantities of material typically not available at the early stages of a program and it is not protein specific. The current imaging technologies depend upon shape-matching algorithms for particle identification. Therefore, a protein-specific detection technique, such as fluorescence, is highly desirable to increase the specificity of this method [67]. Additionally, improvements in sample requirements and reproducibility are needed to further increase the use of MFI technology and suitability of the data in regulatory submissions.

The FDA is looking for information on 2–10 µm particle analysis for inclusion in the BLA as a part of drug product characterization or in the comparability sections for both drug substance and drug product. Considering that the LO method is currently a lot release test for drug product, and is reported in the global license applications, the limitation of this technique to detect the 2–10 µm particles makes it necessary to apply other imaging technologies.

12.6 EVOLVING BIOPHYSICAL TECHNOLOGIES

Methodologies are being continuously developed to meet the increasing requirements to better characterize the structure of therapeutic proteins. In this section, we discuss a few methods that have demonstrated potential use in the development of protein therapeutics.

This discussion is not meant to provide a complete list of evolving biophysical tools, but rather to highlight methodologies that have commercial instrument support and a noted record of publications. Regulatory agencies are well aware of these evolving technologies and encourage their development and use in protein development [68].

12.6.1 Asymmetric Flow Field-Flow Fractionation for Assessing Protein Association State

Field-flow fractionation (FFF) is a solution phase separation technique for characterizing large biopolymers, such as proteins and protein complexes with little or no disruption to the higher order structure [69]. FFF drives macromolecules through an empty channel by laminar longitudinal flow. The separation then occurs through a perpendicular mobile phase flow that moves species with different diffusion coefficients into select regions of the longitudinal flow path. This mechanism gives FFF a considerable advantage because of its ability to directly analyze a protein in its formulated solution. With a growing number of applications in protein drug development, FFF is now considered a promising orthogonal technique to characterize and monitor protein association and aggregation [70]. However, some limitations of FFF has slowed its adoption as a quantitative method in comparability or lot release testing. The use of a force against membranes in FFF can potentially create an interactive binding surface to alter a protein's association state. Additionally, higher molecular weight species such as aggregates have the potential for greater dilution under optimized separation conditions (i.e., greater cross-flow).

Arvinte and coworkers demonstrated the capabilities and limitations of FFF in the assessment of trastuzumab aggregation as well as the basis for administration instructions [71]. FDA and EMEA documentation call for the reconstitution of trastuzumab with 0.9% sodium chloride (saline), while dilution with dextrose is prohibited. The EMEA summary states that dilution of this product in dextrose causes aggregation based on the FFF data. When the antibody samples in saline or 5% dextrose were analyzed using a commercial FFF with saline as mobile phase, no difference in the amount of aggregate was observed between the two solutions. When the same two samples were analyzed with 5% dextrose as mobile phase, sample recovery by UV detection was low. Coupled with in-line MALS detection, the technique confirmed the formation of aggregates in the presence of 5% dextrose and the reversibility of the aggregates by dilution in saline solution. FFF, therefore, demonstrated the nature and control of therapeutic protein aggregation directly from their formulated solution.

12.6.2 Mass Spectrometric-Based Methodologies for Assessing Higher Order Structure

Several mass spectrometry-based methods for characterizing the higher order structures of proteins are moving from discovery applications into the development of protein therapeutics. Modern techniques have expanded the role of mass spectrometry to the study of the folded structure and association states of proteins [72]. Several methods such as combining mass spectrometric detection with chemical cross-linking, ion mobility

(IM), or hydrogen–deuterium exchange (HDX) techniques, which have origins before the birth of the biotechnology industry, have rendered sophisticated approaches for characterizing the higher order structure of proteins.

Chemical cross-linking is one of the oldest approaches to identify and map protein–protein interactions that gained popularity when used to confirm the dimeric structure of insulin [73]. Over the past several decades, the availability of cross-linking reagents and advancements in mass spectrometry has led to the application of this technique in many laboratories. A review of chemical cross-linking techniques illustrates the use of this approach to characterize protein association [74]. A major appeal of protein cross-linking is its potential to capture transient association states of proteins in solution, although the solution conditions needed for the linker reactivity may not match the desired formulated protein conditions. Newer cross-linking chemistries, however, are amenable to near-physiological conditions and have thus been used for discovery work on protein aggregation diseases. Mechanistic studies of protein aggregation observed in Alzheimer, cataract, and Huntington diseases have been studied through cross-linking experiments combined with MS analysis [75–77]. Identification of the cross-linked sites along with the spatial constraints of the linker size is useful for understanding of tertiary and quaternary structures, but the multitude of intra- and intermolecular linked sites presents a tremendous challenge for interpreting the results. While new cross-linking methods and automated algorithms have been successful [78–80], these types of investigations remain atypical or unique for inclusion in protein product licensure. A commercial method has recently been developed to quantify the amount of aggregates in protein therapeutics using cross-linking in combination with high-mass matrix-assisted laser desorption [81].

Another method with great potential to provide high-resolution structural characterization of therapeutic proteins in solution is HDX with mass spectrometric detection [82, 83]. HDX-MS provides structural information by monitoring the amount and rate of exchange of backbone amide protons with deuterons. The amount of HDX is related to the amount of solvent exposure for specific regions of the protein. An increase or decrease in the amount of exchange suggests a corresponding change in the exposed or protected regions of the protein (i.e., protein folding). Global and local structural information can be simultaneously obtained from HDX studies.

Several controlled steps are required to obtain HDX structural information [72, 84]. The studies begin by adding an excess of D_2O to the native protein solution (typically between pH 6.5 and 7.5 for HDX experiments). At select time periods, part of the sample is quenched in a low-pH (2.5), low-temperature solution to maintain the exchange. The amount of global unfolding can be determined by directly analyzing this quenched sample. Localized structural information can be obtained by subsequent proteolytic digestion and LC-MS analysis.

Studies have illustrated the potential of HDX to understand the dynamics and stability of protein therapeutics. One example demonstrates the application of HDX-MS to understand the conformational changes in the therapeutic antiviral protein, interferon (IFN-β1a), upon site-specific alkylation [85]. This work showed that alkylation of one specific cysteine affected the native conformation of the protein and led to reduced activity. Modified forms of therapeutic proteins such as IFN-β1a are often developed to alter the pharmacokinetics or enhance binding to target ligands; therefore, this level of characterization is helpful toward defining the protein product.

Another HDX-MS study demonstrated the unique structural elements of a therapeutic enzyme and the impact of oxidation on its structure [86]. The specificity of HDX-MS showed that remote regions of this protein were destabilized by oxidation under stress conditions, thereby increasing the propensity for aggregation and potentially altered cellular trafficking. Additionally, the long time-course dynamics of this HDX study showed that the change in the structure–function as a result of oxidation was due to changes in the higher order structure remote to the catalytic site. These data supported other biophysical data (i.e., SEC and DSC) to demonstrate that a degradation product, detected as a shoulder peak on SEC, results from global unfolding of the protein under stress conditions.

While the appreciation of HDX-MS as a powerful biophysical tool has been rapidly increasing, incorporation into late-stage product development has been slow. The primary limitation is the complexity of the sample preparation, precision, and data interpretation. This method can also be limited by its lack of sensitivity to discriminate between subtle conformational differences within proteins. The conformational change has to be significant enough to demonstrate approximately 20% change in deuterium exchange. While HDX-MS methodology continues to emerge as a powerful orthogonal method for structural characterization, the current challenges of providing consistent data and detecting subtle structural changes limit its suitability for comparability studies following manufacturing process changes.

IM and ion mobility mass spectrometry (IM-MS) are evolving techniques for studying protein structure in the gas phase [87]. IM is a technique that measures the time it takes for a molecule to transverse through a gas under an electric field. Different conformers are separated in the IM analyzer and then detected at distinct times. When coupled to mass spectrometric detection, quantitative information on the conformation of proteins and protein assemblies can be obtained. The ion detection of each gas phase conformer is related to the relative amount of the corresponding species. This information can also be obtained from the protein of interest even when mixed with other proteins or polymer excipients. The emergence of IM-MS in protein development has been prolonged by the slow materialization of suitable commercial instrumentation and the questionable correlation between gas phase confirmation and the formulated solution phase structure. Since protein samples are introduced into the IM device by an electrospray process, the buffer conditions must be amenable to electrospray. Although the sample solutions are likely different than the formulated protein solution, the standard biophysical measurements would indicate if changes in the protein's buffer have induced structural changes. For larger proteins or protein assemblies, a common approach is to compare the collision cross-sectional areas to X-ray crystal structure measurements. A critical review illustrated that such comparisons must be carefully scrutinized [88]. Another systematic study demonstrated the effect of commercial instrument parameters for the determination of a protein's structure [89]. Based upon this work, the native structure of proteins in solution could be preserved by careful control of several instrument parameters. Another consideration is the lack of resolution to characterize the protein variants in the product, since the partially unfolded proteins or other subtle structural conformers may not be resolved with current capabilities. With the promise of this technique, researchers continue to develop IM-MS instrumentation that may address the needs of commercial protein development [90–93].

12.6.3   Biophysical Functional Assays

The function of protein therapeutics is related to their conformation and higher order structure. While cell-based assay data are required for regulatory filings to demonstrate the desired biological function [94], they are often complex, time consuming, and of lower precision than noncell-based functional assays. Given such limitations of biological assays, biophysical methods can be used to support cell-based bioassays during development of biologics. Surface plasmon resonance (SPR) and bio-layer interferometry (BLI) are two commercially available biophysical tools commonly used for measuring protein–receptor or protein–ligand binding kinetics and stoichiometry. This information can be used throughout the discovery and development phases. For late-stage development, SPR and interferometry can provide supporting functional data for comparability and stability studies.

SPR shows protein affinity or concentration measurements by displaying changes in the refractive indexes (response units (RU)) upon protein–protein binding. The protein of interest or an antibody toward this protein is immobilized on a surface. A solution containing the potential binding partner(s) is passed over the immobilized surface. The binding is determined by the change in refractive index at the surface over time. Using SPR, binding properties for the protein of interest can be determined directly in the cell culture media.

SPR and related optical techniques are commonly used for assessing binding affinity during protein development in addition to discovery efforts. An early application of this method demonstrated the feasibility of monitoring the structural integrity of a human insulin analog (MI3) directly from cell culture media [95]. SPR methodologies have also been validated for high-throughput testing of the immunological response levels of recombinant erythropoietic agents [96]. An SPR method was implemented as part of this development plan due to the reasonable turnaround time, robustness (<20% CV), and limit of detection (0.4 µg/mL). Prior to the use of biophysical functional assays, radiolabeled proteins were commonly used to assess immune response such as studies of reported incidences of neutralizing antibodies to recombinant EPO [97]. In addition to its application toward protein therapeutic development, SPR is emerging as an approach for monitoring the global biophysical attributes during upstream processes [98].

BLI is another label-free optical biosensor technique for monitoring protein–protein interactions during product development [99]. BLI detects changes in the interference pattern from a white light reflection pattern as a function of the thickness of the layer at the biosensor surface. The interference pattern detected by a spectrometer provides a unique spectral profile.

BLI has many of the same advantages and limitations as SPR. Any limitations of using surface biosensor techniques such as SPR and interferometry are dependent upon the protein and the surface binding chemistry. Unstable bound proteins may denature through the regeneration steps required for analysis of replicate samples. Common methods of immobilization can potentially interfere with the desired protein–protein associations. Additionally, high sensitivities are sometimes sacrificed for robustness due to the reduced levels of immobilized ligand density. Such limitations are often overcome through rigorous optimization or the selection of the desired protein fragments to represent the needed binding interface(s).

The development of novel biosensor technologies such as surface acoustic waveform (SAW) devices continues to address current limitations of biosensor technologies by improving the output signal resolution, selectivity, and sensitivity [100]. Additionally, the materials used in SAW biosensors increase the flexibility in the types of immobilization chemistries. The input from a SAW device launches a mechanical wave into a piezoelectrical material that contains the immobilized substrate. Changes in the acoustic wave properties (e.g., amplitude and frequency) are measured as a result of biochemical interactions on the surface [101–103]. By addressing the current limitations of biosensor technologies, biophysical binding assays may become standard industry tools for assessing the structural integrity throughout the therapeutic protein development.

12.6.4 Isothermal Calorimetry

Calorimetry has been previously described as a common biophysical tool for characterizing and comparing the folding of protein therapeutics; however, several biotechnology companies have more recently utilized isothermal titration calorimetry (ITC) and DSC as a means for understanding the effects of buffer constituents upon protein folding and binding [104, 105]. The buffer composition choice for ITC experiments is critical for maximizing the protein–protein binding information. Buffers with negligible enthalpies of ionization maximize the protein–receptor binding data, otherwise measured changes will reflect the buffer composition and interaction with the protein. These optimal experimental conditions, however, may not match the buffer choice for the final protein drug product. The concentrations of protein and its binding partner are also important for obtaining accurate binding data. With these fastidious restrictions, ITC is more commonly used for screening potential protein binding candidates rather than protein development and commercialization [106–108].

12.6.5 Microscale Thermophoresis

Microscale thermophoresis is a new biophysical method for evaluating the interaction between a protein of interest and its ligands or receptors within the relevant biological fluid. This method is the subject of current research to understand the binding kinetics of biomolecules to their target proteins. In this method, biomolecules are transported through fluids by a temperature gradient, a mechanism known as thermophoresis or the Soret effect [109, 110]. In one application of this method, the receptor binding affinity of model proteins such as interferon and calmodulin was shown to decrease significantly in the presence of serum and cell lysates [111].

At this time, we are unaware of any documented use of microscale thermophoresis for protein development. A long maturation period is likely required before robust and interpretable data from such new biophysical tools are demonstrated for practical applications in protein therapeutic development.

12.7 CASE STUDY FOR THE USE OF BIOPHYSICAL METHODS IN THE ELUCIDATION OF STRUCTURE SECTION OF THE LICENSE APPLICATION

Comprehensive biophysical analyses to characterize and demonstrate control of the structural and functional properties of an example therapeutic enzyme during its development and commercialization are summarized in this section. The naturally occurring mutations in this enzyme are known to cause protein misfolding that disrupts the activity and trafficking into the cellular site of action. Therefore, biophysical analyses were used to characterize the secondary, tertiary, and quaternary features of the therapeutic enzyme product. Table 12.5 lists exemplary characteristics and the associated biophysical techniques used in the 2009 license application.

A combination of common, orthogonal, and evolving techniques was used to characterize and monitor the folded structure of this therapeutic enzyme. Common biophysical techniques such as DSC and CD provided information about the global protein structure. Details of the tertiary structure were elucidated by X-ray crystallography. In addition to structural features and disulfide linkages, distinct conformers of the protein were observed within the asymmetric unit cells. The conformational heterogeneity was attributed to certain loop regions surrounding the catalytic site. HDX-MS analysis confirmed similar structures as X-ray crystal analysis as well as added insight into the solution phase dynamics of this protein.

Several biophysical techniques (AUC, SEC-MALS, FFF, and LS) provided an integrated view of the association states of this enzyme. A reversible rapid self-associating system of monomer–dimer was observed in this enzyme by these methods. The same methodologies were also applied to characterize a minor impurity observed in this product. By SEC, an apparent larger molecular weight shoulder to the main peak was observed at less than 3% of the total area, which underwent the same reversible association state as the main peak. SEC-MALS did not show any larger molecular weight species in the shoulder peak than the main peak, which was confirmed by MALDI analysis of the collected shoulder peak. LC-MS analysis of the isolated shoulder peak demonstrated the presence of oxidized residues. In a separate study, oxidative stress conditions resulted in altered conformations identified by HDX analysis. The sample handling instability and the sensitivity of HDX prohibited the use of this technique to confirm the structural attributes of this minor impurity. Nevertheless, the collective data led to the conclusion that a minor degradation product detected as shoulder peak on SEC with an apparently larger hydrodynamic size results from global unfolding of the protein.

Data for receptor binding kinetics and cellular uptake were included in this section as characterization of the biological activity and its relationship to structural characteristics [30]. The data from these studies were as expected from the types of glycan structures on the product that are known to facilitate trafficking to the desired intracellular site for biological action. In this example, the biophysical characterization along with biochemical analyses provided an integrated view of the structure and function of this biologic.

TABLE 12.5. An exemplary set of structural characterization methods used for the development and commercialization of a therapeutic enzyme

Characteristic	Analysis performed	Strategic implications
Secondary structure	Far-UV circular dichroism	Showed specific features of α/β structure, consistent with X-ray crystal structure Applied to comparability analysis
Tertiary structure	X-ray crystallography	Characterized structure at atomic level; was used for comparison to other family members of the enzyme
	Near-UV circular dichroism	Provided global tertiary structural information
	Hydrogen–deuterium exchange mass spectrometry	Provided higher order structural dynamics that were consistent with X-ray crystal structure Characterization of degradation products due to folded structural change
Size distribution	Analytical ultracentrifugation (SV and SE)	Provided orthogonal method for size distribution analysis and confirmed SEC results for aggregate levels The concentration-dependent analysis provided insight into monomer–dimer self-association Applied to comparability analysis
	Light scattering	Provided orthogonal method for size distribution analysis
	SEC-MALS	The concentration-dependent analysis provided insight into monomer–dimer self-association Applied to analysis of minor degradation product of the same size, also showing monomer–dimer association
	Field-flow fractionation (exploratory)	Provided consistent data with SEC-MALS, light scattering, and AUC regarding the number of detected species
Other biophysical properties	UV spectral profile (zero order)	Provided global structural characterization Applied to comparability analysis
	Thermal melt profile by DSC	Provided thermodynamic measurement of thermal unfolding and melting temperature Applied to comparability analysis
Functional properties	Receptor binding kinetics using interferometry biosensor	Provided insight into components contributing to biological activity Applied to comparability analysis
	Cellular uptake bioassay	

12.8 CASE STUDY FOR THE USE OF BIOPHYSICAL METHODS IN THE COMPARABILITY SECTION OF THE LICENSE APPLICATION

Early examples of comparability studies were performed on EPO products to demonstrate consistent identity, purity, potency, and quality during the development and manufacturing processes. One particular EPO product was manufactured through a new proprietary cell culture process that required significant changes to the manufacturing process prior to regulatory approval. Comparability studies were required for process changes and transfer to another manufacturing site. At the time of filing for this product (2001), little biophysical analysis was performed to establish comparability. Circular dichroic spectroscopy and SEC data were the only methods used to characterize this product. The regulatory agencies relied on biological assays at that time to demonstrate adequate structural comparability. A standard biological potency comparison was performed between a representative EPO drug substance and the World Health Organization reference standard.

The modest comparability data used a decade ago contrast sharply with the agency expectations in recent years to use more rigorous characterization assays in comparability studies. Examples of more extensive biophysical studies following process changes are listed in Table 12.5 and in the following section.

12.9 CASE STUDY FOR THE USE OF BIOPHYSICAL METHODS IN THE PHARMACEUTICAL DEVELOPMENT SECTION OF THE LICENSE APPLICATION

Changes in formulation or drug product presentation during development of protein pharmaceuticals require justification in the license application. This is achieved by evaluating the possible impact of such changes on product quality, including the possible impact on structure and function. The impact on stability also requires evaluation through real-time stability studies. Biophysical studies are often used to evaluate potential impacts of changes in formulation and/or drug product presentation on the structure and association states of the protein that might relate to its efficacy and safety.

A recent study utilized a collection of biophysical tools to demonstrate the comparability of therapeutic antibodies after modifications to either the product presentation (from vials to prefilled syringes) or the dosage form (from lyophilized to liquid formulation). Lubiniecki et al. observed that the molecular structure and functional integrity of two antibody-based products were maintained during these modifications. This conclusion was reached based on data from SEC and several biophysical techniques such as CD, DSC, and AUC (SV) [6]. The results indicated that the protein conformation in both drug products was comparable regardless of the applied modifications. The only difference observed was a small but statistically significant difference in the number of subvisible particles (≥ 10 and ≥ 25 µm) between the two drug product presentations. The observation was consistent with other published reports on the number of subvisible particles resulting from the presence of silicon oils in the prefilled syringes [112, 113]. In this report, no adverse clinical effects were observed for the prefilled syringe product.

The increased incidence of neutralizing antibodies observed in patients following changes in formulation and drug product presentation has increased the agencies' expectations for characterizing and monitoring protein aggregation, subvisible particles, and interaction with product container/closures using orthogonal methodologies [114, 115]. Overall, the use of spectroscopic and calorimetric methods to evaluate protein structure needs to be complemented with solution phase assessment of aggregates and subvisible particles to demonstrate the suitability of the product formulation as manufactured and during long-term storage.

12.10   BIOPHYSICAL METHODS FOR EVALUATING PROTEIN—SURFACE/DEVICE INTERACTION

Pharmaceutical development studies on the product interactions with surfaces are important for the selection of contact materials or conditions of use that maintain the activity and quality of the product. Adsorption of proteins to surfaces and the consequent impact on protein structure have been documented [116–128]. Protein–surface interactions may induce global conformational changes in proteins that may lead to the loss of activity [121]. The use of delivery devices is particularly more challenging for some proteins because of long-term contact, mechanical stress, or thermal stress. As such, studies that ensure the maintenance of product quality with the use of devices need to be documented in the license application.

Several solution phase and solid-state biophysical methods, particularly ATR-FTIR, CD, and fluorescence spectroscopy, provide insight into the mechanisms of protein loss or deactivation following interaction with surfaces and delivery devices [125]. Other surface-sensitive absorbance techniques such as optical waveguide light-mode spectroscopy (OWLS) and quartz crystal microbalance have also been used but are less developed than fluorescence and IR spectroscopy [129]. As an example, irreversible inactivation of interleukin 2 in a pump-based delivery device has been reported to be due to the transient surface association of this protein with the catheter tubing as demonstrated by fluorescence spectroscopy, far-UV CD, and ATR-FTIR [125]. The combination of these methods detected conformational changes that the SEC method could not detect [125]. The same protein was shown to undergo conformational changes upon adsorption to the silicon rubber tubing (commonly used as a catheter material) using ATR-FTIR [124]. One example of the application of biophysical tools in surface adsorption studies is the use of ATR-FTIR, CD, and Raman spectroscopy to study fibrinogen structural changes when adsorbed on poly(lactic-*co*-glycolic-acid) films. The employed methods demonstrated that topographical manipulation of the polymer surfaces had a direct effect on the conformational change of fibrinogen as well as its bioactivity [130]. Another example is the study of the structural changes of human and bovine serum albumin when incorporated into the water-in-oil emulsions using DSC, FTIR, and intrinsic tryptophan fluorescence. This investigation demonstrated that the overall secondary structure for these two proteins remained mostly intact while the tertiary structure changed to a larger degree when the proteins were adsorbed to the water-in-oil emulsion [131]. As a final example, the unfolding rate of bovine insulin model at the interface of lipid solution was much faster than for the protein in the bulk solution, and the fibril-like structures formed at the lipid surfaces were different from the ones formed in the bulk of the solution [132].

For drug delivery systems and devices, biophysical methods provide information on surface interactions as related to safety, stability, and device functionality that is needed in the license application.

12.11   IMPLICATION FOR BIOSIMILARS

The structural heterogeneity of protein products presents a challenge for demonstrating structural similarity between biosimilars and innovator products. The complex process of manufacturing proteins makes it difficult to duplicate the resulting heterogeneity even with minor changes to an existing process. The premise that the process defines the product creates a challenge to demonstrate the "sameness" of a follow-on product that is manufactured by an alternative process.

One of the requirements for approval of biosimilars is a thorough state-of-the art characterization of the biosimilar and reference product to provide assurance that the materials will have a comparable biological effect. Regulatory guidelines suggest that human clinical testing may be reduced or even eliminated based upon the results of such cutting-edge characterization [133]. Biophysical analyses are thus used to support efforts to demonstrate structural similarity between a biosimilar and the innovator product. As an example, Zarzio® is a biosimilar recombinant human granulocyte colony-stimulating factor that has been approved in the EU. The comparison study between Zarzio® and its reference product Neupogen® [38] demonstrated structural comparability through biophysical analyses ranging from more common (e.g., CD) to more elaborate (e.g., NMR) technologies. The biophysical comparability was consistent with the similar *in vitro* bioactivity of Zarzio® and Neupogen®. Ultimately, comparable safety and efficacy was established through a clinical trial with healthy and chemotherapy-naïve breast cancer patients [134].

Regulatory agencies have recently placed greater emphasis on the control and characterization of folding, association states, and aggregation of protein therapeutics to assess the potential risk of immunogenicity [135–137]. The development of EPO biosimilars demonstrates the need for complete structural characterization with attention to immunogenic implications. The antibody-mediated pure red cell aplasia (PRCA) in patients treated with Eprex®, a biosimilar form of EPO, is a frequently cited case of an immune response subsequent to process and structural changes [138]. An increase in aggregation of Eprex® during storage compared to reference EPO alfa was suspected to contribute to the observed immunogenic responses. This increase in aggregation and immunogenic response coincided with a change in the formulation excipients and product contact materials. A subsequent biophysical comparison showed that the biosimilar was not structurally the same as the original recombinant EPO alfa [4].

The requirements for biosimilar approval are set on a case-by-case basis due to the complexity of protein therapeutics and the lack of a single path for commercialization of biosimilars [139–141]. Based upon lessons learned from follow-on proteins such as Eprex®, current and future development and commercialization of protein biosimilars will require extensive structural characterization. As the requirements for approval of biosimilars evolve, the expectations for biophysical characterization would likely increase. While biophysical analyses are useful, their link to biological effect is often speculative and difficult to prove.

12.12 CONCLUDING REMARKS

Continuous evolution of biophysical technologies mirrors the regulatory agencies' increasing expectations to apply such technologies to the evaluation of therapeutic protein structure. In this chapter we have described the biophysical tool box and its use during development to commercialization of protein therapeutics, taking into consideration the limitations and complexity of the technologies.

The available and emerging advanced biophysical methodologies generate a knowledge base that helps with the development of controlled and effective products. Such detailed knowledge puts a constraint on the development of biosimilars or future new protein therapeutics. The innovator's analytical methodologies, as reflected in their respective license application, set precedence for developers of subsequent protein therapeutics. The significant additions and advances to the biophysical tool box over the past two decades have also created a corresponding need for increased expertise with the instrumentation, data analysis, data interpretation, and data application. As such, companies have chosen to apply certain methodologies based on their experience with their products' history, their internal capabilities, and the phase of development. Overall, biotechnology companies face the challenge of how to best apply their resources in implementing as many methodologies as practical while maintaining the ability to interpret the complex data in relation to safety, efficacy, and consistency of product quality throughout the product's lifecycle.

ACKNOWLEDGMENTS

The authors thank Drs Katherine Taylor and Philip Savickas (Rare Business Unit, Shire) for their insightful suggestions during the writing of this chapter. The authors also thank Prof. David B. Volkin (University of Kansas) for referring them to his work in reference 6.

REFERENCES

1. Davern, S., et al. (2011) Effect of lysine modifications on the stability and cellular binding of human amyloidogenic light chains. *Biochim. Biophys. Acta*, **1812**, 32–40.

2. Young, D.S., et al. (2009) Effect of enzymatic deimination on the conformation of recombinant prion protein. *Biochim. Biophys. Acta*, **1794**, 1123–1133.

3. Wang, S., et al. (2010) Separation of post-translational modifications in monoclonal antibodies by exploiting subtle conformational changes under mildly acidic conditions. *J. Chromatogr. A*, **1217**, 6496–6502.

4. Deechongkit, S., Aoki, K.H., Park, S.S., and Kerwin, B.A. (2006) Biophysical comparability of the same protein from different manufacturers: a case study using Epoetin alfa from Epogen and Eprex. *J. Pharm. Sci.*, **95**, 1931–1943.

5. Chirino, A.J. and Mire-Sluis, A. (2004) Characterizing biological products and assessing comparability following manufacturing changes. *Nat. Biotechnol.*, **22**, 1383–1391.

6. Lubiniecki, A., et al. (2011) Comparability assessments of process and product changes made during development of two different monoclonal antibodies. *Biologicals*, **39**, 9–22.

7. Bloemendal, M. (2005) Circular dichroism spectroscopy. In: Jiskoot, W. and Crommelin, D. editors. *Methods for Structural Analysis of Protein Pharmaceuticals*. AAPS Press, pp. 83–130.

8. Alford, J.R., Kendrick, B.S., Carpenter, J.F., and Randolph T.W. (2008) High concentration formulations of recombinant human interleukin-1 receptor antagonist: II. Aggregation kinetics. *J. Pharm. Sci.*, **97**, 3005–3021.

9. Kong, J. and Yu, S. (2007) Fourier transform infrared spectroscopic analysis of protein secondary structures. *Acta Biochim. Biophys. Sin. (Shanghai)*, **39**, 549–559.

10. Toft, J., et al. (1996) Resolution of overlapping mid-infrared spectra using SIMPLISMA and a second-derivative approach. *Vib. Spectrosc.*, **10**, 125–138.

11. Dong, A. and Caughey, W.S. (1994) Infrared methods for study of hemoglobin reactions and structures. *Methods Enzymol.*, **232**, 139–175.

12. Carpenter, J.F., Prestrelski, S.J., and Dong, A. (1998) Application of infrared spectroscopy to development of stable lyophilized protein formulations. *Eur. J. Pharm. Biopharm.*, **45**, 231–238.

13. Chittur, K.K. (1998) FTIR/ATR for protein adsorption to biomaterial surfaces. *Biomaterials*, **19**, 357–369.

14. Dole, M.N., Patel, P.A., Sawant, S.D., and Shedpure, P.S. (2011) Advance applications of Fourier transform infrared spectroscopy. *Int. J. Pharm. Sci. Rev. Res.*, **7**, 159–166.

15. Van De Weert, M., Hering, J.A., and Haris, P.I. (2005) Fourier transform infrared spectroscopy. In: Jiskoot, W. and Crommelin, D. editors. *Methods for Structural Analysis of Protein Pharmaceuticals*. AAPS Press, pp. 131–166.

16. Nilsson, M.R., Driscoll, M., and Raleigh, D.P. (2002) Low levels of asparagine deamidation can have a dramatic effect on aggregation of amyloidogenic peptides: implications for the study of amyloid formation. *Protein Sci.*, **11**, 342–349.

17. Wen, Z.Q. (2007) Raman spectroscopy of protein pharmaceuticals. *J. Pharm. Sci.*, **96**, 2861–2878.

18. Wen, Z.Q., Cao, X., and Vance, A. (2008) Conformation and side chains environments of recombinant human interleukin-1 receptor antagonist (rh-IL-1ra) probed by Raman, Raman optical activity, and UV-resonance Raman spectroscopy. *J. Pharm. Sci.*, **97**, 2228–2241.

19. Chi, Z., Chen, X.G., Holtz, J.S., and Asher, S.A. (1998) UV resonance Raman-selective amide vibrational enhancement: quantitative methodology for determining protein secondary structure. *Biochemistry*, **37**, 2854–2864.

20. Ozdemir, A., Lednev, I.K., and Asher, S.A. (2002) Comparison between UV Raman and circular dichroism detection of short alpha helices in bombolitin III. *Biochemistry*, **41**, 1893–1896.

21. Simpson, J.V., et al. (2011) Pre-processing of ultraviolet resonance Raman spectra. *Analyst*, **136**, 1239–1247.

22. Blundell, T.L. and Patel S. (2004) High-throughput X-ray crystallography for drug discovery. *Curr. Opin. Pharmacol.*, **4**, 490–496.

23. Jhoti, H. (2007) Fragment-based drug discovery using rational design. *Ernst Schering Found. Symp. Proc.*, **3**, 69–185.

24. Tickle, I., et al. (2004) High-throughput protein crystallography and drug discovery. *Chem. Soc. Rev.*, **33**, 558–565.

25. Read, R.J. (1996) As MAD as can be. *Structure*, **4**, 11–14.

26. Bhattacharya, S. and Osman, H. (2009) Novel targets for anti-retroviral therapy. *J. Infect.*, **59**, 377–386.

27. Walter, M.R., et al. (1998) Review of recent developments in the molecular characterization of recombinant alfa interferons on the 40th anniversary of the discovery of interferon. *Cancer Biother. Radiopharm.*, **13**, 143–154.

28. Dvir, H., et al. (2003) X-ray structure of human acid-beta-glucosidase, the defective enzyme in Gaucher disease. *Eur. Mol. Biol. Org. Rep.*, **4**, 704–709.

29. Roeber, D., et al. (2003) Crystallization and preliminary X-ray analysis of recombinant human acid beta-glucocerebrosidase, a treatment for Gaucher's disease. *Acta Crystallogr. D Biol. Crystallogr.*, **59**, 343–344.

30. Brumshtein, B., et al. (2010) Characterization of gene-activated human acid-beta-glucosidase: crystal structure, glycan composition, and internalization into macrophages. *Glycobiology*, **20**, 24–32.

31. Peterson, B., et al. (2010) Comparison of in-vitro cellular uptake of velaglucerase alfa to that of imiglucerase: effect of chemical parameters and receptor specific inhibition, in *9th International Meeting of the European Working Group on Gaucher Disease*, Cologne, Germany.

32. Yang, Y., et al. (2010) An Achilles' heel in an amyloidogenic protein and its repair: insulin fibrillation and therapeutic design. *J. Biol. Chem.*, **285**, 10806–10821.

33. Jacoby, E., et al. (1996) Structure and dynamics of a protein assembly. 1H-NMR studies of the 36 kDa R6 insulin hexamer. *J. Mol. Biol.*, **258**, 136–157.

34. Williamson, K.L. and Williams, R.J. (1979) Conformational analysis by nuclear magnetic resonance: insulin. *Biochemistry*, **18**, 5966–5972.

35. EMEA (2006) Scientific discussion: alternative treatment for diabetes mellitus. Available at http://www.ema.europa.eu/docs/en_GB/document_library/EPAR_-_Scientific_Discussion/human/000698/WC500051842.pdf.

36. Hudson, F.M. and Andersen, N.H. (2004) Exenatide: NMR/CD evaluation of the medium dependence of conformation and aggregation state. *Biopolymers*, **76**, 298–308.

37. Cheetham, J.C., et al. (1998) NMR structure of human erythropoietin and a comparison with its receptor bound conformation. *Nat. Struct. Biol.*, **5**, 861–866.

38. Sorgel, F., Lerch, H., and Lauber, T. (2010) Physicochemical and biologic comparability of a biosimilar granulocyte colony-stimulating factor with its reference product. *BioDrugs*, **24**, 347–357.

39. Freedberg, D.I. (2005) Using magnetic resonance spectroscopy to characterize biologicals. In: Mire-Sluis, A.R. editor. *State of the Art Analytical Methods for the Characterization of Biological Products and Assessment of Comparability.* International Association for Biologicals, Geneva, pp. 77–83.

40. Robertson, A.D. and Murphy, K.P. (1997) Protein structure and the energetics of protein stability. *Chem. Rev.*, **97**, 1251–1268.

41. Sharma, V.K. and Kalonia, D.S. (2003) Temperature- and pH-induced multiple partially unfolded states of recombinant human interferon-alpha2a: possible implications in protein stability. *Pharm. Res.*, **20**, 1721–1729.

42. Bushueva, T.L., Busel, E.P., and Burstein, E.A. (1978) Relationship of thermal quenching of protein fluorescence to intramolecular structural mobility. *Biochim. Biophys. Acta Protein Struct.*, **534**, 141–152.

43. Permyakov, E.A. and Burstein, E.A. (1984) Some aspects of studies of thermal transitions in proteins by means of their intrinsic fluorescence. *Biophys. Chem.*, **19**, 265–271.

44. Erickson, U.B., et al. (2006) Thermofluor-based high-throughput stability optimization of proteins for structural studies. *Anal. Biochem.*, **357**, 289–298.

45. He, F., et al. (2011) Screening of monoclonal antibody formulations based on high-throughput thermostability and viscosity measurements: design of experiment and statistical analysis. *J. Pharm. Sci.*, **100**, 1330–1340.

46. Vedadi, M., et al. (2010) Biophysical characterization of recombinant proteins: a key to higher structural genomics success. *J. Struct. Biol.*, **172**, 107–119.

47. Sharma, V.K. and Kalonia, D.S. (2003) Steady-state tryptophan fluorescence spectroscopy study to probe tertiary structure of proteins in solid powders. *J. Pharm. Sci.*, **92**, 890–899.

48. He, F., et al. (2010) High throughput thermostability screening of monoclonal antibody formulations. *J. Pharm. Sci.*, **99**, 1707–1720.

49. Niesen, F.H., Berglund, H., and Vedadi, M. (2007) The use of differential scanning fluorimetry to detect ligand interactions that promotes protein stability. *Nat. Protoc.*, **2**, 2212–2221.

50. Hawe, A., Filipe, V., and Jiskoot, W., (2010) Fluorescent molecular rotors as dyes to characterize polysorbate-containing IgG formulations. *Pharm. Res.*, **27**, 314–326.

51. Lindgren, M., Sorgjerd, K., and Hammarstrom, P., (2005) Detection and characterization of aggregates, prefibrillar amyloidogenic oligomers, and protofibrils using fluorescence spectroscopy. *Biophys. J.*, **88**, 4200–4212.

52. Haidekker, M.A. and Theodorakis, E.A. (2010) Environment-sensitive behavior of fluorescent molecular rotors. *J. Biol. Eng.*, 4, 11. Available online at: http://www.jbioleng .org/content/pdf/1754–1611–4–11.pdf

53. Sutharsan, J., et al. (2010) Molecular rotors: synthesis and evaluation as viscosity sensors. *Tetrahedron*, **66**, 2582–2588.

54. Mach, H., Volkin, D.B., Burke, C.J., and Middaugh, C.R. (1995) Ultraviolet absorption spectroscopy. In: Shirley, B.A. editor. *Methods in Molecular Biology*. Humana Press, Totowa, NJ, pp. 91–114.

55. Kueltzo, L.A., Ersoy, B., Ralston, J.P., and Middaugh, C.R. (2003) Derivative absorbance spectroscopy and protein phase diagrams as tools for comprehensive protein characterization: a bGCSF case study. *J. Pharm. Sci.*, **92**, 1805–1820.

56. Kueltzo, L.A. and Middaugh, C.R. (2003) Structural characterization of bovine granulocyte colony stimulating factor: effect of temperature and pH. *J. Pharm. Sci.*, **92**, 1793–1804.

57. Philo, J.S. (2006) Is any measurement method optimal for all aggregate sizes and types? *Am. Assoc. Pharm. Sci. J.*, **8**, E564–E571.

58. Schuck, P. (2000) Size-distribution analysis of macromolecules by sedimentation velocity ultracentrifugation and Lamm equation modeling. *Biophys. J.*, **78**, 1606–1619.

59. Demeule, B., Shire, S.J., and Liu, J. (2009) A therapeutic antibody and its antigen form different complexes in serum than in phosphate-buffered saline: a study by analytical ultra-centrifugation. *Anal. Biochem.*, **388**, 279–287.

60. Demeester, J., De Smedt, S.S., Sanders, N.N., and Haustraete, J. (2005) Light scattering. In: Jiskoot, W. and Crommelin, D. editors. *Methods for Structural Analysis of Protein Pharmaceuticals.* AAPS, Arlington, VA, pp. 245–275.

61. Huang, C.T., Sharma, D., Oma, P., and Krishnamurthy, R., (2009) Quantitation of protein particles in parenteral solutions using micro-flow imaging. *J. Pharm. Sci.*, **98**, 3058–3071.

62. Narhi, L.O., et al. (2009) A critical review of analytical methods for subvisible and visible particles. *Curr. Pharm. Biotechnol.*, **10**, 373–381.

63. Singh, S.K., et al. (2010) An industry perspective on the monitoring of subvisible particles as a quality attribute for protein therapeutics. *J. Pharm. Sci.*, **99**, 3302–3321.

64. Zölls, S., et al. (2011) Particles in therapeutic protein formulations, part 1: overview of analytical methods. *J. Pharm. Sci.*, **101**, 914–935.

65. Sharma, D.K., King, D., Oma, P., and Merchant, C. (2010) Micro-flow imaging: flow microscopy applied to sub-visible particulate analysis in protein formulations. *Am. Assoc. Pharm. Sci. J.*, **12**, 455–464.

66. Strehl, R., et al. (2011) Discrimination between silicone oil droplets and protein aggregates in biopharmaceuticals: a novel multiparametric image filter for sub-visible particles in microflow imaging analysis. *Pharm. Res.*, **29**, 594–602.

67. Mach, H., et al. (2011) The use of flow cytometry for the detection of subvisible particles in therapeutic protein formulations. *J. Pharm. Sci.*, **100**, 1671–1678.

68. Shacter, E. (2011) Regulatory expectations for application of new and advanced technologies for the analysis of protein products. In: *2nd Annual Characterization and Comparability for Biologics.* Cambridge Health Institute, Bethesda, MD.

69. Reschiglian, P., et al. (2005) Field-flow fractionation and biotechnology. *Trends Biotechnol.*, **23**, 475–483.

70. Rambaldi, D.C., Reschiglian, P., and Zattoni, A. (2011) Flow field-flow fractionation: recent trends in protein analysis. *Anal. Bioanal. Chem.*, **399**, 1439–1447.

71. Demeule, B., et al. (2009) New methods allowing the detection of protein aggregates: a case study on trastuzumab. *mAbs*, **1**, 142–150.

72. Kaltashov, I.A. and Eyles, S.J. (2005) Mass spectrometry in biophysics. Desiderio, D.M. and Nibbering, N.M. editors. *Wiley-Interscience Series on Mass Spectrometry.* John Wiley and Sons, Inc., Hoboken, NJ, pp. 458.

73. Zahn, H. and Meienhofer, J. (1958) Reactions of 1,5-difluoro-2,4-dinitrobenzene with insulin. II. Experiments with insulin. *Makromol. Chem.*, **26**, 126–152.

74. Trakselis, M.A., Alley, S.C., and Ishmael, F.T. (2005) Identification and mapping of protein-protein interactions by a combination of cross-linking, cleavage, and proteomics. *Bioconjug. Chem.*, **16**, 741–750.

75. Dehay, B., Weber, C., Trottier, Y., and Bertolotti, A. (2007) Mapping of the epitope of monoclonal antibody 2B4 to the proline-rich region of human huntingtin, a region critical for aggregation and toxicity. *Biotechnol. J.*, **2**, 559–564.

76. Du, J. and Murphy, R.M. (2010) Characterization of the interaction of beta-amyloid with transthyretin monomers and tetramers. *Biochemistry*, **49**, 8276–8289.

77. Wang, S.S., Wu, J.W., Yamamoto, S., and Liu, H.S. (2008) Diseases of protein aggregation and the hunt for potential pharmacological agents. *Biotechnol. J.*, **3**, 165–92.

78. Back, J.W., de Jong, L., Muijsers, A.O., and de Koster, C.G. (2003) Chemical cross-linking and mass spectrometry for protein structural modeling. *J. Mol. Biol.*, **331**, 303–313.

79. Singh, P., Panchaud, A., and Goodlett, D.R., (2010) Chemical cross-linking and mass spectrometry as a low-resolution protein structure determination technique. *Anal. Chem.*, **82**, 2636–2642.

80. Young, M.M., et al. (2000) High throughput protein fold identification by using experimental constraints derived from intramolecular cross-links and mass spectrometry. *Proc. Natl. Acad. Sci. U.S.A.*, **97**, 5802–5806.

81. Bich, C., Nazabal, A., Wenzel, R.J., and Zenobi, R. (2010) Cross-linking reagents for molecular interactions analysis of proteins and/or other compounds bearing amino functions. WO patent 2010136539 (December 2, 2010) and US patent 20110177617 (July 21, 2011).

82. Houde, D., Berkowitz, S.A., and Engen, J.R. (2011) The utility of hydrogen/deuterium exchange mass spectrometry in biopharmaceutical comparability studies. *J. Pharm. Sci.*, **100**, 2071–2086.

83. Kaltashov, I.A., et al. (2010) Conformation and dynamics of biopharmaceuticals: transition of mass spectrometry-based tools from academe to industry. *J. Am. Soc. Mass Spectrom.*, **21**, 323–337.

84. Engen, J.R., (2009) Analysis of protein conformation and dynamics by hydrogen/deuterium exchange MS. *Anal. Chem.*, **81**, 7870–7875.

85. Bobst, C.E., et al. (2008) Detection and characterization of altered conformations of protein pharmaceuticals using complementary mass spectrometry-based approaches. *Anal. Chem.*, **80**, 7473–7481.

86. Bobst, C.E., et al. (2010) Impact of oxidation on protein therapeutics: conformational dynamics of intact and oxidized acid-beta-glucocerebrosidase at near-physiological pH. *Protein Sci.*, **19**, 2366–2378.

87. Chen, G., et al. (2011) Characterization of protein therapeutics by mass spectrometry: recent developments and future directions. *Drug Discov. Today*, **16**, 58–64.

88. Jurneczko, E. and Barran, P.E. (2011) How useful is ion mobility mass spectrometry for structural biology? The relationship between protein crystal structures and their collision cross sections in the gas phase. *Analyst*, **136**, 20–28.

89. Michaelevski, I., Eisenstein, M., and Sharon, M. (2010) Gas-phase compaction and unfolding of protein structures. *Anal. Chem.*, **82**, 9484–9491.

90. Bleiholder, C., Dupuis, N.F., Wyttenbach, T., and Bowers, M.T. (2011) Ion mobility-mass spectrometry reveals a conformational conversion from random assembly to beta-sheet in amyloid fibril formation. *Nat. Chem.*, **3**, 172–177.

91. Bohrer, B.C., et al. (2008) Biomolecule analysis by ion mobility spectrometry. *Annu. Rev. Anal. Chem. (Palo Alto Calif.)*, **1**, 293–327.

92. Hogan, C.J., et al. (2011) Tandem Differential mobility analysis-mass spectrometry reveals partial gas-phase collapse of the GroEL complex. *J. Phys. Chem. B*, **115**, 3614–3621.

93. Kurulugama, R.T., Valentine, S.J., Sowell, R.A., and Clemmer, D.E. (2008) Development of a high-throughput IMS-IMS-MS approach for analyzing mixtures of biomolecules. *J. Proteomics*, **71**, 318–331.

94. Goldman, N.D. (1999) *Federal Registry Notices: International Conference on Harmonisation; Guidance on Specifications: Test Procedures and Acceptance Criteria for Biotechnological/Biological Products.* H.A.H. Services, editor. Federal Register, Washington, DC, pp. 44928–44935.

95. Disley, D.M., Morrill, P.R., Sproule, K., and Lowe, C.R. (1999) An optical biosensor for monitoring recombinant proteins in process media. *Biosens. Bioelectron.*, **14**, 481–493.

96. Mason, S., et al. (2003) Validation of the BIACORE 3000 platform for detection of antibodies against erythropoietic agents in human serum samples. *Curr. Med. Res. Opin.*, **19**, 651–659.

97. Casadevall, N., et al. (2002) Pure red-cell aplasia and antierythropoietin antibodies in patients treated with recombinant erythropoietin. *N. Engl. J. Med.*, **346**, 469–475.

98. Thillaivinayagalingam, P., et al. (2010) Biopharmaceutical production: applications of surface plasmon resonance biosensors. *J. Chromatogr. B Analyt. Technol. Biomed. Life Sci.*, **878**, 149–153.

99. Moore, J.D., et al. (2011) Chemical and biological characterisation of a sensor surface for bioprocess monitoring. *Biosens. Bioelectron.*, **26**, 2940–2947.

100. Sankaranarayanan, S.K.R.S., Singh, R., and Venkat, R. (2010) Acoustic streaming induced elimination of nonspecifically bound proteins from a surface acoustic wave biosensor: mechanism prediction using fluid-structure interaction models. *J. Appl. Phys.*, **108**, 104507–1045018.

101. Cooper, M.A. (2007) Resonant acoustic profiling (RAP) and rupture event scanning. In: Wolfbeis, O.S., Steinem, C., and Janshoff, A. editors. *Piezoelectric Sensors*. Springer Series on Chemical Sensors and Biosensors. Vol. **5**, pp. 449–479.

102. Gizeli, E., et al. (2003) Sensitivity of the acoustic waveguide biosensor to protein binding as a function of the waveguide properties. *Biosens. Bioelectron.*, **18**, 1399–1406.

103. Rocha-Gaso, M.I., March-Iborra, C.M., Montoya-Baides, A., and Arnau-Vives, A. (2009) Surface generated acoustic wave biosensors for the detection of pathogens: a review. *Sensors*, **9**, 5740–5769.

104. Demarest, S.J., et al. (2006) An intermediate pH unfolding transition abrogates the ability of IgE to interact with its high affinity receptor Fc epsilon RI alpha. *J. Biol. Chem.*, **281**, 30755–30767.

105. Garidel, P., Hoffmann, C., and Blume, A. (2009) A thermodynamic analysis of the binding interaction between polysorbate 20 and 80 with human serum albumins and immunoglobulins: a contribution to understand colloidal protein stabilisation. *Biophys. Chem.*, **143**, 70–78.

106. Betzi, S., et al. (2007) Protein-protein interaction inhibition (2P2I) combining high throughput and virtual screening: application to the HIV-1 Nef protein. *Proc. Natl. Acad. Sci. U.S.A.*, **104**, 19256–19261.

107. Langelaan, D.N., Ngweniform, P., and Rainey, J.K. (2011) Biophysical characterization of G-protein coupled receptor-peptide ligand binding. *Biochem. Cell Biol.*, **89**, 98–105.

108. Pierce, M.M., Raman, C.S., and Nall, B.T. (1999) Isothermal titration calorimetry of protein-protein interactions. *Methods*, **19**, 213–221.

109. Villain-Guillot, S. and Wurger, A. (2011) Thermal diffusion in a binary liquid due to rectified molecular fluctuations. *Phys. Rev. E Stat. Nonlin. Soft Matter Phys.*, **83**, 030501/1–030501/4.

110. Weinert, F.M., Mast, C.B., and Braun, D. (2011) Optical fluid and biomolecule transport with thermal fields. *Phys. Chem. Chem. Phys.*, **13**, 9918–9928.

111. Wienken, C.J., et al. (2010) Protein-binding assays in biological liquids using microscale thermophoresis. *Nat. Commun.*, **1**, 100.

112. Jones, L.S., Kaufmann, A., and Middaugh, C.R. (2005) Silicone oil induced aggregation of proteins. *J. Pharm. Sci.*, **94**, 918–927.

113. Sun, L., et al. (1997) Protein denaturation induced by cyclic silicone. *Biomaterials*, **18**, 1593–1597.

114. Bennett, C.L., et al. (2004) Pure red-cell aplasia and epoetin therapy. *N. Engl. J. Med.*, **351**, 1403–1408.

115. Boven, K., et al. (2005) The increased incidence of pure red cell aplasia with an eprex formulation in uncoated rubber stopper syringes. *Kidney Int.*, **67**, 2346–2353.

116. Antia, M., et al. (2006) Single molecule fluorescence studies of surface-adsorbed fibronectin. *Biomaterials*, **27**, 679–690.

117. Castillo, E.J., Koenig, J.L., Anderson, J.M., and Lo, J. (1984) Characterization of protein adsorption on soft contact lenses. I. Conformational changes of adsorbed human serum albumin. *Biomaterials*, **5**, 319–325.

118. Drobny, G.P., et al. (2003) Structural studies of biomaterials using double-quantum solid-state NMR spectroscopy. *Annu. Rev. Phys. Chem.*, **54**, 531–571.

119. Engel, M.F., van Mierlo, C.P., and Visser, A.J. (2002) Kinetic and structural characterization of adsorption-induced unfolding of bovine alpha-lactalbumin. *J. Biol. Chem.*, **277**, 10922–10930.

120. Hu, H.Y., Li, Q., Cheng, H.C., and Du, H.N. (2001) Beta-sheet structure formation of proteins in solid state as revealed by circular dichroism spectroscopy. *Biopolymers*, **62**, 15–21.

121. Sethuraman, A., et al. (2004) Protein unfolding at interfaces: slow dynamics of alpha-helix to beta-sheet transition. *Proteins Struct. Funct. Bioinform.*, **56**, 669–678.

122. Thurow, H. and Geisen, K. (1984) Stabilisation of dissolved proteins against denaturation at hydrophobic interfaces. *Diabetologia*, **27**, 212–8.

123. Tsai, D.H., et al. (2011) Adsorption and conformation of serum albumin protein on gold nanoparticles investigated using dimensional measurements and in situ spectroscopic methods. *Langmuir*, **27**, 2464–2477.

124. Tzannis, S.T., Hrushesky, J.M., Wood, P.A., and Przybycien, T.M. (1997) Adsorption of a formulated protein on a drug delivery device surface. *J. Colloid Interface Sci.*, **189**, 216–228.

125. Tzannis, S.T., Hrushesky, W.J., Wood, P.A., and Przybycien, T.M. (1996) Irreversible inactivation of interleukin 2 in a pump-based delivery environment. *Proc. Natl. Acad. Sci. U.S.A.*, **93**, 5460–5465.

126. Weidner, T., et al. (2010) Sum frequency generation and solid-state NMR study of the structure, orientation, and dynamics of polystyrene-adsorbed peptides. *Proc. Natl. Acad. Sci. U.S.A.*, **107**, 13288–13293.

127. Welinder, A.C., Zhang, J., Steensgaard, D.B., and Ulstrup, J. (2010) Adsorption of human insulin on single-crystal gold surfaces investigated by in situ scanning tunnelling microscopy and electrochemistry. *Phys. Chem. Chem. Phys.*, **12**, 9999–10011.

128. Zhai, J., et al. (2010) Changes in beta-lactoglobulin conformation at the oil/water interface of emulsions studied by synchrotron radiation circular dichroism spectroscopy. *Biomacromolecules*, **11**, 2136–2142.

129. Höök, F., et al. (2002) A comparative study of protein adsorption on titanium oxide surfaces using in situ ellipsometry, optical waveguide lightmode spectroscopy, and quartz crystal microbalance/dissipation. *Colloids Surf. B*, **24**, 155–170.

130. Koh, L.B., Rodriguez, I., and Venkatraman, S.S. (2010) Conformational behavior of fibrinogen on topographically modified polymer surfaces. *Phys. Chem. Chem. Phys.*, **12**, 10301–10308.

131. Jorgensen, L., et al. (2004) Probing structural changes of proteins incorporated into water-in-oil emulsions. *J. Pharm. Sci.*, **93**, 1847–1859.

132. Sharp, J.S., Forrest, J.A., and Jones, R.A. (2002) Surface denaturation and amyloid fibril formation of insulin at model lipid-water interfaces. *Biochemistry*, **41**, 15810–15819.

133. Johnston, G. (2005) Generic biopharmaceuticals: towards a science-based abbreviated process. *J. Generic Med.*, **2**, 118–124.

134. EMEA. (2008) *CHMP Assessment Report for Zarzio*. European Medicines Agency, London, p. 33.

135. Chow, S.C. and Liu, J.P. (2010) Statistical assessment of biosimilar products. *J. Biopharm. Stat.*, **20**, 10–30.

136. EMEA. (2007) *Guideline on Immunogenicity Assessment of Biotechnology-Derived Therapeutic Proteins*. EMEA, London.

137. Roger, S.D. (2010) Biosimilars: current status and future directions. *Expert Opin. Biol. Ther.*, **10**, 1011–1018.

138. Schellekens, H. (2009) Biosimilar therapeutics–what do we need to consider? *NDT Plus*, **2**, i27–i36.

139. Schellekens, H. and Moors, E. (2010) Clinical comparability and European biosimilar regulations. *Nat. Biotechnol.*, **28**, 28–31.

140. Hennessy, S., Leonard, C.E., and Platt, R. (2010) Assessing the safety and comparative effectiveness of follow-on biologics (biosimilars) in the United States. *Clin. Pharmacol. Ther.*, **87**, 157–159.

141. Hirsch, B.R. and Lyman, G.H. (2011) Biosimilars: are they ready for primetime in the United States? *J. Natl. Compr. Canc. Netw.*, **98**, 934–942.

INDEX

Biophysical Methods for Biotherapeutics: Discovery and Development Applications, First Edition. Edited by Tapan K. Das.
© 2014 John Wiley & Sons, Inc. Published 2014 by John Wiley & Sons, Inc.